The Boreal Owl

Ecology, Behaviour and Conservation of a Forest-Dwelling Predator

Widespread in North American forest regions including the Rocky Mountains, the boreal owl (*Aegolius funereus*) was once the most numerous predatory bird in Eurasian boreal forests. Synthesising the results of unique long-term studies of boreal owls, this book explores hunting modes, habitats and foods, prey interactions, mating and parental care, reproduction, dispersal, survival and mortality, population regulation and conservation in boreal forests.

Providing a detailed introduction to the species, the authors study the complex interactions of boreal owls with their prey species. They examine the intersexual tug-of-war over parental care, and the behavioural and demographic adaptations to environmental conditions that markedly and predictably fluctuate both seasonally and multiannually. They also question whether boreal owls are able to time their reproductive effort to maximise lifetime reproductive success. Discussing the effect of modern forestry practices on owl populations, the book also examines how boreal owls could be managed to sustain viable populations.

Erkki Korpimäki is Professor of Animal Ecology in the Department of Biology at the University of Turku, Finland. His long-term research questions have focused on how predators are adapted to the large spatio-temporal fluctuations in their main prey densities, the apparent impacts of predators on prey populations and the effects of human-induced changes in the environment on viability of populations.

Harri Hakkarainen is Adjunct Professor of Animal Ecology in the Department of Biology at the University of Turku, Finland. His main fields of research have included landscape ecology, environmental ecology, life-history and demography, habitat and diet selection, sexual selection and dispersal of individuals.

The Boreal Owl

Ecology, Behaviour and Conservation of a Forest-Dwelling Predator

ERKKI KORPIMÄKI AND HARRI HAKKARAINEN

Section of Ecology, Department of Biology, University of Turku, Finland

CAMBRIDGE
UNIVERSITY PRESS

University Printing House, Cambridge CB2 8BS, United Kingdom

Published in the United States of America by Cambridge University Press, New York

Cambridge University Press is part of the University of Cambridge.

It furthers the University's mission by disseminating knowledge in the pursuit of education, learning and research at the highest international levels of excellence.

www.cambridge.org
Information on this title: www.cambridge.org/9781107425323

First published 2012
First paperback edition 2014

A catalogue record for this publication is available from the British Library

Library of Congress Cataloguing in Publication data
Korpimäki, Erkki.
The boreal owl : ecology, behaviour, and conservation of a forest-dwelling predator / Erkki Korpimäki, Harri Hakkarainen.
    pages   cm
Includes bibliographical references and index.
ISBN 978-0-521-11371-7
1. Aegolius funereus – Ecology.   2. Aegolius funereus – Behavior.   3. Aegolius funereus – Conservation.   I. Hakkarainen, Harri.   II. Title.
QL696.S83K67   2012
333.95'897–dc23

                                                                  2012004205

ISBN 978-0-521-11371-7 Hardback
ISBN 978-1-107-42532-3 Paperback

Contents

The color plates will be found between pages 242 and 243.

Preface

The senior author

It is early one sunny but chilly morning in late March 1966. The black woodpecker, *Dryocopus martius*, is drumming in the distance. It is easy to ski, because the surface snow layer is still hard enough to bear the weight of two teenage boys. The senior author of this book (EK, born in 1952 in Kauhava, South Ostrobothnia, western Finland) and Reijo Passinen, another enthusiastic young local birdwatcher, are checking some nest-boxes and natural cavities of boreal owls (*Aegolius funereus*) that they had set up or found during the previous autumn and winter. Pine-dominated coniferous forest gradually changes to denser old-growth spruce-dominated forest, where big aspen trees also grow sparsely. On the trunk of a tall aspen, eight metres above the ground, there is a natural cavity made by a black woodpecker. In the vicinity of this tree cavity, a male boreal owl has been hooting eagerly during calm evening nights from early February onwards, although the ambient temperature has often been below −15°C. EK lightly scrapes the aspen trunk, and the facial disc of a female boreal owl suddenly emerges from the entrance hole of the cavity. She has a surprised look, and only her yellow and black eyes can easily be seen looking out from the greyish-white facial disc against the background of the aspen trunk. It was their first visit to the nest of a boreal owl. When they went back, they found yet another boreal owl nest, this time in the new nest-box that they had set up the previous autumn.

This expedition remains vivid in EK's memory and engendered a lifelong fascination in predators, in particular in secretive nocturnal owls, which are really fascinating creatures. It was also the starting point of a period of explorative study, and later a professional research project, that he never imagined would still be going strong more than 40 years later.

When he was 13 years old, EK began some explorative studies on boreal owls in the natural-history clubs of the elementary and high schools at Kauhava and erected the first nest-boxes for these owls in autumn 1965, together with Reijo Passinen. The data collected during the explorative period from 1966 onwards in the Kauhava region included the number of nests, clutch size, breeding success and diet composition of boreal owls (Table 1).

During the three summers of the years in high school (1968–1970), EK worked as a research assistant on the northernmost tip of Finnish Lapland (Karigasniemi, Utsjoki) in a

Table 1. Number of available nest-boxes and nest-holes, total number of boreal owl nesting attempts (at least one egg laid; NE), number of clutches that were taken by pine martens (PM), number of clutches that failed for other reasons (mostly were deserted because of probable food scarcity; FA), percentage of boxes and holes with owl nests, number of female and male parents ringed or re-trapped at nests, and number of boxes with hooting bachelor males in the Kauhava region, western Finland during 1966–2010 (number of hooting males only in 1978–93). Data from Korpimäki (1987e), Korpimäki et al. (2008) and unpublished.

Year	Number of boxes	No. of nests			Percent nested	Number trapped		Number of bachelors[b]
		NE	PM	FA		Female[a]	Male	
1966	35	4	0	0	11.4			
1967	56	11	0	4	19.6			
1968	61	2	0	0	3.3			
1969	63	4	0	0	6.3			
1970	65	5	0	0	7.7			
1971	83	5	0	0	6.0			
1972	90	2	0	0	2.2			
1973	99	18	1	4	18.2			
1974	122	13	1	0	10.7			
1975	133	3	0	1	2.3			
1976	235	19	1	4	8.1	7	0	
1977	290	63	1	12	21.7	29	0	
1978	335	14	0	3	4.2	6	0	5
1979	355	37	2	8	10.4	22	11	6
1980	395	24	2	9	6.1	25	9	16
1981	395	10	0	0	2.5	7	6	18
1982	415	33	1	3	8.0	36	24	14
1983	450	26	1	13	5.8	25	19	30
1984	450	8	2	2	1.7	10	7	4
1985	450	50	4	8	11.1	35	36	11
1986	450	85	4	16	18.9	81	76	13
1987	450	11	0	4	2.4	12	10	8
1988	500	99	2	15	19.8	93	92	22
1989	500	147	1	43	29.4	133	129	40
1990	500	19	0	6	3.8	16	15	2
1991	500	115	1	35	23.0	95	91	12
1992	500	163	2	64	32.6	123	121	13
1993	500	13	0	5	2.6	14	12	4
1994	500	49	0	17	9.8	44	40	
1995	500	29	0	14	5.8	27	26	
1996	500	38	0	15	7.6	39	30	
1997	500	21	0	13	4.2	18	11	
1998	500	4	0	1	0.8	3	1	
1999	500	24	0	1	4.8	23	17	
2000	490	26	0	16	5.3	29	25	
2001	490	2	0	0	0.4	2	1	
2002	490	40	0	4	8.2	25	10	
2003	490	138	1	14	28.2	90	19	
2004	490	10	1	0	2.0	5	4	
2005	490	76	1	19	15.5	29	18	
2006	490	20	0	10	4.1	17	11	

Table 1. (cont.)

| Year | Number of boxes | No. of nests | | | | Number trapped | | |
		NE	PM	FA	Percent nested	Female[a]	Male	Number of bachelors[b]
2007	490	14	0	1	2.9	13	12	
2008	490	60	0	11	12.2	50	41	
2009	490	50	2	10	10.2	41	33	
2010	490	3	0	1	0.6	2	2	

[a] In some years, the number of ringed or re-trapped females was higher than the number of nests, because some females were also ringed in the areas surrounding the main owl study area in the Kauhava region.
[b] Number of hooting males was monitored using the point stop method.

project on the behaviour, breeding density, breeding performance and success of peatland bog and tundra birds. This project was led by the late Olavi Hildén (Department of Zoology, University of Helsinki, Finland). Docent Hildén was one of the most celebrated field ornithologists in Finland in the 1950s to 1990s who, among other things, made long-term studies on archipelago birds (Hildén 1966, Hildén et al. 1995) and published a widely cited review on habitat selection in birds (Hildén 1965). Hildén led the search for the hidden nests of the secretive arctic wader and passerine species such as the broad-billed sandpiper *Limicola falcinellus*, the spotted redshank *Tringa erythropus*, the jack snipe *Lymnocryptes minimus*, the red-necked phalarope *Phalaropus lobatus*, the Lapland bunting *Calcarius lapponicus* and the red-throated pipit *Anthus cervinus*. Docent Hildén also researched specific topics in behavioural ecology from the 1960s onwards, including the mating and breeding systems of waders (e.g. Hildén 1975, 1978). EK learned how to study the behaviour and breeding performance of individuals in bird populations at Karigasniemi.

EK embarked on his studies in biology at the Department of Zoology, University of Oulu, Finland in September 1971. There he was privileged to benefit from the wide experience, enthusiastic teaching and deep knowledge of professors Lauri Siivonen and Seppo Sulkava. The late Professor Siivonen was internationally renowned for his studies and reviews on the population cycles of forest grouse and other game animals (Siivonen 1954, 1948). Professor Sulkava is a pioneer investigator of the diet composition and predator–prey interactions of north European birds of prey, including the goshawk *Accipiter gentilis*, the peregrine *Falco peregrinus*, the golden eagle *Aquila chrysaetos*, the white-tailed sea-eagle *Haliaetus albicilla*, the eagle owl *Bubo bubo*, and the great grey owl (see, e.g., Sulkava S. 1966, 1968, Sulkava S. and Huhtala 1997, Sulkava S. et al. 1984, 1997, 1998, 2008). His tuition was of particular importance to EK, because he and his brother, Phil. Lic. Pertti Sulkava, had performed the first study on regional and between-year variations in the diet composition of boreal owls in Finland (Sulkava P. and Sulkava S. 1971). Seppo Sulkava's wide knowledge of ecology and nature in general – and of birds and mammals in particular – greatly benefited EK in planning the study designs and data collection for his Ph.D. thesis on breeding perform-ance and predator–prey interactions in boreal owls. Under his supervision, EK also began to collect data on between-year variations in breeding density and breeding performance in relation to fluctuations in main food abundance, diet composition and prey selection of boreal owls, as well as on the daily activity patterns of parents and growth of nestlings. These data were analysed and summarised in his Ph.D. thesis (Korpimäki 1981).

Later on, EK's research projects were extended to include population dynamics, reproductive success and behavioural ecology of many other birds of prey including, for example, the short-eared owl *Asio flammeus*, the long-eared owl *A. otus*, the hawk owl *Surnia ulula*, the Ural owl *Strix uralensis*, the eagle owl *Bubo bubo*, the pygmy owl *Glaucidium passerinum*, the Eurasian kestrel *Falco tinnunculus*, the hen harrier *Circus cyaneus*, the common buzzard *Buteo buteo*, and the goshawk *Accipiter gentilis* (see Korpimäki 1984, 1986c, 1987c, 1992a, Korpimäki and Huhtala 1986, Huhtala et al. 1987, Korpimäki et al. 1990, Korpimäki and Norrdahl 1991a, 1991b, Reif et al. 2004, Tornberg et al. 2006, Suhonen et al. 2007). The research team headed by EK has also studied the responses of mammalian predators (the least weasel *Mustela nivalis*, the stoat *M. erminea*, the American mink *M. vison*, and the red fox *Vulpes vulpes*) to the population cycles of small mammals (e.g. Korpimäki et al. 1991, Klemola et al. 1999, Dell'Arte et al. 2007). Finally, the impacts of avian and mammalian predators on the fluctuating small mammal and game animal populations have also been studied (e.g. Norrdahl and Korpimäki 1995b, 1996, Korpimäki and Norrdahl 1998, Korpimäki et al. 2002, 2005a; and reviews in Korpimäki and Krebs 1996, Korpimäki et al. 2004, Salo et al. 2010). These two main long-term research topics have thus dealt with predator–prey interactions from both sides of the fence. First, how predators are adapted to the large spatio-temporal fluctuations in their main prey densities and, second, what are the apparent impacts of predators on prey populations? A third, more applied, research question is: how have human-induced changes in the environment, such as modern forestry practices, changes in the intensity of agriculture, and invasion of alien predators, altered the composition of predator assemblages, individual-level performance and survival of predators and their impact on prey populations (e.g. Nordström et al. 2002, 2003, Nordström and Korpimäki 2004, Salo et al. 2007, 2008).

The junior author

The junior author, HH, was born in Jyväskylä in central Finland in late March 1966, perhaps on the very same sunny but chilly morning that EK met his first boreal owl at Kauhava. HH started his studies at the Department of Biology, University of Jyväskylä, central Finland in 1987. Two years later, he joined EK's boreal owl research team, and was the first student on the project to complete a master's thesis on boreal owls. At that time, HH had a keen interest in solving the puzzle of reversed sexual size dimorphism in boreal owls and other birds of prey, and subsequently worked in the same area of study for 10 years. He first collected the data for his Ph.D. thesis on reproductive effort, body size and their fitness consequences in boreal owls (Hakkarainen 1994), which consists of six chapters (Hakkarainen and Korpimäki 1991, Hakkarainen and Korpimäki 1991,

1994a, 1994b, 1994c, 1995), and then performed postdoctoral research, mostly funded by the Academy of Finland.

HH's main fields of interest have included a wide spectrum of topics in ecology, such as landscape ecology, environmental stress, environmental ecology, life-history ecology, reproductive success, survival, habitat and diet selection, sexual selection, and dispersal of individuals. Today his main aim is to study whether changes in forest structure owing to modern forestry techniques have had effects on individuals' reproductive success, survival, physiological stress, food resources and morphological characteristics. In addition, intra- and interspecific interactions, predation and competition have been examined in differently fragmented forest landscapes. He has approached the problem of forest fragmentation by collecting long-term data on forest-dwelling species, such as the Eurasian treecreeper (*Certhia familiaris*), wood ants (*Formica* spp.) and the three bird of prey species: the boreal owl, the goshawk and the common buzzard.

During the boreal owl studies, HH has been taught by EK how to perform very efficient field studies in population ecology. He soon noticed that owl studies, as such, are energetically demanding; in the 1991 field season, with 120 owl nests to monitor, HH lost 7 kilos in weight, due to hundreds of climbs to the nest-boxes and natural cavities, and staying awake for long periods at night. But perhaps the most memorable experience occurred during the 1992 field season (a record year – with 163 nests altogether), when our new Ph.D. student, Vesa Koivunen, introduced 'the ladder technique' to the project. Previously the nest-box trees had been climbed without any instruments. 'The ladder technique' increased our daily record for checking nest-boxes considerably, from 80 to 120. This new technique, unfortunately, also turned out to increase HH's body mass, and these days, in fact, ladders are obligatory when checking nest-boxes.

During the last 25 years, our research on boreal owls and other birds of prey has also included an experimental approach along with the observational approach. In the experiments, we have attempted to test the predictions of hypotheses that have been derived on the basis of observational results. We hope that these studies will continue in the future, as the value of long-term data in animal population ecology is always increasing.

In conclusion, it has been our pleasure and privilege to perform long-term studies of the boreal owl – perhaps the most fascinating bird in the coniferous forests of the world.

Acknowledgements

This almost lifelong field project has been fortunate to enjoy the field assistance of many voluntary local bird-watchers and field technicians. We mention just a few, who have assisted during more than one field season, including Mikko Hast, Ossi Hemminki, Timo Hyrsky, Mikko Hänninen, the late Sakari Ikola, Reijo Passinen, Jorma Nurmi, Stefan Siivonen and Rauno Varjonen. Members of the Ornithological Society of Suomenselkä, in particular Jussi Ryssy, Erkki Rautiainen, Mauri Korpi, Jaakko Härkönen, Risto Saarinen, Tarmo Myntti, Kari Myntti, Pertti Sulkava and Risto Sulkava, have played a crucial role in collecting the data on breeding densities and reproductive success of boreal owls in the Seinäjoki, Ähtäri and Keuruu regions. They all deserve our special thanks for their heroic field efforts and long-lasting friendship.

Many Ph.D. students and postdoctoral researchers have worked on our boreal owl population studies and collected data for their theses and publications. We thank Petteri Ilmonen, Vesa Koivunen, Toni K. Laaksonen, Juan A. Fargallo, Michael Griesser, Robert L. Thomson and Markéta Zárybnická for their contributions to the long-term data files and publications that have been written on the basis of these data. We acknowledge Serge Sorbi (Belgium), Jochen Wiesner (Germany), Wilhelm Meyer (Germany), Markéta Zárybnická (Czech Republic), Pierre-Alain Ravussin (Switzerland), Pierre Henrioux (Switzerland), Ted Swem (Alaska, USA) and Jackson S. Whitman (Alaska, USA), who willingly sent us their unpublished data from boreal owl populations, unpublished manuscripts, reprints of publications, and pictures for the book.

Our studies on the population dynamics and breeding success of boreal owls and other birds of prey were supported financially by the Emil Aaltonen Foundation, the Otro Seppä Memorial Foundation of the Finnish Cultural Foundation, the Jenny and Antti Wihuri Foundation, and the Academy of Finland. Three grants from the South Ostrobothnia Foundation of the Finnish Cultural Foundation in the last three years, to compile the old data in the files, to analyse these data, and to write this book were crucial for the completion of this long-term book project. Rauno Varjonen helped in putting old data in the files and carefully assisted in drawing figures and creating tables for the book. We also are very grateful to nature photographers Benjam Pöntinen (Lapua, Finland; http://www.pontinen.fi/) and Pertti Malinen (Vaasa, Finland) for the many magnificent photographs that they kindly provided for this book.

Dr. Jari Valkama, the head of the Ringing Centre, Natural History Museum (Helsinki, Finland) was very helpful in rapidly responding to our many queries on nationwide ring recoveries of boreal owls, and also made valuable comments on many chapters of the book. The manuscript also benefited from the indispensible comments of Dr. Toni K. Laaksonen and Dr. Alexandre Villers. The Section of Ecology, Department of Biology, University of Turku (Finland) provided great working facilities and an inspiring working atmosphere. The final chapters of the book were written during the study visit of EK to Estacion Biologica de Donana (CSIC), Seville, Spain. EK would like to thank Dr. Fabrizio Sergio and Dr. Juan Jose Negro for good working facilities.

Finally, we would like to thank our families for their endless patience and encouragement during this – perhaps too long-lasting – book project. EK would like to warmly thank his son Teemu and his daughters Heli and Hanna, and his life companion Maike, for their support and love. HH would like to thank his family: wife Terhi, daughter Riina and son Roni for their support during the long writing process.

1 Introduction

Vertebrate predators, particularly large charismatic apex predators, have traditionally elicited much esteem and deference, because they have been depicted as being beautiful, majestic and strong. Predator–prey interactions are also an intriguing topic in ecology and evolution. Most people would be impressed by the skilful hovering of Eurasian kestrels in the sky, which then swoop down towards a vole hidden in the undergrowth, or by the auditory location of voles by the great grey owl (*Strix nebulosa*) underneath a 60-cm-deep layer of compacted snow, through which the owl dives and often successfully captures the voles. It is also intriguing that snowy owls '*Bubo scandiaca*' in the northern tundra are able to track peak densities of lemmings thousands of kilometres apart, and find and settle in areas with high lemming densities to rear broods of ten owlets. Territories of birds of prey, particularly the large top avian predators such as golden eagles and eagle owls, are expected to have a high biodiversity value (Sergio et al. 2005, 2006), which suggests that conservation efforts focusing on avian predators can be ecologically justified.

This book is intended for general readers interested in nature, and in particular for ecology students, professional ecologists, and other scientists who are interested in life-history evolution, parental care and mating systems, population dynamics and predator–prey relationships of vertebrate animals. In addition, we expect that bird-watchers, naturalists, gamekeepers, and farmers, as well as forest managers, would also be interested in this book, so we will attempt to write in a style that is easily readable both for members of the public and for professionals and scholars. Many keen bird-watchers in temperate areas are making increased efforts to watch northern species, and are particularly interested in birds of prey. At present, there are not many relevant books available, because most books about avian predators are concerned with species that mainly live in temperate areas, or have a community-wide approach and cannot therefore concentrate on individual predator species in great detail. The overriding interest of farmers and forest managers might be based on the fact that owls are widely considered as biological agents controlling small mammal densities. In the peak phase of the population cycle, voles can cause serious damage to forestry and agriculture, and the annual cost can run into many millions of euros, as in Finland (Huitu et al. 2009).

Most books on the ecology and behaviour of birds and other organisms deal with species that live in temperate, subtropical and tropical regions with no marked between-year fluctuations in main environmental characteristics. In the Eurasian and North American boreal and arctic areas, there are marked between-year cyclic population fluctuations of many animal species, including some insects, voles, lemmings, shrews

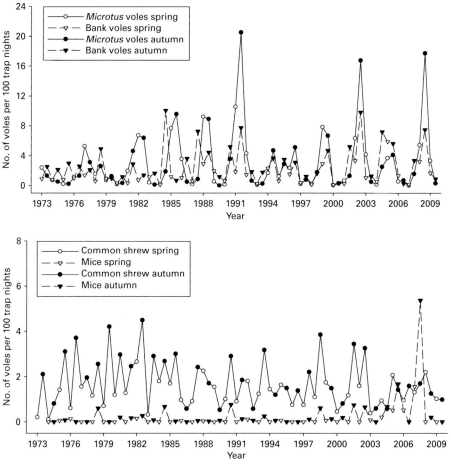

Fig. 1.1.   Abundance estimates (no. of individuals snap-trapped per 100 trap-nights) of *Microtus* voles (circles) and bank voles (triangles) (upper panel), and common shrews (circles) and pooled harvest and house mice (triangles) (lower panel, note different scale of *y*-axis) in the Kauhava region, western Finland in spring (May; open symbols) and autumn (September; closed symbols) during 1973–2009 (Korpimäki et al. 2002, 2005a and unpublished data).

and small game animals (e.g. hare and forest grouse), and their avian and mammalian predators. Of these, the most regular cyclic fluctuations have been found in small rodents in Eurasia and in snow-shoe hares in North America, and in their avian and mammalian predators in both continents (Boutin et al. 1995, Korpimäki and Krebs 1996, Klemola et al. 2002, Korpimäki et al. 2004). These multi-annual cyclic fluctuations have fascinated ecologists, naturalists and local people living in these areas for a long time (e.g. Collett 1878, Collin 1886, Elton 1942, Elton and Nicholson 1942, Hagen 1952, Siivonen 1954, 1957). Although some researchers have recently claimed that these regular high-amplitude population cycles of small mammals and their predators in northern areas have faded out (Hörnfeldt et al. 2005, Ims et al. 2008), possibly because of climate change, long-term data from our study area (Fig. 1.1; Korpimäki et al. 2005a, 2008) and elsewhere in northern Europe (Aunapuu and Oksanen T. 2003, Oksanen T. et al. 2008, Brommer et al. 2010, Lehikoinen et al. 2011) show that the heart of the well-known and fascinating north European vole and owl cycle is still beating.

The densities of most vole populations in central and western Europe fluctuate only seasonally and thus exhibit low densities in spring and higher densities in autumn, after the reproductive season of voles (see Hansson and Henttonen 1988). In addition, there are also some local multi-annual low-amplitude cyclic fluctuations of voles of the genus *Microtus* and their avian predators in central and western Europe (figure 14.1 in Taylor I. 1994, Lambin et al. 2000, Salamolard et al. 2000, Tkadlec and Stenseth 2001). However, there are also at least five distinct differences in the population dynamics of voles between west-central Europe and northern Europe.

1. The northern multi-annual vole cycles cover hundreds of thousands of square kilometres in pristine boreal and arctic ecosystems and in human-made forest plantations and agricultural fields (Kalela 1962), whereas the 3-year population cycle of, for example, field voles *Microtus agrestis* in Kielder Forest, Northumberland, England, is a smaller-scale phenomenon covering approximately 600 km^2 in man-made forest plantations (Lambin et al. 2000).

2. The low phases of northern vole cycles reach densities 1 to 2 orders of magnitude lower than in temperate Europe (for example, <1 per hectare in northern Europe vs. 25–50 per hectare in Kielder Forest) (Henttonen et al. 1987, Oksanen T. et al. 1999, Lambin et al. 2000).

3. Northern vole cycles are high-amplitude fluctuations, characterised by 50- to 500-fold (sometimes up to 1000-fold) differences between peak and minimum densities (Hanski et al. 1991, Oksanen L. and Oksanen T. 1992), whereas in temperate Europe differences between peaks and lows are only about 10-fold (Lambin et al. 2000).

4. The cycles of Kielder Forest are spatially synchronous at a small scale only (8–20 km; Lambin et al. 1998), whereas the northern vole cycles are synchronous at an essentially larger spatial scale (70–600 km; Henttonen et al. 1987, Hanski et al. 1991, Huitu et al. 2003b, Sundell et al. 2004).

5. In northern Europe, populations of field voles fluctuate in close temporal synchrony with those of other herbivorous voles of the genera *Microtus* and *Myodes* (earlier *Clethrionomys*) species, and even with insectivorous shrews (*Sorex* spp.) (Fig. 1.1). All of them show their lowest population densities simultaneously (Henttonen et al. 1987, 1989, Hanski and Henttonen 1996), whereas interspecific synchrony with the aforementioned species or other small mammals has not been documented in temperate Europe.

These five differences in population dynamics between temperate and northern Europe indicate clear differences in the underlying mechanisms. In particular, these five characteristics are also crucial for the predators that subsist on voles as their main foods.

This book tells the story of the boreal owl (*Aegolius funereus* L.; Tengmalm's owl in Europe): its hunting modes, habitats and foods, its interactions with prey, parental care and mating systems, reproduction, dispersal, survival and mortality, family planning, population regulation and conservation in boreal forests. The boreal owl used to be the most numerous predatory bird in Eurasian boreal coniferous forests, and it is also widespread in North American coniferous forest regions including the Rocky Mountains. Here we use the American English species name, the boreal owl, rather than the British English species name, Tengmalm's owl, which originates from the Swedish physician and naturalist, Peter Gustaf Tengmalm (1754–1803). P. G. Tengmalm was interested in owls and improved upon Linnaeus' owl classification in a paper submitted to the Swedish Academy of Sciences. Johann Friedrich Gmelin named an owl after him (*Strix tengmalmi*) in 1788, but believed mistakenly that P. G. Tengmalm had been the first to describe this species for science. We think that 'boreal owl' describes this species better than 'Tengmalm's owl', because this owl species occupies boreal (taiga) forests in Eurasia and North America. In addition, at least in the Finnish language, human names are not normally used as the names of animal species.

Boreal owls subsist on small mammals (voles, mice, lemmings and shrews), most of which show cyclic population fluctuations in Eurasian boreal areas with a cycle length of 3–4 years (see Fig. 1.1 for an example of a 3-year vole cycle in western Finland). In the course of the 3–4-year cycle, vole populations show 50- to 250-fold fluctuations in their densities. Therefore, vole-eating predators such as boreal owls may experience 'lean' periods of 1–2 years in length and 'fat' periods of 1–2 years in length during a single vole cycle. These owls need to adjust their main behavioural and life-history traits to these cyclically varying 'fat' and 'lean' periods. They should be able to survive 'lean' periods either by long-distance dispersal movements, because vole cycles are synchronous over large areas (see above), and/or by shifting to alternative prey, such as small birds, because the densities of all main small mammal populations are synchronously low in northern areas. On the other hand, owls should also be able to adjust their reproductive effort so that they can take full advantage of an abundance of food, for example by producing large clutches and broods with high-quality offspring during 'fat' periods. Our own long-term studies (>40 years) and studies by our colleagues elsewhere in Europe and in North America show that boreal owls are the avian predators that are apparently best adapted to these cyclically varying environmental conditions. In addition, they have to cope with seasonally varying environmental conditions, because the distribution of these owls covers regions where long snowy winters are, or at least have been, regular and predictable. In boreal regions, long-lasting deep snow cover protects small mammals against aerial predators, and boreal owls are too small to penetrate a deep snow layer and to catch voles hiding beneath the snow.

The main purposes of our book are first to give a short introduction to the species, and then to describe and discuss the complex interactions of boreal owls with their main prey species. Next, we aim to examine the behavioural and life-history (demographic) adaptations of boreal owls to environmental conditions that markedly fluctuate both seasonally and multi-annually, cyclically in a predictable manner, in boreal forests. We investigate whether boreal owls are able to time their reproductive effort so as to maximise reproductive success; in other words to 'plan' their families. Fourth, we explore the factors that might regulate boreal owl populations. Fifth, we aim to reveal the possible detrimental effects of loss of boreal coniferous forests due to modern forestry practices on boreal owl populations. Finally, we intend to explore how boreal owl populations could be managed in order to sustain viable populations.

2 The boreal (or Tengmalm's) owl: in brief

Source: Drawn by Marke Raatikainen

2.1.   Plumage and morphological characteristics

The boreal owl is a quite small fieldfare-sized long-winged avian predator in Eurasian and North American coniferous (taiga) forests. This almost entirely night-active (nocturnal) owl is very inconspicuous in the daytime but is sometimes seen perched on a branch of a dense conifer, usually a spruce tree. It stands motionless in an upright position near the trunk in the manner of a long-eared owl. Therefore, humans are not often able to observe boreal owls, but their loud primary song, 'pu' or 'po', has been heard by many people who live and work in the countryside (see Chapter 2.3). A long time ago, laymen and hunters in the Finnish countryside believed that this vocalisation was made by the arctic hare *Lepus timidus*. In the boreal owl, the sexes are alike but females are substantially larger than males.

In comparison to the similar-sized little owl, *Athene noctua*, boreal owls have a disproportionately large head, longer wings and a very distinct facial disc. The greyish-white facial disc is framed by a brown-black border and highlighted by raised white eyebrows, which give the owl a surprised expression (Fig. 2.1). This expression is

Fig. 2.1.   The greyish-white facial disc of the boreal owl is framed by a brown-black border and highlighted by raised white eyebrows. The upper part is dark brown with white spotting, and the under-part is creamy white, broadly streaked with dark brown or russet. Photo: Pertti Malinen. (For colour version, see colour plate.)

Fig. 2.2. Fledglings of boreal owls are dark brown with very distinct white 'eyebrow' markings and some white spots on the scapulars and wing coverts. Photo: Benjam Pöntinen. (For colour version, see colour plate.)

complemented by the large bright yellow-black eyes. The upper part of the plumage is dark brown with white spotting, which gives the species its Finnish name '*helmipöllö*' and its Swedish name '*pärluggla*' ('pearl owl'). The under-part is creamy white, broadly streaked with dark brown or russet. This streaking is especially dense on the breast but less so on the lower belly. Between-individual colour variation in adult owls is substantial: some are more greyish, whereas others have a more reddish-brown (russet) colour on their upper parts. Yearling owls are probably darker than older owls. The short tail is brown above, with narrow white cross-bars. As in most other northern owl species, the legs and feet are densely covered with white-grey feathers, which is also a striking difference from little owls. The claws are blackish-brown and the bill is wax-yellow. The closest allospecific, the saw-whet owl *Aegolius acadicus*, is smaller than the boreal owl, has chestnut rather than greyish-brown streaking on the breast, and the general tone of its upper part is reddish-brown rather than greyish-brown.

The coloration of young boreal owls during the late nestling, fledging and post-fledging periods differs distinctly from that of adults. Young owls are dark brown overall, even darker and sootier on the facial disc, with very distinct white 'eyebrow' markings (Fig. 2.2). They have some white spots on the scapulars and wing coverts. Young saw-whet owls have a paler two-tone breast and belly than young boreal owls.

2.2.   Moulting, ageing and sexing

Some European owls moult all their primary feathers each year, whereas boreal owls take more than a year to replace the primaries of the juvenile plumage (Altmüller and

Fig. 2.3.   The wings of yearling (2nd calendar year) boreal owls have only dark brown and unworn primary and secondary feathers. Photo: Rauno Varjonen. (For colour version, see colour plate.)

Kondrazki 1976, Glutz von Blotzheim and Bauer 1980). In the post-juvenile moult during the post-fledging and first independence periods of the young, boreal owls moult their first adult plumage, including all the primary and secondary feathers of the wings. Therefore, in their first breeding season at the age of 1 year (i.e. as yearlings), parent boreal owls have only one age class of primary and secondary feathers, which are dark brown and unworn (Fig. 2.3). From late June–July up to September of the second calendar year of their life, the owls moult between one and six (on average four) outermost primaries. Therefore, in their second autumn and breeding season the next spring, 2-year-old owls have two age classes of primaries on their wings: the outermost dark-brown unworn 1–6 primary feathers and the light-brown worn innermost 4–9 primaries (Fig. 2.4). In the third calendar year of life, the owls moult a middle group of their primaries, usually 2–4 primaries. Therefore, 3-year-old owls have three age classes of primaries on their wings (Fig. 2.5). At the age of 4 years, the wave of primary moulting proceeds towards the innermost primaries; the ninth and tenth innermost primaries are first moulted only at the age of 5 years (Fig. 2.6).

To age non-ringed boreal owls trapped at the nest or outside the breeding season, this moult pattern method has been tested by using known ages of ringed individuals in German (Schwerdtfeger 1984, 1991), Finnish (Korpimäki and Hongell 1986, Lagerström and Korpimäki 1988) and Swedish (Hörnfeldt et al. 1988) populations. It has been found that it is reasonable to age the owls into at least three age classes: 1-year, 2-year and +2-year-old individuals. Hörnfeldt et al. (1988) concluded that 90% of females and >80% of males can also be aged as 3-year-old individuals. They also gave detailed descriptions of the duration and patterns of moult of boreal owls (see also Erkinaro 1975, Glutz von Blotzheim and Bauer 1980). All in all, this age determination method has proved to be very valuable in individual-level population studies of boreal owls, because most unknown-age adult owls can be reliably aged into these three age

Fig. 2.4. Two-year-old (3rd calendar year) boreal owls have the outermost dark-brown unworn 1–6 (in this case 5) primary feathers and the light-brown worn innermost 4–9 (in this case 5) primaries. Photo: Rauno Varjonen. (For colour version, see colour plate.)

Fig. 2.5. Three-year-old (4th calendar year) boreal owls have three age classes of primaries on their wings. This male has moulted the fourth to sixth primary feathers in the previous autumn, the first three primaries in the autumn before that, and the four innermost primaries are still unmoulted. Photo: Rauno Varjonen. (For colour version, see colour plate.)

classes when trapped and ringed for the first time. Nestlings can be reliably aged by measuring wing length (Fig. 2.7): the accuracy of age estimation appears to be 1 day up to the age of 3 weeks, and thereafter it is 2 days up to the age of 4–5 weeks. The body mass is apparently a reliable estimate of age of the young only up to the age of 2 weeks (Fig. 2.8), because the body mass of nestlings and fledglings varies in relation to main food

Fig. 2.6.   At the age of 4 years, the wave of primary moulting proceeds towards the innermost primaries; the 9th and 10th innermost primaries are first moulted only at the age of 5 or more years. This >7-year-old male has not yet moulted the 10th innermost primary. Photo: Rauno Varjonen. (For colour version, see colour plate.)

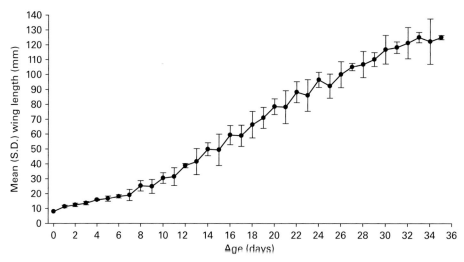

Fig. 2.7.   Mean (s.d.) wing length (mm) of boreal owl chicks from hatching to the age of 35 days. Pooled data from chicks from nine nests during 1974–5, 1977–9 and 1980 (Korpimäki 1981 and unpublished data).

abundance, and there is also hatching-order and intersexual variation in body mass (Kuhk 1969, 1970, Korpimäki 1981, Carlsson B.-G. and Hörnfeldt 1994, Schwerdtfeger 2000, Valkama et al. 2002, Suopajärvi P. and Suopajärvi M. 2003).

Sexing of breeding owls is easy, because females, but not males, have very distinct brood patch; only females incubate eggs and brood the young (Korpimäki 1981, Mikkola

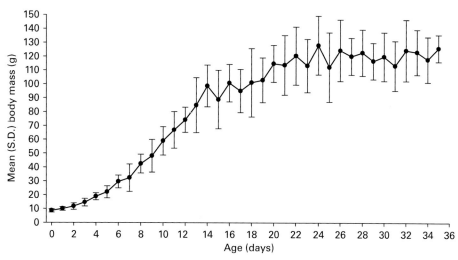

Fig 2.8. Mean (s.d.) body mass (g) of boreal owl chicks from hatching to the age of 35 days. Pooled data from chicks from nine nests during 1974–5, 1977–9 and 1980 (Korpimäki 1981 and unpublished data).

1983, Hayward and Hayward 1993). Adult females are much heavier (approx. 50%) and also have longer wings (approx. 5%) and tails (approx. 5%) than males (Fig. 2.9). There is also some age-related difference in body mass as well as wing and tail length, with yearlings being lighter and having slightly shorter wings and tails than +1-year-old owls (Figs. 2.9 and 2.10). The range of wing and tail lengths and other body dimensions of both sexes somewhat overlap. Therefore, only about 90% of females and males can be reliably sexed on the basis of body dimensions (Korpimäki and Hongell 1986, Hayward and Hayward 1991, Hipkiss 2002). No reliable sexing method of young in the nest or in their juvenile plumage on the basis of body dimensions or external plumage characters has been found (Korpimäki 1981, Hayward and Hayward 1993, Carlsson B.-G. and Hörnfeldt 1994, Schwerdtfeger 2000). To sex owlets in the nest and non-breeding adult owls, one needs to sample a drop of blood and perform polymerase chain reaction (PCR) tests in a molecular laboratory (Fridolfsson and Ellegren 1999, Hörnfeldt et al. 2000, Hipkiss et al. 2002b).

2.3.   Vocalisations

The vocal array of boreal owls consists of at least eight vocalisations: the primary territorial or 'staccato' song, a prolonged staccato song ('stutter song', *sensu* König and Weick 2008), delivery call, screech ('skiew'), peeping call ('cheep'), weak call, 'chuuk' call, and hiss. These vocalisations have been described in detail, for example by März (1968), Glutz von Blotzheim and Bauer (1980), Hayward and Hayward (1993), and König and Weick (2008). Therefore, only descriptions of the most important vocalisations are given here. The primary song of the male boreal owls consists of

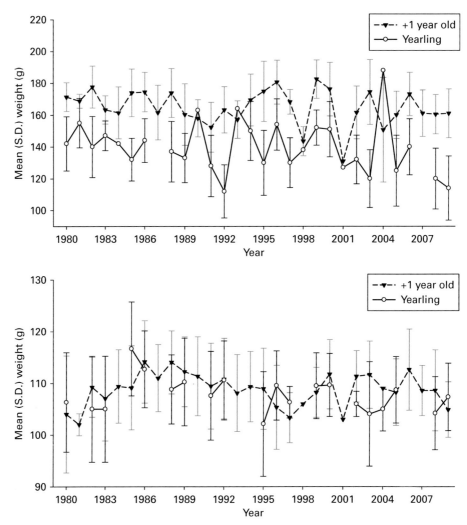

Fig. 2.9.   Yearly mean (s.d.) body mass (g) of yearling and adult (+1-year-old) boreal owl females (upper panel) and males (lower panel) during 1980–2009 (data from Korpimäki 1990a and unpublished).

a quick succession of 'bu' or 'po' notes or hoots that usually increase in volume during a trill. The number of hoots is usually from four to nine (up to 20) hoots and the silent interval between trills is 1–5 seconds but varies with the intensity of excitement. The individual notes usually begin rather dim and mellow, then increase to full volume and finally break off suddenly. Singing bouts frequently last 20 minutes but may continue for 2–3 hours, with some pauses of several minutes. Wide between-individual differences are apparent in the length of the trill, number of hoots, mean length of hoots and mean inter-hoot length (König 1968b), but there appear to be no different regional dialects (König and Weick 2008). In addition, some males also have peculiar characteristics in their song. Therefore, by analysing and comparing the sonograms of males

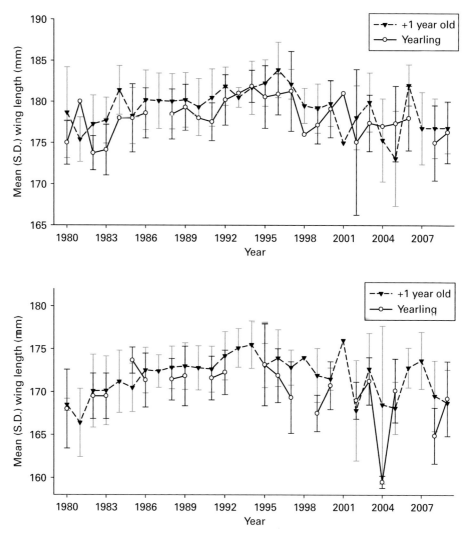

Fig. 2.10. Yearly mean (s.d.) wing length (mm) of yearling and adult (+1-year-old) boreal owl females (upper panel) and males (lower panel) during 1980–2009 (data from Korpimäki 1990a and unpublished).

uttering at different cavities and boxes, individual identification can probably be made (Carlsson B.-G. 1991). This individual identification of hooting males has already been successfully used for the other owl species; for example, the pygmy owl (Galeotti et al. 1993), the tawny owl (Galeotti 1998) and the scops owl *Otus scops* (Galeotti and Sacchi 2001).

The loudness of the primary song also varies considerably between individuals. This is suggested by the audibility of the song to humans on calm or fairly calm nights with no precipitation: from 700 m to 3.5 km in our study area, from 700 m up to 3 km in Sweden (Holmberg 1979), and from 1.5 km up to 3.5 km in Idaho, in the USA (Hayward and

Hayward 1993). Primary song usually begins about 30–60 minutes after sunset, peaks before midnight, and may continue throughout the night until dawn (Kuhk 1953, König 1968b, Korpimäki 1981, Hayward and Hayward 1993), but sometimes males (probably unpaired) also hoot during the daytime. In northern Europe, the first males usually sing at the turn of the year (late December to early January): for example, in our study area, the first hooting males were recorded on 1–10 January 1975, 24–26 December 1975, 22 December 1976 to 15 January 1977, 15 December 1977 and 5–11 January 1979 (Korpimäki 1981). The hooting activity peaks in late February and early March, whereas in late April to May it is mostly unpaired males that sing (Lundin 1961, Korpimäki 1981, Mikkola 1983).

Possibly the peak of singing activity is later in central Europe, because during 1939–1946 the pioneer investigator of boreal owls, Dr. Rudolf Kuhk, in Lüneburger Heide (Lüneburg Heath), Germany, recorded the peaks in singing activity in late March to mid-May (Kuhk 1953). He was also the first to record hooting of males in autumn: late September to November. In addition, the singing activity peaked only in April–May in southern Bohemia, the Czech Republic and eastern Slovakia, and a similar peak was again recorded in October and November (Kloubec and Pacenovsky 1996). In Idaho, USA, singing started about 20 January, reached peak intensity by late March and was uncommon by late April (Hayward et al. 1993). In Colorado, USA, under the harsh conditions of the Rocky Mountains, singing was recorded from 18 February to 21 June, with peaks in late April and again in late May–June (Palmer 1987).

There is some disagreement about the effects of environmental conditions on hooting activity: heavy snowfall, rain and wind can reduce calling, while temperature, cloud cover and lunar phase appear not to affect the number of hooting owls. In Finland, boreal owls respond to playbacks of their primary song throughout the year, but responsiveness peaks in March–April and again in late August–November. The same is also true in the Italian Alps, but the responses to playbacks have not been recorded in July–August (Mezzavilla et al. 1994). Playing back a recording of the primary song of males to attract boreal owls to mist nets for ringing purposes is a useful method that has been used by bird-ringing observatories for a long time. In autumn, this playback method seems to be attractive to both males and females, with no intersexual differences in attraction (Hipkiss et al. 2002b).

The primary song of males is usually uttered from within 100 m, and frequently within 10 m, of a suitable nest-cavity (Hayward and Hayward 1993). It has been suggested that males cease their primary song when they attract a mate (e.g. König 1968b, Holmberg 1979, Mikkola 1983), or at least when their female occupies a nest-hole some days before the laying of the first egg (Hayward and Hayward 1993). Detailed observations of individual males showed, however, that males were singing at up to five different nest-holes before attracting a mate (Carlsson B.-G. 1991, Hayward and Hayward 1993). They sung mostly at the nest-hole of what became the primary breeding site and with decreasing frequency at more distant holes (Carlsson B.-G. 1991). All early-mating males continued to sing at secondary nest-holes after attracting primary (first) females. Secondary singing locations were on average 800 m (range 300 m to 3.2 km) from the nest-holes of primary females, although closer locations were used for singing before mating for the first time (Carlsson B.-G. 1991). The wide audibility of loud primary song helps to find night-active, secretive, forest-dwelling boreal owls, which also readily respond to playback and imitation. Because males can hoot at different nest-holes before attraction of a mate, already-mated males continue to sing in many new distant locations, and unpaired males are the most eager in hooting, one has to be cautious when estimating the number of breeding owl pairs on the basis of singing locations only. Records of individual primary songs and their detailed analyses are badly needed in order to reveal the various singing locations of individual breeding males and unpaired males.

Males utter prolonged 'staccato' song from the initiation of courtship to the early incubation period. This 'stutter song' consists of a long trill of notes similar to primary song, but lasts up to 1 minute, and notes are more irregularly spaced than in the primary song. This 'stutter song' normally leads into a long, mellow trill consisting of up to 350, but normally fewer, rapid notes. A male singing the primary song changes to prolonged song when flying to a potential nest-hole and continues prolonged song from the entrance to the nest-hole. Males may also repeat the flights between female and cavity while singing prolonged song. The function of this call is presumably to do with pair formation and bonding.

Delivery call is a brief trill of 4–10 notes with same note and inter-note length as prolonged song. Males utter a delivery call prior to transfer of prey to the female or young in the nest, or to fledglings when they have left the nest-hole. Screech ('skiew') is a loud rough call by males and females and is often followed by bill clapping. Humans often hear screech in response to playback of primary song. Peeping call ('cheep') is a soft call of the female used throughout the breeding period, particularly as a response to the prolonged song or delivery call of males. The rough hiss call is usually accompanied by repeated bill clapping and is uttered by the female when the nest is disturbed, or sometimes by both sexes when being handled by humans.

2.4. Distribution, variation and population estimates

The boreal owl has circumpolar Holarctic distribution, occupying Eurasian and North American boreal coniferous (taiga) forests and also high-elevation subalpine forests further south (Figs. 2.11 and 2.12). Distribution in Europe corresponds well with the natural distribution of the Norway spruce *Picea abies* (Glutz von Blotzheim and Bauer 1980). Older research recognises seven subspecies: one in North America and six in Eurasia (März 1968, Glutz von Blotzheim and Bauer 1980), but more recent work acknowledges only five (König and Weick 2008). Of these later-mentioned five subspecies, four are found across the boreal forests of Eurasia (*Aegolius funereus funereus*, *A. f. magnus*, *A. f. pallens* and *A. f. caucasicus*). Distinctions between these four Eurasian subspecies are still quite unclear, although it has been suggested that inter-subspecies boundaries appear to be in the Ural Mountains in western Russia, and along the Kolyma River in north-east Russia (Mysterud 1970). It has also been suggested that boreal owls may represent a clinal variation in size and coloration across the northern boreal forest. Throughout Eurasia, owls seem to vary from dark brown and smaller in south-west and northern Europe to paler

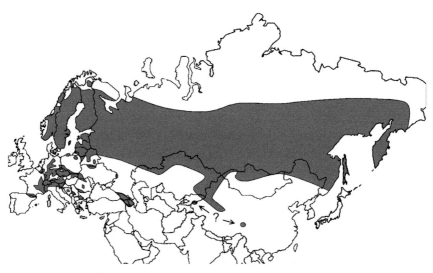

Fig. 2.11.   Distribution map of boreal owls in Europe and Asia.

Fig. 2.12.   Distribution map of boreal owls in North America.

or less rufous, more profusely white-spotted and larger in north-east Russia (März 1968, Mysterud 1970, Mikkola 1983). Therefore, boreal owls appear to follow a general rule that animals in warm and humid areas are more heavily pigmented than those in cool and dry areas. This colour variation may be attributable to cryptic coloration and/or reduced thermoregulatory costs in colder climate regions (Norberg 1987). However, possible inter-areal differences in coloration and body size throughout Eurasia have never been quantitatively analysed. Therefore, this suggestion remains to be tested adequately with comparable data from breeding boreal owls in Siberia.

According to recent research (König and Weick 2008), the nominate boreal owl *A. f. funereus* ranges from northern Europe to the Pyrenees in the west, to the Alps and Carpathians in the south, and from the Baltic Republics to the north of the Caspian Sea, and Russia in the east. *A. f. magnus* (König and Weick 2008 included *A. f. jakutorum* as a synonym) ranges from north-east Siberia to Kamchatka. Its upper parts are ashy grey-brown with heavy white spotting, and the dark markings on the under-parts are rather faint. It is generally paler and larger than the nominate boreal owl. The range of *A. f. pallens* (König and Weick 2008 included the taxon *A. f. sibiricus* as a synonym) covers western Siberia, Tian-Shan, south Siberia, south through north Mongolia to north-east China and east Russia including Sakhalin. The subspecies *A. f. caucasicus* has a patchy distribution in high elevation coniferous forests in the Caucasus (König and Weick 2008 included the taxon *A. f. beickianus* as a synonym) (Fig. 2.11).

In northern Europe, the breeding range of boreal owls (*A. f. funereus*) is distributed in a fairly regular area, covering northernmost Finland, Sweden and Norway below the tree line in the north, and southern Norway and Sweden as well as the Baltic States and Belarus in the south (Fig. 2.11). More or less isolated local populations occur in high-elevation coniferous forests in Germany, Switzerland, Austria, Czech Republic, Slovakia and Rumania. In western Europe, the distribution extends to the eastern parts of Belgium, the Netherlands and France, to the Pyrenees in southern France and northern Spain, and in the south to the north Italian Alps, to Slovenia, Croatia, Serbia, Greece and Bulgaria (Fig. 2.11).

The subspecies in North America, *A. f. richardsoni*, ranges throughout boreal forests in Canada and Alaska, but its distribution also continues further south in high-elevation subalpine forests of the Rocky Mountains, Blue Mountains and Cascade ranges (Fig. 2.12; see also Hayward et al. 1987, and figure 1 in Hayward and Hayward 1993). There has been some debate that boreal owls have expanded their range into or within the western USA since 1963, but Stahlecker and Duncan (1996) concluded, based on paleontological and archaeological specimens and autumn observations during the late nineteenth and early twentieth centuries, that boreal owls have been present in Colorado and New Mexico for centuries (see also Hayward et al. 1987, Ryder et al. 1987). Boreal owls remained largely undocumented, however, because their high-elevation subalpine conifer forest habitat is snowbound during their most active hooting period. This sub-species is probably markedly larger than conspecifics of *A. f. funereus* in Finland (Table 2.1, see also Fig. 2.13).

Using seven microsatellite markers, genetic samples were analysed from 275 boreal owls in different parts of the North American range, from 36 boreal owls in southern

Table 2.1. Mean (s.e. = standard error, range, *n*) body mass and wing length of breeding male and female boreal owls in the Kauhava region, Finland (Korpimäki 1990a and unpublished data) and in Idaho, USA (Hayward and Hayward 1991).

Variable	Kauhava				Idaho			
	Mean	s.e.	Range	*n*	Mean	s.e.	Range	*n*
Males								
Body mass (g)	109.6	0.25	86–137	953	117.3	1.39	93–139	50
Wing length (mm)[a]	171.9	0.12	159–183	953	172.5	0.58	163–179	41
Females								
Body mass (g)	165.6	0.52	112–216	1153	166.8	2.46	132–215	53
Wing length (mm)[a]	179.1	0.12	164–190	1173	183.6	0.72	174–198	49

[a] Cord from the carpal joint tip to the tip of the longest primary feather, with flattening for the data from the Kauhava region and without flattening for the data from Idaho.

Fig. 2.13.   The North American subspecies of the boreal owl, *Aegolius funereus richardsoni*. Photo: Jackson S. Whitman. (For colour version, see colour plate.)

Norway and 5 owls in north-eastern Russia (Koopman et al. 2005, 2007a). There was no detectable genetic differentiation between Norwegian and Russian owls, but this result appears to be hampered by the small sample size of Russian owls. Similarly, no genetic differentiation was found between boreal owls sampled in Manitoba, Canada and Alaska, USA, situated almost 3500 km apart in the boreal forests (Koopman et al. 2007a). In contrast, notable genetic differentiation was documented between Eurasian and North American specimens, suggesting a dispersal barrier caused by the Bering Strait and, in particular, the North Atlantic Ocean (Koopman et al. 2005). The finding of low intra-continental genetic differentiation was expected, in the sense that it has already been documented many times that juveniles of both sexes and adult female owls show high rates of long-distance dispersal (i.e. they are nomadic) (Wallin and Andersson 1981, Löfgren et al. 1986, Korpimäki et al. 1987, Sonerud et al. 1988, Schwerdtfeger 1993, 1996). This probably results in a large amount of gene flow between different local owl populations and raises the important question of whether the recognition of six different subspecies throughout Eurasia is really justified.

Congeners of boreal owls include the saw-whet owl, the unspotted saw-whet owl *Aegolius ridgwayi*, and the buff-fronted owl *A. harrisii*. All of these are smaller than boreal owls and occur in the New World only. Saw-whet owls range throughout boreal and temperate forest belts in Canada and Alaska, but their range is generally more southern than that of boreal owls (Johnsgard 1988, Cannings 1993, Hayward and Hayward 1993). These two owl species of the genus *Aegolius* also co-exist in wide areas, mainly around the border zone of Canada and the USA, and in the Rocky Mountains of, for example, Idaho and Colorado, USA. In areas of co-existence, boreal owls usually occupy higher-elevation forests than the smaller saw-whet owls (Hayward and Garton 1984, Palmer 1987, Hayward and Garton 1988). Unspotted saw-whet owls occupy Central America from south Mexico to west Panama, and buff-fronted owls are found in South America, including the Andes of Venezuela, Colombia, Peru, Bolivia, north-western Argentina and eastern parts of Brazil (Duncan 2003, König and Weick 2008). Therefore, their ranges do not appear to overlap with boreal owls.

The boreal owl was probably the most common bird of prey in the coniferous forests of Finland and Sweden in the 1950s and 1960s (Merikallio 1958, Ulfstrand and Högstedt 1976, Mikkola 1983), although it is fairly difficult to estimate the population size of nocturnal species with wide between-year fluctuations in breeding populations. The European-wide Bird Atlas estimated that the total European population in the late 1980s and early 1990s averaged 48 000 (range 37 000–71 000) with the largest breeding populations in areas west of the Ural Mountains, Russia (mean 31 000, range 10 000–100 000), Sweden (20 000, 10 000–40 000), Finland (12 000, 8000–20 000), Belarus (4000, 3000–5000) and Norway (3000, 2000–10 000) (Korpimäki 1997).

3   Study areas and research methods

3.1.   Main study area in the 'Wild West' of Finland

Our main study area of boreal owls is situated on the border between the south- and mid-boreal vegetation zones (Ahti et al. 1968) in the Kauhava region (approx. 63°N, 23°E) of the province of South Ostrobothnia, the 'Wild West' of Finland (Fig. 3.1). South Ostrobothnia is the flat area on the east coast of the Gulf of Bothnia, Baltic Sea. At the beginning of the explorative study period, the study area was approx. 35 km^2 in the village of Ruotsala, in the southern part of Kauhava, but later it was gradually extended almost every year, reaching approx. 200 km^2 in 1973 and including 99 nest-boxes and natural cavities (Table 1). The increase in the number of nest-boxes and the extension of the study area continued up to 1983, when the number of nest-boxes and natural cavities totalled 450 in an area of 1100 km^2. The final increase in the number of nest-boxes and natural cavities happened in 1988, with up to 500 nest-sites in an area of 1300 km^2, including the whole township of Kauhava and also the surrounding areas of the townships of Lapua, Lappajärvi and Kortesjärvi (Fig. 3.2). From 1983 onwards, the number of natural cavities found and annually checked in the area has ranged between 20 and 30. During the 2000s, the number of natural cavities somewhat decreased due to clear-cutting, and therefore the number of available nest-sites for boreal owls has been approx. 490 from 2000 onwards (see Table 1).

Only three main roads cross the study area: the west-to-east road (Vaasa–Jyväskylä) in the southern part of the study area, the south-to-north road (Seinäjoki–Uusikaarlepyy) in the western part of the study area, and the west-to-north-east road (Kauhava–Ylivieska–Oulu). Boreal owls sometimes breed in natural cavities very close (minimum only 5 m) to the main roads (Fig. 3.3), and all the traffic kills of these owls have been recorded on these main roads (see Chapter 10.4 on mortality factors) so it is advisable not to set up nest-boxes of owls close to main roads.

There is a dense network of minor roads (<3 m wide) in our study area that have mainly been built for farming and forestry access. Nest-boxes of boreal owls have been set up quite close (50–200 m) to these minor roads to facilitate relatively fast inspection. The boxes were hung some 4–7 m (mean 5 m) above ground, mainly on spruce trees, so their entrance holes were visible to owls at a distance when they were flying or perching in the forest (Fig. 3.4). During the explorative study period, most boxes were hung much higher on the trees (7–15 m), but boreal owls willingly accept nest-boxes that are only 4–5 m above ground. The density of nest-boxes in our study area (approx. 1 nest-box per 2 km^2)

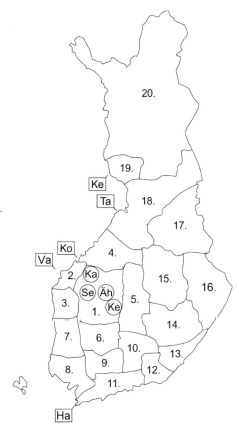

Fig. 3.1.   Map showing the main study area of the Kauhava region (Ka), the Seinäjoki region (Se), the Ähtäri region (Äh) and the Keuruu region (Ke) in the district of Ornithological Society of Suomenselkä (1). The map also shows other local ornithological societies in Finland (2 = Quarken, 3 = Suupohja, 4 = Central Ostrobothnia, 5 = Central Finland, 6 = Pirkanmaa, 7 = Satakunta, 8 = Varsinais-Suomi, 9 = Kanta-Häme, 10 = Päijät-Häme, 11 = Uusimaa, 12 = Kymenlaakso, 13 = South Karelia, 14 = South Savonia, 15 = North Savonia, 16 = North Karelia, 17 = Kainuu, 18 = North Ostrobothnia, 19 = Kemi-Tornio, 20 = Lapland). Permanent autumn capturing and ringing localities of irruptive boreal owls in Finland are: Ke = Kemi, Ta = Tauvo, Ko = Kokkola, Va = Valassaaret, and Ha = Hanko.

does not substantially differ from natural conditions, because the density of suitable natural cavities in pristine coniferous forests of southern Finland is 0.5–1.5/km^2 (Pouttu 1985, Virkkala et al. 1994). In addition, in coniferous forests managed for forestry in central Sweden, the density of natural cavities made by black woodpeckers is 0.3–0.4/ km^2 (Johnsson et al. 1993).

The weather station of the Finnish Meteorological Institute at Kauhava airport is situated in the middle of the study area (50 m asl). The warmest month is July and the coldest is February (Fig. 3.5). Most of the precipitation occurs in July–August and the snow layer is deepest in February–March.

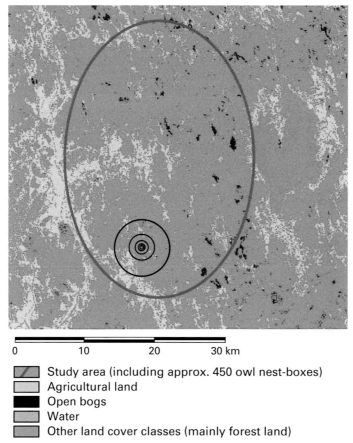

0   10   20   30 km

▨ Study area (including approx. 450 owl nest-boxes)
▢ Agricultural land
■ Open bogs
▢ Water
▨ Other land cover classes (mainly forest land)

Fig. 3.2.   Landscape composition of the study area in the Kauhava region, western Finland. Lappajärvi is the large lake on the eastern border of the study area. Concentric black circles show the five landscape scales (250 m, 500 m, 1000 m, 2000 m, 4000 m) around a nest-box (see Hakkarainen et al. 2003). (For colour version, see colour plate.)

3.2.   Habitats in the main study area

The south-western part of our study area is only 30–40 m asl, and the terrain gradually elevates, reaching 100–120 m asl in the north-eastern parts of the study area, close to the western shore of Lake Lappajärvi (Fig. 3.2). About 61% of the study area is covered by different-aged forests: 39% young forests (growing stock volume <101 m^3/ha), 11% middle-aged forests (101–151 m^3/ha), and 11% old-growth forests (>151 m^3/ha) (Hakkarainen et al. 2003). Practically all forests in the study area are managed for forestry purposes, which means that forests are first harvested by thinning when trees are 30–40 years old, and then clear-cutting is done at intervals of 60 up to 100 years. Thereafter, clear-cut areas are seeded or saplings of spruce or pine are planted. Therefore, old-growth forests nowadays comprise less than 1% of the area. In 1990, this proportion was still somewhat higher (2%, Hakkarainen et al. 2003), and when our studies

Fig. 3.3.   Boreal owls sometimes bred in natural cavities only 5 m from a main road. Natural cavity in aspen tree in Ritamäki, Lapua in 1985. Photo: Erkki Korpimäki.

Fig. 3.4.   The boxes of boreal owls were hung some 4–7 m (mean 5 m) above ground, mainly on spruce trees. Photo: Erkki Korpimäki.

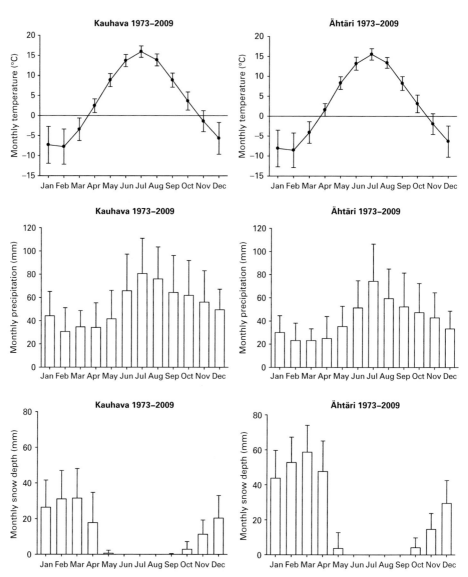

Fig. 3.5.   Mean monthly ambient temperature (°C, upper panels), precipitation (mm, middle panels) and snow depth (mm, lower panels) in the main study area (Kauhava, left) and Ähtäri, 100 km south-east of the main study area (right) (Finnish Meteorological Institute 1994 and unpublished data).

were initiated in the late 1960s and early 1970s, the proportion of old-growth forest could be estimated to be as high as 10% (Fig. 3.6). Agricultural farmland (arable fields and pasture) covers 25% of the area (Fig. 3.7), clear-cut and sapling areas 6% (Fig. 3.8), peatland bogs 2% (growing stock volume <12 m³/ha; Fig. 3.9), other areas (settlements, roads, peat production, etc.) 3%, and water (lakes, rivers, creeks, etc.) 2% of the area (Fig. 3.10).

    Farmland areas predominate in the western part of the study area, whereas the eastern parts are forested, with scattered open peatland bogs (Fig. 3.2). The main crops cultivated in the fields are oats, barley and hay (mostly timothy, *Phleum pratense*) for cattle (mainly

Fig. 3.6.   Old-growth forests nowadays comprise less than 1% of the study area. The forest nature reserve at Passinmäki, Kauhava covers some 40 ha and is restricted to clear-cut areas on the eastern side. Photo: Erkki Korpimäki. (For colour version, see colour plate.)

Fig. 3.7.   Main crops cultivated on agricultural fields are oats, barley, potatoes and hay. Hay fields are good habitats for field and sibling voles. Photo: Erkki Korpimäki. (For colour version, see colour plate.)

dairy cows) and potatoes for humans (Fig. 3.7). Although most farms in the area have been traditionally small in comparison with, for example, those in southern Sweden or Denmark, agricultural practices have gradually intensified in Finland, including our study area, from the 1960s onwards. The main intensification happened, however, when Finland joined the European Union in 1995. As a result, the proportion of

Fig. 3.8.   Clear-cut and sapling areas cover 6% of the study area. Recent clear-cut area in the picture. Photo: Pertti Malinen. (For colour version, see colour plate.)

Fig. 3.9.   Wet peatland bog with small pine trees is called '*räme*', and treeless even wetter peatland bog is called '*neva*' in Finnish. Ympyriäisneva, Kauhava. Photo: Erkki Korpimäki. (For colour version, see colour plate.)

abandoned fields has decreased substantially, and most fields are now cultivated with oats, barley, potatoes and sometimes wheat, while the proportion of hay fields and pastures has decreased, along with the decreasing number of dairy farms. Most fields are also drained, which decreases winter habitats for voles, which in agricultural fields are nowadays largely confined to narrow grassy strips in the vicinity of ditches. The study area shows a productivity gradient from the western agriculture-predominant areas to

Fig. 3.10.   Lakes, rivers and creeks cover 2% of the study area. The creek Mustalamminluoma, Kauhava is flooding in April. Photo: Erkki Korpimäki. (For colour version, see colour plate.)

Fig. 3.11.   Pine (*Pinus silvestris*) forests cover approx. 65% of our study area. These pine forests mainly grow on dry, less productive acid soils. Photo: Erkki Korpimäki. (For colour version, see colour plate.)

barren hinterlands distant from farmland. Human settlement is mainly scattered close to the farmland areas and the density of settlement is highest in the western farmland areas. Overall, approx. 15 000 people are living within the study area.

The majority (approx. 65%) of the forests in our study area are pine-dominated areas (Fig. 3.11). These pine (*Pinus silvestris*) forests mainly grow on dry, less productive acid soils or in the surroundings of open peatland bogs, called '*neva*' in Finnish (Fig. 3.9),

Fig. 3.12.   Spruce (*Picea abies*) forest with some small deciduous trees, birch (*Betula* spp.) and aspen
(*Populus tremula*), scattered in between the spruces. Photo: Erkki Korpimäki. (For colour version,
see colour plate.)

which are too wet for the growth of trees – only the pine tree is able to grow in these less
wet surroundings, although very slowly. This kind of wet peatland bog with small slow-
growing pine trees is called '*räme*' in Finnish (Fig. 3.9). The rest of the forests are spruce
(*Picea abies*) forests (>30%), and in some areas deciduous trees (birches *Betula* spp. and
aspen *Populus tremula*) are scattered in between the spruces, sometimes even forming
small birch- or aspen-dominated patches (Fig. 3.12). Spruce, birch and aspen forests
grow on luxuriant, less dry soils.

   In our study area, boreal owls avoided breeding in boxes that were mainly located in
extensive, uniform forests dominated by pine (Table 3.1). Moreover the proportion of
agricultural land was low and that of peatland bogs high in the vicinity of rarely occupied
boxes. Territories of boreal owls were graded according to the frequency of nesting attempts
during the first 10-year period (originally 1977–86; see Korpimäki 1988d): grade 0 = no
nests, 1 = one nest, 2 = two nests, 3 = three nests, 4 = four nests, and 5 = between five and
nine nests. This method was first used by Newton and Marquiss (1976) to grade territories
occupied by sparrowhawks, *Accipiter nisus*, in Scotland, and has since been used for a
number of other birds of prey (see review in Sergio and Newton 2003). This territory
grading method appears not to be entirely independent of the occupant quality, because
high-quality individuals tend to occupy better territories than lower-quality individuals do.
Therefore, the suitability of this method was analysed for boreal owls in our study area. The
percentages of various habitats within one square kilometre of all the nest-boxes used in
the territory were calculated from landscape maps (scale 1:20 000) and from personal
records in the field. Thereafter, the mean percentages of different habitats were calculated
for each territory, and these means were used when analysing the habitat distribution of
territory grades (Korpimäki 1988d). Six main habitat categories were differentiated: pine
forest, spruce forest, agricultural land, peatland bog, water-body and inhabited area. The

Table 3.1. Habitat distribution in boreal owl territories of different grades. The figures show the mean percentages (± s.d. = standard deviation) of different habitat categories within 1 square kilometre of the nest-boxes of the territories. Pooled data from 1977–86 (Korpimäki 1988d).

	Grade of territory[a]					
	0	1	2	3	4	5
Pine forest[b]	65.7 ± 13.7	55.4 ± 18.5	50.4 ± 22.3	50.3 ± 17.3	27.7 ± 17.3	24.4 ± 17.6
Spruce forest[b]	20.5 ± 13.2	17.9 ± 11.3	20.3 ± 16.2	21.5 ± 11.8	32.0 ± 15.1	35.6 ± 15.0
Agricultural land[b]	7.6 ± 9.6	23.2 ± 13.2	26.9 ± 20.8	25.7 ± 16.6	39.4 ± 17.0	38.3 ± 16.8
Peatland bog	4.8 ± 9.3	2.3 ± 7.8	0.4 ± 1.3	0.7 ± 1.5	0.6 ± 2.2	0.8 ± 2.2
Water-body	1.4 ± 3.8	0.6 ± 2.9	1.6 ± 5.0	1.0 ± 3.4	0.0 ± 0.0	1.0 ± 4.0
Inhabited area	0.0 ± 0.0	0.6 ± 1.8	0.4 ± 1.3	0.7 ± 1.9	0.3 ± 1.1	0.0 ± 0.0
No. of territories	29	33	18	24	13	16

[a] Territories were graded according to frequency of use in 1977–86: grade 0 = not used for nesting, 1 = used in 1 year, 2 = used in 2 years, 3 = used in 3 years, 4 = used in 4 years, and 5 = used for nesting in 5–9 years.
[b] For pine forest, spruce forest and agricultural land, the trend of values in different territory grades was significant using Spearman rank correlation: $r_s = -0.53$ for pine forest, $r_s = 0.34$ for spruce forest and $r_s = 0.57$ for agricultural land (df = 131 and $p < 0.001$ in each case).

analyses revealed that there were also some significant differences in the habitat distributions of territory grades 1–5. As expected, the proportion of pine forest decreased, and the proportions of spruce forest and agricultural land increased, with increasing grade of territories (Table 3.1). In contrast, there were no obvious trends in the small proportions of peatland bogs, water-bodies and inhabited areas between territory grades 1–5.

3.3. Foods in the main study area

The most important prey groups in the diet of boreal owls in our study area are voles of the genera *Microtus* (the field vole *M. agrestis* and the sibling vole *M. rossiaemeridionalis*) and *Myodes* (the bank vole *M. glareolus*) and shrews (mainly the common shrew *Sorex araneus* and, to a smaller extent, the lesser shrew *S. minutus*) and small birds (see Sulkava P. and Sulkava S. 1971, Korpimäki 1981, 1988b) (Fig. 3.13). The two *Microtus* species mainly inhabit fields, whereas the bank vole and common shrew are the most common small mammals in woodlands of northern Europe (Henttonen et al. 1977, 1989, Korpimäki 1981, Hansson 1983). Both spring and autumn densities of bank voles are about three times higher in spruce forest than in pine forest, and a similar – but smaller – difference is also evident in the densities of common shrews (Korpimäki 1981). The mean densities of small birds are also higher in spruce forests than in pine forests (331 versus 260 breeding pairs/km^2; Korpimäki 1981). In addition, dense spruce forests may offer better protection than sparsely growing pine forests against predators such as pine martens (*Martes martes*) and larger birds of prey. Thus, spruce forest provides a more productive, and possibly safer, nesting habitat for boreal owls than pine forest. The high percentage of spruce forests and low percentage of pine forests in good territories is

Fig. 3.13.   Bank voles and common shrews mainly occupy woodland in our study area. Male boreal owl about to deliver a bank vole (upper panel) and a common shrew (lower panel) to the nest-box. Photo: Benjam Pöntinen. (For colour version, see colour plate.)

thus not unexpected. Our results are consistent with those on sparrowhawks, in which the occupancy of nesting territories and the nest spacing are positively related to the densities of the available bird prey, which is in turn partly dependent on soil-productivity and elevation (Newton et al. 1986).

Because *Microtus* voles are about twice as heavy as bank voles, and about four times as heavy as common shrews, they were the preferred prey of boreal owls in our study area (Korpimäki 1981) (Fig. 3.14). These voles mainly occupy open habitats (i.e. agricultural land and clear-cuts) where snow melts earlier than in woodland. Breeding starts when snow is still melting (Korpimäki 1987f). At that time, boreal owls mainly hunt at the edge zones of fields and clear-cut areas, where the availability of voles is better than in woodland. Therefore it is not surprising that there was a positive correlation between the proportion of agricultural land and the occupancy of boreal owl territories. On the

Fig. 3.14.   Two vole species of the genus *Microtus*: the field vole (illustrated) and the sibling vole, occupy open habitats (i.e. agricultural land and clear-cut areas). They are the preferred prey of boreal owls in our study area. Photo: Benjam Pöntinen.

edges of woodland, the densities of small mammals and birds are generally higher than in the middle of woodland (see e.g. Hansson 1982, Helle 1986). The length of edge habitat in territories increases with the proportion of agricultural land, and this further improves the positive correlation between the territory grade and the proportion of farmland. Because the fields probably lie in the most productive places in the local landscape, the proportions of agricultural land and spruce forest in the owl territories may be partly intercorrelated. In peatland bogs, the densities of breeding small birds are much lower than in woodland (Korpimäki and Rajala 1985), and the same is true for voles (Henttonen et al. 1977, Heino 1985). The food supply of boreal owls is poor in peatland bogs, which explains its higher proportion in areas identified as unsuitable for breeding.

3.4.   Explorative studies, breeding density and performance

At the beginning of the explorative studies, we were not sure how to carry out intensive, and possibly disturbing, investigations at boreal owl nests. Because all the future studies would have been worthless if the owls had responded adversely to disturbance, we were careful during the first years to ensure that we did not induce any nest desertions. The boxes were not visited during the expected egg-laying periods, and only in the late incubation period were the numbers of eggs counted. Only after the explorative study period during 1966–9, did we also start to visit nests during the egg-laying and early incubation periods and also began to ring the young and parent females in the nests. At that time we discovered that incubating females can be caught by hand, and the number of eggs and prey items in the owls' food stores can also be identified without any risk of nest desertion. We also found out from other experienced Finnish owl ringers that female parents can be captured and ringed during the early incubation period, and became aware that boreal owls are extremely good study objects in the sense that they are not particularly sensitive to disturbance at their nests.

We collected data on breeding density, breeding performance and breeding success of boreal owls by inspecting the boxes and all known natural cavities each year. The first inspection round was usually made in late March to late April, depending on the timing of breeding of owls, and the second inspection round was performed in mid-May to early June. One checking round of approx. 500 boxes and cavities of owls usually took about 7–10 days, with a two-man patrol in a four-wheel-drive car working for at least 12 hours a day, because each box and cavity needed to be checked by climbing the tree. The roofs of our nest-boxes can be opened, and this facilitates a closer look at the contents of the box. Environmental conditions in early spring are usually very variable and often also quite harsh. The ground can be snow-covered, mostly with soft wet snow, but sometimes hard frosted snow. Alternatively, melting snow produces widespread flooding in the flat countryside, making the roads muddy and the terrain so wet that the forest floor will not bear the weight of a man and even a 4-wd car often gets stuck. In late March and early April of the 1960s to the late 1980s, there was normally so much snow in our study area that it was faster to ski than to walk to the boxes, but over the last 20 years the use of skis by box-check patrols has largely decreased. (In Finland, and elsewhere in northern Europe, cross-country skiing has been the traditional way to move around forests, for example when hunting forest grouse and fur-bearing animals in winter.) In the context of box inspections, we have also made an attempt to find new natural cavities made by woodpeckers; particularly at those sites where the calls of black woodpeckers have been heard.

During 1966–2010, we found 1605 nests of boreal owls in the main study area (Table 1). A nest was defined as when at least one egg was laid in the box. Found nests were checked during the incubation period, and at the time of estimated hatching, as well as in the mid and late nestling periods, to record the final number of eggs laid (i.e. the clutch size), as well as the number of young hatched, ringed and fledged. The final inspection of the nests was made after the breeding season to observe any of the young that had died in the nest after ringing. The date of laying of the first egg was obtained mainly by back-dating from hatching dates using 29 days as the incubation period for the first-laid egg (Korpimäki 1981). If young were already hatched during the second inspection of the nest, their wing lengths were measured and hatching dates could then be estimated on the basis of growth curves of wing lengths of known-age young (Fig. 2.7). Usually there was only one nesting attempt of boreal owls per box each season, but a few re-nesting attempts were recorded in cases where the first clutch had been preyed upon before the second was laid. Clutches were classified as preyed upon when broken eggs, shells, killed nestlings or adults, or freshly cut feathers were found. Therefore, clutches that were taken before the first checking of a nest were also recorded, minimising underestimation of nest predation by failure to detect nests preyed upon in the early phase of a breeding attempt (Sonerud 1985a). A nesting attempt was defined as successful if it produced at least one fledgling.

To maintain the network of approx. 500 boxes for boreal owls in our study area, a two-man patrol had to work for one week each autumn to replace some 20–40 boxes that were lost in clear-cutting or were cracked and decayed. All in all, it can be estimated that we visited nests and boxes of boreal owls approx. >41 000 times during 1966–2009. Using an average of 15 minutes per box visit means >10 000 hours fieldwork, equalling >250 forty-hour weeks and totalling approximately 5 years of fieldwork altogether!

3.5.   Individual-level population studies and radio-tracking

Females of 1224 nests during 1976–2009 and males of 957 nests during 1979–2009 were trapped and ringed in our study area (Table 1). During 1980–2001 and 2005–9 most (80–100%) of the breeding females of the study population were ringed or re-trapped, and during 1981–96, 1999–2000 and 2007–9, between 60% and 100% of the breeding males in the study population were ringed or re-trapped. Females were caught inside their nest-boxes by closing the entrance hole by hand or rod while they were incubating eggs or brooding young. Males were trapped by fixing a box trap with a swing door to the entrance hole of the nest-box when the oldest owlet was 2–3 weeks old (Korpimäki 1981). This swing-door trap was modified from the trap developed for trapping male Ural and tawny owls (Saurola 1987) and has also been successfully used for other box-nesting birds of prey including Eurasian kestrels (Korpimäki 1988e, Korpimäki and Wiehn 1998) and saw-whet owls (Hinam and St. Clair 2008). About 1–2% of males were reluctant to enter the box traps, and were therefore trapped using the bow net trap developed and described by Altmüller and Kondrazki (1976). Parent owls were ringed with metal rings or, if already ringed, their ring number was recorded, the owls were aged by checking the moult of primary and secondary feathers (see Chapter 2.2), their body mass was taken to an accuracy of 1 g, and wing, tail, bill and tarsus lengths were measured (Fig. 3.15). In addition, in the years 1969–72 some of the fledglings, and from 1973 onwards all the fledglings, were ringed, normally at the age of 24–25 days of the oldest owlet of the brood (a total of 4158 fledglings during 1970–2009). From 1984 onwards, body mass and wing length of most of the fledglings were measured at the time of ringing.

Home range size was determined by tracking breeding male boreal owls using radio-telemetry. Fortunately, boreal owls are robust to trapping, handling and radio-tagging (Korpimäki 1981, Hayward et al. 1993, Eldegard and Sonerud 2009), making them suitable for individual-level population and telemetry studies. Eleven males in the breeding

Fig. 3.15.   The wing length was taken from carpal joint tip to tip of longest primary feather with flattening. Photo: Pertti Malinen.

season of 2005 and 15 males in 2009 were captured when the oldest nestling was at least 13 days old, ringed and fitted with tail-mount PD-2 transmitters (Holohil Systems; weight = 2 g, about 2% of the male mass). After release, males were given one full night to acclimate to the transmitter and were then tracked for three focal bouts: two focal bouts during the nestling and one during the early fledgling period (within 2 weeks after fledging). Each focal bout began after sunset and lasted a minimum of 2 hours. Within this period, a single observer continuously followed an owl using a three-element 'Yagi' antenna plugged into a hand-held VHF receiver (Biotrack Ltd.) to record the male's movements. Owls were followed mostly on foot, or with the aid of a car along forest tracks and minor secondary roads when necessary, in order to get as close as possible to the individual followed. Locations (fixes) were recorded every 8 minutes on average. Individual locations were estimated in the field and marked on base maps of the top-ography (1:20 000; provided by the National Land Survey of Finland) using a combination of signal strength, knowledge of the terrain and observer experience (O'Donnell 2000). Locations were confirmed afterwards based on direction (taken with a compass to the nearest degree), observer location (marked with a Garmin GPS), and the linear distance to the owl estimated from signal strength, based on previous experimental calibrations in the field (Bernard and Fenton 2003). Location accuracy was ±100 m. If the observer felt that there was poor resolution of a fix (signal faint, i.e. high gain, direction uncertain or signal bounce) then the fix was omitted from the analysis.

Home range size of males was estimated using Ranges VI software (Bernard and Fenton 2003) with two methods. First, nightly home ranges of hunting males were estimated with the 100% minimum convex polygon (MCP) method (Kenward 2001). Second, fixes pooled from the locations of hunting males at night and daytime roosts during the breeding period were used to compute a kernel estimator of 95% of the fixes (Seaman and Powell 1996). We used groups of auto-correlated night-time fixes pooled with independent daytime roost fixes in order to retain more information on individual movements (Hinam and St. Clair 2008, Santangeli et al. 2012). A smoothing parameter for the home range contours was calculated using the least squares cross-validation method (Seaman and Powell 1996). As the value obtained by this method depends on home range size and number of fixes (Kenward 2001), which ranged from 32 to 60 fixes per owl in our data from 2009, we first calculated a smoothing parameter for each individual and then the median of these values. The obtained median value was used as the smoothing parameter to estimate the home ranges, which were then comparable between males (Kenward 2001, Hinam and St. Clair 2008, Santangeli et al. 2012).

3.6.   Prey number and diet composition

We used two methods to study the number of prey delivered to nests and the diet composition of boreal owls: (1) identification of prey items stored in the nests, and (2) identification of prey remains found in the pellets and collected from nest-boxes. During nest-box inspections in the explorative study period, we found that the nest-holes of boreal owls regularly contain a food store of prey animals. Therefore, the number, species and state (whole vs. partially eaten) of prey animals in the food stores was regularly recorded on visits to all the nests of our study population from 1973 onwards (a total of 12 554 prey items identified in food stores of nests during 1973–2009; Korpimäki 1988b and unpublished data). To avoid counting any individual in a food store twice in nests that were visited frequently, the tail of small mammals or a claw of birds was cut off. Boreal owls always begin to eat their prey from the head (Scherzinger 1971, Korpimäki 1981, 1988b). This method is useful in the sense that, besides the prey species, one can obtain information of the body mass and other measures, sex, age, reproductive status and body condition of prey items killed by boreal owls (Korpimäki 1981, Koivunen et al. 1996a, 1996b, 1998a, 1998b).

A layer of pellets and other prey remnants accumulates at the bottom of the nest-hole of boreal owls during the nestling period (Sulkava P. and Sulkava S. 1971, Korpimäki 1981). Each food sample collected consisted of pellets and other prey remains from a nest-hole. It is important to collect prey detritus layers from as many nests or roosts as possible, because prey items found in a nest do not form an independent observation unit necessary to run most statistical tests. Because earlier experience on hunting in the same territory can probably affect prey selection, and thus prey items delivered to the nest, an independent observation unit for statistical tests would be a pair of owls or an individual owl (Korpimäki et al. 1994).

The number of nests where food samples (prey detritus layers) were collected totalled 24 during 1966–72 (a total of 1560 prey items), and 392 food samples containing 27 759 prey items identified from pellets and other prey remains during 1973–2009 (Sulkava P. and Sulkava S. 1971, Korpimäki 1981, 1988b and unpublished data). In 'poor vole years', when the number of nests was low, the food samples were collected from the majority of nests of the population, whereas in 'good vole years' the sampled nests were randomly chosen. These food samples contained mainly prey items brought by the parent owls to the nest during the last 2 weeks of the nestling period. Before that, the females remove prey items from the nests (Korpimäki 1981). Most of the prey is brought by the males, because many females do not take part in feeding at all, or take part in feeding only at a low rate for less than 2 weeks at the end of the nestling period (Korpimäki 1981, Zárybnická 2009b, Eldegard and Sonerud 2010). The mass of these prey detritus layers was closely positively correlated with the number of fledglings present in the nest (linear regression, $r^2 = 0.934$; Whitman 2008). In addition, the number of prey animals delivered by males to their nests during a 4-hour nightly observation period when the young were 2 weeks old was closely positively related to the total number of prey items found in prey detritus layers after the young had fledged (Hakkarainen and Korpimäki 1994a). In addition, boreal owls regularly use boxes, or shallow cavities in trees, for resting in winter, and pellets and other food remains accumulate at the bottom of the boxes and cavities. We collected food samples from 117 roosting holes and nest-boxes during 1970–89 to study diet composition in winter (a total of 819 prey items; Korpimäki 1981, 1986e and unpublished data).

Pellets and prey remains in food samples were dried, and bones, feathers and scales were then separated. Hairs from large samples were dissolved in sodium hydroxide (according to Degn 1978). Small mammals were identified according to Siivonen (1974; see also Uttendörfer 1952, Sulkava P. and Sulkava S. 1971). The numbers of individuals were counted on the basis of the mandibles. The field and sibling voles were sometimes difficult to differentiate because the joint branch of the mandible was broken.

Therefore, not all these individuals could be identified to species level. The identification of water voles *Arvicola terrestris* and other larger mammals was mostly based on leg bones and reference material in the Zoological Museum, University of Oulu, Finland. Birds were usually identified by comparing the humeri and other larger bones, and beaks and feathers, with reference material derived from road kills and other carcasses of known bird species. The identification of insects was as described by Itämies and Mikkola (1972) and Itämies and Korpimäki (1987).

An analysis of pellets and prey remnants collected from and in the vicinity of nests and roosts is a traditional, well-developed and tested method to study the diet composition of birds of prey. It has been widely used for more than 60 years after the pioneering investigations of Errington (1930, 1932) and Craighead and Craighead (1956) in North America, and Uttendörfer (1952), Sulkava S. (1964) and Sulkava P. (1965) in Europe. In general, pellet analyses are more reliable for dietary investigations of owls, but less reliable for diurnal raptors because many of the latter species dismember their prey prior to swallowing and might not ingest all portions (Craighead and Craighead 1956). Diurnal raptors also digest bones to a greater extent than owls do (Duke et al. 1975). It was earlier believed that only young owls digested bones to any great extent (Errington 1932), but there are also some reports of noticeable bone loss attributed to digestion in adult owls (Raczynski and Ruprecht 1974, Lowe 1980). Nevertheless, Mikkola (1983) found a very close correlation between food eaten and prey remains found in pellets. In addition, März (1968), Marti (1987) and Marti et al. (2007) have written comprehensive reviews on the methods, and their possible strengths and weaknesses, for studying the diet composition of birds of prey in general. For example, Herrera and Hiraldo (1976), Jaksic and Jiménez (1986), Mikkola (1983), Marti et al. (1993a, 1993b), and Korpimäki and Marti (1995) analysed and summarised the food niche metrics and trophic structures of owls and diurnal raptors in Europe, North America and South America.

We also wanted to discover any possible weaknesses in the data for diet composition of boreal owls analysed on the basis of pellets and other prey remains found in nests and roosting sites.

First, we tested this method by comparing the diet compositions determined by pellet analyses with the prey items identified and counted in the food stores of owls in nest-boxes. These two methods give quite a similar picture of the diet composition and its between-year variation. However, most prey remains and pellets accumulate at the bottom of the nest-boxes only during the last week of the nestling period, when the female no longer broods the young and cleans the nest (Korpimäki 1981). On the other hand, most prey items in the food stores were found during the egg-laying, incubation, hatching and early nestling periods (Korpimäki 1987d). Therefore, the diet determined by pellet analyses reflects diet composition at the end of the nestling period, and the diet determined by prey items found in food stores reflects the diet composition 2–6 weeks earlier, which largely explains the inter-method differences found in diet composition (Korpimäki 1981, 1988b). Results of earlier studies also indicate that pellet analysis is an admirable method for studying the diet composition of small and medium-sized owls, including, in addition to boreal owls, Eurasian pygmy owls (Kellomäki 1977), barn owls *Tyto alba* (Taylor I. 1994), long-eared owls (Korpimäki 1992a) and short-eared owls (Korpimäki and Norrdahl 1991b).

Second, working in partnership with technical genius Reijo Passinen, we designed, built and employed a camera recording apparatus to monitor three nests of boreal owls (one nest in each year from 1977 to 1979; Korpimäki 1981). There was a wooden box on the back wall of the nest-box at the level of the entrance hole in the front wall, in which we placed an 8-mm film camera taking single pictures and a flash light. A clock originating from a car was attached onto the front wall. The camera was wired to a light-sensitive cell (1977-8) or mechanical switch (1979) mounted at the entrance hole of the nest-box. The latter was found to be better because it needed current only when releasing the camera and flash light. The current for the camera was taken from a 6-volt accumulator, and that for the watch, flash light and releasing mechanism from a 12-volt accumulator. This camera recording apparatus was one of the earliest pieces of equipment employed to record food provision rates and food items provided by parent birds; particularly by parent birds of prey to their offspring (Korpimäki 1981; see Reif and Tornberg 2006 for a review).

In the pictures taken with this apparatus, the facial disc of the owl and the clock were visible. The prey delivered to the nest was also recorded, because although boreal owls carry the prey to the vicinity of the nest in their claws, they then transfer the prey item to their bill in order to deliver it to the nest (Norberg 1964, 1970, Korpimäki 1981, Hakkarainen and Korpimäki 1994a). The prey species (a total of 370 during 1977–9) were determined from the pictures on the film after using apparatus designed to cut 8-mm film into single frames. Comparisons of the diet composition determined by pellet analyses and film recordings from the same three nests revealed that overall differences in the diet composition determined by the two methods were relatively small. Pellet analyses tended to underestimate the proportion of small mammals in the diet by an average of 8%, while the proportion of birds tended to be overestimated by 8%. Among the mammalian prey, pellet analyses tended to slightly underestimate the proportion of *Microtus* voles (4%) and shrews (2%), whereas the opposite result was found for bank voles (a 2% overestimate) (Korpimäki 1981). Comparisons of the two methods also revealed that in the barn owl, contents of pellets and prey delivered to nests as recorded by continual photographic monitoring during the same period gave a very similar picture of the diet composition (Taylor I. 1994).

During the last 10 years, the available techniques for recording prey delivery rates and other behaviours of day-active birds at their nests have developed considerably, including the use of digital video recording (reviewed by Reif and Tornberg 2006); nevertheless, recording techniques are confronted with a difficult challenge in the darkness of night. One of the most recent developments in this technique has been used for studying boreal owls (Zárybnická 2008, Zárybnická et al. 2009). The equipment consists of a digital camera, a chip reader device, a movement data-logger, a movement infrared detector, and infrared lighting. The camera was installed inside the nest-box opposite the entrance hole and the shutter was released by the infrared detector, which was sensitive to movements at the entrance hole to the box. The time of detection was recorded by the movement data-logger and one to three photos were taken for each event. At night the entrance hole was illuminated by infrared diodes at the moment when photos were being taken by the camera. Parent owls and owlets were marked with chip rings. A chip reader device fixed on the entrance hole detected and archived all movements of chips in the entrance hole of the nest-box. Using this equipment, it was possible to record visits of male and female parents to the nest-box and to determine whether the parents arrived with or without prey. This equipment was used at 12 nests in the Ore Mountains, north-western Czech Republic (6 nests in 2004 and 6 nests in 2006) and at 9 nests in our study area in 2005 (Zárybnická 2009a, 2009b, Zárybnická et al. 2009). Pellet analysis appeared to under-estimate the proportion of field and bank voles and overestimate the proportion of wood mice in the diet (Zárybnická et al. 2011), but this difference might also be explained by the fact that pellets and prey remains accumulate at the bottom of the nest-boxes during the last 2 weeks of the nestling period, whereas prey deliveries were recorded during the 4 weeks of the nestling period.

3.7.   Estimation of the available food

To estimate small mammal abundance, we conducted snap-trappings in early May and mid-September at two sites within the boreal owl study area. The study plot at Ruotsala is situated in the core of the main study area, where we began to erect nest-boxes in the mid-1960s (Fig. 3.2), and was snap-trapped during 1973–2009. The study plot at Alajoki is situated in the south-western part of our owl study area and is about 14 km from Ruotsala. There the snap-trappings were done during 1977–2009. We sampled the main habitat types used by foraging boreal and other owls (a cultivated field, an abandoned field, a spruce forest and a pine forest) at both sites by means of 50–100 metal mouse-type snap-traps per plot (as commonly used in Finland and suitable for *Microtus* voles, bank voles, harvest mouse *Micromys minutus*, house mouse *Mus musculus*, and shrews including the common shrew, lesser shrew and water shrew *Neomys fodiens*) (Korpimäki and Norrdahl 1991a, 1991b). The traps were set at intervals of 10 m apart along runways of small mammals and were checked once a day for 3–4 days. The mouse snap-traps were baited with mixed-grain bread (i.e. bread made of a mixture of rye and wheat flours), which has been shown to be an appropriate bait for voles, mice and shrews (see Koivunen et al. 1996a, 1996b). In addition, 20–50 Finnish metal rat-type snap-traps, which are suitable for trapping larger water voles and brown rats *Rattus norvegicus* (Korpimäki et al. 1991, 2005a), were used at both Ruotsala and Alajoki during 1981–2009. The rat snap-traps were baited with dried apple and mixed-flour bread, which are both appropriate baits for water voles (Myllymäki et al. 1971). We pooled the results from the 3–4-night trapping periods for each species separately and standardised them to the number of animals caught per 100 rat-trap nights for water voles and brown rats, and per 100 mouse trap nights for all other small mammal species. As the synchrony of trap indices between Ruotsala and Alajoki, which are only 14 km apart, was high (cross-correlation coefficient with lag 0 for *Microtus* voles = 0.68, for bank voles = 0.62; see also figure 1 in Korpimäki and Wiehn 1998), the species-wise density indices used in all the analyses in this book and presented in Fig. 1.1 are the means of the two sites in the study area, apart from 1973–6 when we trapped small mammals at the Ruotsala study plot only.

   According to long-term snap-trapping data from our study area, the mean body mass of both *Microtus* species voles is approximately 26 g (field vole 28 g, sibling vole 25 g; Norrdahl and Korpimäki 2002a). Bank voles (18 g), mice (harvest mouse 7 g and house mouse 18 g) and shrews (common shrew 8 g, lesser shrew 3.5 g and water shrew 14.5 g) are substantially smaller than the main prey (Norrdahl and Korpimäki 2002a). Water voles (adults 200 g, young 100 g; Norrdahl and Korpimäki 2002a), in turn, are much larger than the main prey of avian predators in our study area, and boreal owls are able to kill and carry only the young water voles to their nests.

To study the assemblage and density of breeding birds available for hunting boreal owls at the Ruotsala study plot, we carried out 122-km-long line censuses in late May to mid-June during 1974–8 (Korpimäki 1981). Censuses were carried out each year on the same lines situated in spruce-dominated and pine-dominated forests and in agricultural fields, and covered the main habitat types of the study area in similar proportions. The breadth of the main census belt was 50 m, but for estimating the densities of less frequent bird species, 100-m and 200-m-wide lines were also used. The censuses were held from sunrise (at approx. 03:00 hours) until 07:30 hours. According to different studies, 25–>80% of the bird numbers revealed by more accurate mapping methods are recorded in single-line censuses, and their efficiency is dependent on the habitat (Merikallio 1958, Järvinen O. 1979, Helle 1986). Therefore, the results do not show the actual densities of breeding birds, but are probably sufficient to reveal the assemblage of bird species available for hunting boreal owls in our study area (Korpimäki 1981). A similar census method has also been used to study the assemblage and density of breeding birds available for hunting pygmy owls (Kellomäki 1977).

3.8.   Point stop counts of hooting owls

Point stop counts of hooting owls with playbacks have traditionally been used in attempts to survey population densities of secretive night-active owls (e.g. Lundin 1961, Lundberg 1978, Korpimäki 1981, Hayward et al. 1993, Kloubec and Pacenovsky 1996, Lane et al. 2001, Duncan et al. 2009), but the reliability of this method in monitoring breeding and non-breeding male owls has not often been tested against field data. Non-breeding floaters are a hidden, and largely unstudied, proportion of populations of birds of prey, including both night-active owls (e.g. Rohner 1996) and day-active raptors (e.g. Kenward et al. 2000). Therefore, we attempted to find out whether listening with the point stop method and playback technique could be an appropriate method to study the numbers of non-breeders in the boreal owl populations.

We modified the point stop method with playbacks suggested by Lundberg (1978) and Holmberg (1979) for boreal owls and conducted listening surveys during 14 springs (1979–92) using standard methods. We attempted to locate occupied nest-sites by listening to display calls on calm rainless nights in mid-February to late March, when male boreal owls advertise their primary nest-holes (Carlsson B.-G. 1991, Korpimäki 1991b). During these nights, ambient temperatures were usually less than −10°C, because snowfall, rain and low temperatures decrease the hooting activity of boreal owl males (Holmberg 1979, Hongell 1986). We drove slowly by car to each point (i.e. within 100–300 m of a nest-site), stopped, and opened and closed the door as quietly as possible. At each point, we first listened to the display calls of owls for 2.5 minutes, then we played a recording of staccato song of male boreal owls for 30 seconds. Finally, we listened for another 2.5 minutes and made a note of whether owls hooted before or after the playback. Thereafter, the point was left and we drove slowly to the next point. If a hooting owl was still audible at the next point, it was not re-counted as a new owl. Between-individual differences in the hoots also helped to avoid re-counting of the same individual owls (see Chapter 2.3).

The dense network of minor roads in our main study area made it possible to screen most of the study area once. The audibility of the primary song of boreal owls to humans is usually 1.5–2.0 km (minimum 0.7–1 km in the worst conditions; see also Holmberg 1979, Lane et al. 2001). Therefore, the point stops were arranged so that the distance between them and each nest-hole of the area was <1 km. We also avoided surveying the same route more than once a season, because all the early-breeding males (67% of total number of males) in a good vole year continued to sing at secondary nest-holes after attracting and installing primary females. In addition, all bigynous males were also singing at other nest-holes later, in order to attract a third mate (Carlsson B.-G. 1991). In fact, polygynous males are polyterritorial, with two nests – on average 1.3 km apart – and empty nest-holes, or even other breeding males in between (Korpimäki 1991b; see also Carlsson B.-G. et al. 1987). Therefore, many males singing at nest-holes from mid-April onwards are probably already paired and breeding males, which could then be surveyed twice more in later point stop counts. We initiated point stop counts 1–2 hours after sunset and continued until the early hours of the morning, sometimes up to 2 hours before sunrise, because boreal owls usually start active hooting 1 hour after sunset and continue until sunrise in northern Europe (Wahlstedt 1959, Lundin 1961, Hongell 1986).

A total of 394 male boreal owls were recorded on 3680 stops at nest-holes on routes extending 7546 km during 1979–92 (Table 3.2). Later nest-box inspections revealed that 53% of these hooting males with suitable nest-holes remained bachelors, i.e. did not attract a mate and start to breed (Korpimäki 1991b). Therefore, the point stop method with playbacks emerged as a particularly appropriate method for revealing the number of bachelor boreal owl males in the study area. Of the hooting males, 73% were recorded to hoot without playback, and only a minority (27%) responded to playback only (Table 3.2). It has been reported that 76% of males also hooted spontaneously in the Sumava Mountains, Czech Republic (Kloubec and Pacenovsky 1996). Therefore, although playbacks substantially increased the efficiency of point stop counts, one can still question whether the use of playbacks is necessary, considering that they might also attract enemies of boreal owls. Sometimes, Ural owls responded to the playbacks of boreal owl primary song, and there is also at least one record in Finland where an Ural owl attacked the magneto-phone playing the song of pygmy owls (Hämäläinen 1979). In addition, boreal owls sometimes respond to the song of the smaller pygmy owls. Because boreal owls are food competitors and enemies of pygmy owls (Korpimäki 1981, Suhonen et al. 2007), the latter may be intimidated by the playback hoots of boreal owls. Therefore, our suggestion is that the use of playback should be limited to scientific owl studies in which it is really the only option in order to gather important information on the occurrence of floaters which will be used to manage and conserve owl populations.

Table 3.2. Number and length (km) of point stop census routes, number of boreal owl nest-boxes surveyed with the point stop method, total number of boreal owl males recorded (before/after the playback), number of males recorded per 10 nest-boxes, and number of bachelor males in the Kauhava region, western Finland, during 1979–90.

Year	No./length of routes	No. of boxes	No. of males	No. of males per 10 boxes	No. of bachelors
1979	26/262	145	7 (6/1)	0.48	6
1980	29/411	221	17 (16/1)	0.77	16
1981	16/424	168	18 (16/2)	1.07	18
1982	28/530	267	12 (8/4)	0.45	12
1983	20/776	334	38 (31/7)	1.14	30
1984	10/363	184	4 (2/2)	0.22	4
1985	7/ 414	185	12 (10/2)	0.65	11
1986	16/595	267	31 (29/2)	1.16	13
1987	19/814	313	11 (7/4)	0.35	8
1988	11/678	344	53 (34/19)	1.54	22
1989	8/ 470	277	90 (58/32)	3.25	40
1990	9/ 595	314	1 (1/0)	0.03	2
1991	7/ 565	313	34 (24/10)	1.09	12
1992	8/ 739	348	66 (44/22)	1.90	13
Total	214/7546	3680	394(286/108)	1.07	209

There was a large (108-fold) between-year variation in the number of male boreal owls recorded per 10 nest-boxes of playback surveys during 1979–92 (range 0.03–3.25, mean 1.07; Table 3.2). The number of calling boreal owls heard per survey kilometre also showed a 12-fold variation in a 4-year study in Idaho, USA (range 0.02–0.12; Hayward et al. 1993). The yearly number of hooting males per 10 boxes was significantly positively correlated with the abundance index of bank voles in the current spring (Spearman rank correlation, $r_s = 0.72$, two-tailed $p = 0.003$, $n = 14$), but not with the abundance index of *Microtus* voles in the current spring ($r_s = 0.45$, $p = 0.11$, $n = 14$). Unexpectedly, the abundance indices of these voles in the previous autumn appeared not to have an obvious association with the relative number of hooting males recorded. However, in Kluane boreal forest, Yukon, Canada, the number of calling boreal owl males peaked at 11 per 24 km^2 of road transect during 1990–6, and numbers were positively related to pooled densities of *Microtus* and *Myodes* voles in the previous summer (Doyle and Smith J. 2001). Instead, the number of hooting males per 10 boxes was a good predictor of the future breeding density of owls, because it was closely positively correlated to the number of boreal owl nests found per 100 nest-boxes in the subsequent nest-box inspections (Fig. 3.16).

3.9.   Long-term study areas of boreal owls in Finland

There are four additional long-term (>35 years) study areas of boreal owls in Finland, where a network of nest-boxes has been maintained, and data on the number of owl nests, number of nest-boxes and nest-holes, clutch sizes and brood sizes have been collected each year for >30 years. Three of these areas (the Seinäjoki, Ähtäri and Keuruu regions)

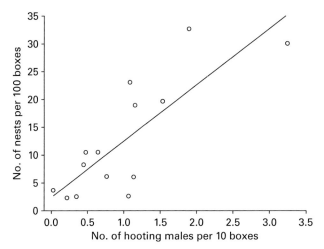

Fig. 3.16.   Number of boreal owl nests found in 100 boxes plotted against the number of hooting males
recorded per 10 nest-boxes during 1979–92 (Pearson correlation, $r = 0.81$, $p < 0.01$, $n = 14$).

have been maintained by the members of the Ornithological Society in Suomenselkä
(Fig. 3.1).

The Seinäjoki region in South Ostrobothnia (approx. 50 km south of our study area;
see Fig. 3.1) consists of flat lowlands where the coverage of forests, agricultural land and
peatland bogs is similar to our study area (Table 3.3a). There has been a network of
approx. 80–120 boxes suitable for boreal owls during 1973–7, and 220–500 boxes
during 1978–2009. Erkki Rautiainen, Jussi Ryssy and Mauri Korpi have managed and
annually inspected the owl boxes.

The Ähtäri region (approx. 100 km south-east of our study area; Fig. 3.1) is mainly
forest-covered area with more lakes and less agricultural fields than in the Kauhava and
Seinäjoki regions (Table 3.3a). Jussi Ryssy, Jaakko Härkönen, Risto Saarinen and co-
workers have maintained a network of 180–340 boreal owl boxes and inspected the nest-
boxes annually during 1973–2009.

The Keuruu region (approx. 180 km south-east of our study area; Fig. 3.1) is also
mainly forest-covered terrain with fewer agricultural fields and peatland bogs than the
other three areas (Table 3.3a). Tarmo Myntti, Kari Myntti, Pertti Sulkava, Risto Sulkava
and co-workers have maintained and inspected 60–120 boreal owl boxes annually during
1973–9; 170–250 owl boxes during 1980–95; and 100–190 boxes during 1996–2009.

In the regions of Ähtäri and Keuruu, the first nest-boxes for boreal owls had already
been erected in the field in 1957. A total of 5 nests of boreal owls were found in 1958, 4 in
1959, 1 in 1960, 5 in 1961, 33 in 1962, 5 in 1963, 6 in 1964, and 26 in 1965 (Korpimäki
1985c). In the Kauhava region, the first nest-boxes for boreal owls were set up in autumn
1965 and have been checked from 1966 onwards (Table 1). In the Seinäjoki region, the
first nest-boxes were erected in 1968 and have been checked from 1969 onwards
(Korpimäki 1981).

There are also marked altitude-related differences in the snow conditions between the
Kauhava–Seinäjoki and Ähtäri–Keuruu regions, along with fundamental differences in

Table 3.3. (a) The coverage of main habitat types in our study area (Kauhava) and in three other long-term study areas (Seinäjoki, Ähtäri and Keuruu) in western Finland. (b) The proportions of various owl species for pooled number of nests and territories recorded in Kauhava, Seinäjoki, Ähtäri and Keuruu regions during 1976–84 (data from Korpimäki 1987b).

(a)	Kauhava	Seinäjoki	Ähtäri	Keuruu
Forest	46%	51%	62%	71%
Agricultural field	26%	28%	10%	6%
Peatland bog	26%	21%	19%	12%
Lake, river, etc.	1%	1%	9%	11%
Altitude	30–120 m	40–100 m	120–200 m	200–300 m

(b)	Kauhava	Seinäjoki	Ähtäri	Keuruu
Boreal owl	42.8%	39.9%	28.0%	27.3%
Eagle owl	13.7%	16.6%	22.7%	5.0%
Ural owl	5.0%	7.2%	32.5%	42.3%
Great grey owl	0.0%	0.3%	0.0%	0.0%
Tawny owl	0.0%	0.0%	1.7%	6.8%
Short-eared owl	25.3%	6.6%	2.6%	0.7%
Long-eared owl	11.4%	8.8%	3.8%	3.8%
Hawk owl	0.8%	0.0%	0.0%	0.4%
Pygmy owl	1.0%	10.7%	8.7%	13.8%
Mean no. of nests per year	53.3	25.2	36.8	30.8
Mean no. of territories per year	39.3	16.2	27.4	31.4

landscape characteristics (Fig. 3.5). In the former (western) regions the mean monthly snow depth in December–March was 23 cm (range 11–32 cm; Fig. 3.5) during 1961–90, whereas in the latter (eastern) regions the corresponding snow depth was substantially higher (mean 39 cm, range 20–51 cm; Fig. 3.5) (data from Solantie 1975, 1977, 2000, Solantie et al. 1996). Even in April, during the breeding season of boreal owls, there was on average a 28 cm snow layer in the two eastern regions, but only a 7 cm snow layer in the two western areas. During November–April of 1977–88, the mean number of days with snow cover was 27.4 per month in the east but only 24.5 per month in the west. The mean monthly ambient temperature in winter (December–February) was −8.3°C in the west and −8.8°C in the east, in spring (March–May) it was +2°C in the west and +1.5°C in the east, and in summer (June–August) it was +14.2°C in the west and +13.8°C in the east (data from the Finnish Meteorological Institute 1994).The amplitude of vole cycles was essentially the same in the Kauhava–Seinäjoki and Ähtäri–Keuruu regions. Populations of *Microtus* voles in agricultural field habitats exhibited a higher degree of spatial synchrony in the two western study areas with more continuous agricultural landscape than in the two eastern study areas with forest-dominated landscape (Huitu et al. 2003b). In contrast, bank vole populations in agricultural fields were more synchronised within the eastern study areas with forest-dominated landscape (Huitu et al. 2003b). This suggests that landscape composition may affect the temporal and spatial dynamics of vole populations, which in turn could modify the population dynamics of vole-eating predators, including boreal owls.

Indeed, the assemblage of owl communities in these four long-term study areas differed substantially during 1976–84 (Korpimäki 1987b). In the lowland study areas of Kauhava and Seinäjoki, the boreal owl was clearly the most frequent breeding species, whereas in the higher-altitude study areas of Ähtäri and Keuruu, the Ural owl was the most numerous breeding species (Table 3.3b). The largest member of the owl assemblage, the eagle owl, was a more frequent breeder than the Ural owl in the Kauhava and Seinäjoki areas, and also relatively numerous in the Ähtäri region. Of the forest-dwelling species, the pygmy owl was a relatively common breeder in the regions of Seinäjoki, Ähtäri and Keuruu, but did not usually breed in the Kauhava area up until 2002, when a network of nest-boxes was established for pygmy owls. During 2003–9 the number of pygmy owl nests ranged from 5 to 25 in the Kauhava region (E. Korpimäki, unpublished data). Short-eared and long-eared owls hunting in open country were numerous breeders in the Kauhava and Seinäjoki regions but much less numerous in the regions of Ähtäri and Keuruu. Of the southern species, the tawny owl was only an occasional breeder in the Ähtäri and Keuruu regions, whereas of the northern species, the hawk owl and the great grey owl occasionally bred in the Kauhava and Keuruu regions (the former) and in the Seinäjoki region (the latter). It is noteworthy that all these four study areas included an excess of nest-boxes for Ural, tawny and boreal owls during 1976–84 and that no obvious differences existed in the searching traditions and methods that could explain these differences found in the composition of owl communities (Korpimäki 1987b). To conclude, the intra-guild position of boreal owls was essentially different in these four areas, because the worst intra-guild enemy, the Ural owl, which substantially decreases the habitat quality for boreal owls (Hakkarainen and Korpimäki 1996), was not common in our study area and in the Seinäjoki region. In addition, the forest-dwelling pygmy owl, which is probably one of the most serious food competitors of boreal owls, particularly in winter (Korpimäki 1981, 1986e), was a less common breeder in our study area than in the other three areas up to 2002.

The fifth long-term study area of boreal and other owls is located in the vicinity of Tampere, in the province of Pirkanmaa, 230 km south of our study area (Fig. 3.1). One of the pioneer investigators of Finnish bird populations, Pentti Linkola initiated studies on owls and diurnal raptors in an area east of Tampere covering approx. 3500 km^2 (Linkola and Myllymäki 1969). With his co-workers, he studied breeding density, clutch size and production of young, of boreal, tawny, Ural and long-eared owls, as well as Eurasian kestrels and common buzzards during 1952–66 (0 nests of boreal owls in 1952, 10 nests in 1953, 24 in 1954, 3 in 1955, 4 in 1956, 7 in 1957, 12 in 1958, 8 in 1959, 6 in 1960, 19 in 1961, 28 in 1962, 13 in 1963, 34 in 1964, 56 in 1965, and 59 in 1966; Linkola and Myllymäki 1969). Martti Lagerström, one of the Finnish pioneers in boreal owl studies, initiated his investigations in an area covering approx. 7000 km^2 in the vicinity of Tampere in 1961 (13 nests in 1961, 13 in 1962, and 5 in 1963). With several voluntary bird-ringers and bird-watchers, he organised a group to erect a large number of nest-boxes. Thereafter 191–669 nest-boxes and holes were inspected annually in his study area during 1964–9, and 733–1394 nest-boxes during 1970–80. The annual number of boreal owl nests found varied from 5 to 134 during the study period and the total number of nests found during 1961–80 was 902 (Lagerström and Häkkinen 1978, Lagerström 1980, 1982, 1983a, 1983b, 1985a, 1985b, 1991, Korpimäki and Lagerström 1988). The study areas surrounding Tampere are quite similar in habitat composition to the Keuruu region mentioned above: the coverage of coniferous forests is approx. 70% and this area also includes some big lakes.

Nationwide monitoring of the population dynamics and breeding success of owls in Finland was organised by local ornithological associations (see map in Fig. 3.1), was initiated in 1979 (Forsman et al. 1980) and has continued since then (e.g. Jokinen et al. 1982, 1983, Saurola 1985, 1986, 2008, Haapala and Saurola 1989, Haapala et al. 1992, 1996a, 1996b, 1997, Honkala and Saurola 2006). Unfortunately, there was a gap in nationwide reports of the results during 1982–5, but for most areas of the 20 local ornithological societies of Finland, we have been able to gather information on number of nests found and nest-boxes/-holes of boreal owls inspected during this period. Therefore, there are long-term data (>20 years) available about variations in the relative breeding densities of boreal owls in Finland in 20 large study areas (Table 3.4).

3.10.   Study areas elsewhere in Europe and North America

One of the few long-term study areas of boreal owls in other north European countries is located near Umeå, in the province of Västerbotten, northern Sweden (64°N, 20°E) (see Löfgren et al. 1986, Carlsson B.-G. et al. 1987, Hörnfeldt et al. 1990, 2005, Hörnfeldt and Nyholm 1996, Hipkiss and Hörnfeldt 2004). The area belongs to the middle boreal vegetation zone and is dominated by managed coniferous forests with small areas of low-intensity agriculture. During 1980–2003, about 200–500 nest-boxes for boreal owls (base 20 × 20 cm, entrance hole diameter 8.5 cm), set up along roads at approx. 1-km intervals, have been annually inspected to find nests of boreal owls and to collect data on breeding performance and success as well as brood sex ratio, nest-site tenacity and factors affecting the occurrence of polygyny, etc. In addition, Professor Geir A. Sonerud (1985a, 1989, 1993) has studied reproductive success – in particular nest predation of boreal owls by pine martens – in two areas in Hedmark county, south-eastern Norway (61°N, 11°E) during 1970–83. Both areas, situated in the northern boreal zone, consist of coniferous forests mixed with peatland bogs, and contain no permanent human settlements apart from the occasional small summer dairy farm with pastures. A total of 44 boreal owl nests were found in the smaller study area (40 km^2) and 97 nests in the larger area (400 km^2) during 1971–83. Later studies in this area have concentrated on parental care of boreal owls (Eldegard and Sonerud 2009, 2010).

One of the few long-term study areas of boreal owls in central Europe is located in western Harz Mountain in Germany (51°N, 10°E), where Dr. Ortwin Schwerdtfeger has maintained a network of 200 nest-boxes in an area of approx. 200 km^2. This area covers spruce-dominated coniferous forests on mountain slopes at 400–880 m elevation. He studied breeding density and reproductive success of boreal owls during the years 1979–2008 (e.g. Schwerdtfeger 1984, 1988, 2008). Most of the female and male parents and fledglings of boreal owl nests have also been trapped and ringed, which facilitates the studies on dispersal and age composition of the breeding population (Schwerdtfeger 1990, 1991, 1996, 1997). There also are study sites for population dynamics and diet composition of boreal owls in other areas of Germany (e.g. König 1969, Franz et al.

Table 3.4. Study areas, number of nest-boxes, number of nests per year and cyclicity indices (CV) of boreal owl populations in Europe and North America.

Country, locality	Years	No. of nest-boxes			No. of nests per year				Source
		Mean	s.d.	Range	Mean	s.d.	Range	CV	
Finland, Kauhava region, 63°N, 23°E	1977–2009	372.2	170.7	35–500	35.4	41.1	2–163	116.0	1
Finland, Seinäjoki region, 62.5°N, 22.5°E	1977–2009	304.7	115.1	83–507	19.8	20.4	0–84	103.1	1
Finland, Ähtäri region, 62.4°N, 22°E	1977–2009	310.4	49.5	183–449	16.3	12.0	0–58	73.9	1
Finland, Keuruu region, 62.2°N, 24.4°E	1977–2009	176.1	64.2	60–283	7.8	6.9	0–31	87.7	1
Finland, Quarken, 63°N, 21.5°E	1979–2009	432.2	185.6	100–672	24.3	27.2	1–118	111.8	2
Finland, Suupohja, 62.3°N, 21.5°E	1977–2009	783.9	382.2	70–1280	36.8	40.8	2–205	110.9	2
Finland, Central Ostrobothnia, 64°N, 23°E	1979–2009	722.3	324.2	159–1355	89.4	84.7	9–341	94.7	2
Finland, Central Finland, 62.5°N, 25.5°E	1980–2009	679.2	256.9	115–1017	33.3	34.3	1–165	102.8	2
Finland, Pirkanmaa, 61.5°N, 24°E	1977–2009	1595.3	754.3	191–2847	66.3	57.9	2–225	87.4	2
Finland, Satakunta, 61.5°N, 21.5°E	1979–2009	340.7	169.3	36–550	12.4	12.9	0–46	103.8	2
Finland, Varsinais-Suomi, 60.5°N, 22.5°E	1979–2009	712.9	261.5	139–1068	19.9	32.9	0–150	165.6	2
Finland, Kanta-Häme, 61°N, 24.3°E	1979–2009	717.8	252.0	60–1097	20.8	28.5	0–110	137.1	2
Finland, Päijät-Häme, 61°N, 25.3°E	1977–2009	416.4	138.4	154–631	11.9	10.3	0–37	86.7	2
Finland, Uusimaa, 60.2°N, 25°E	1979–2009	1520.3	759.8	100–2669	24.0	42.2	0–163	175.6	2
Finland, Kymenlaakso, 60.7°N, 27°E	1978–2009	1253.7	446.8	550–1897	47.7	51.4	0–204	107.7	2
Finland, South Karelia, 61°N, 28°E	1986–2009	476.4	101.7	204–642	17.7	15.0	0–59	84.4	2
Finland, South Savonia, 62°N, 29°E	1986–2009	127.5	30.7	88–206	6.5	10.7	0–48	165.4	2
Finland, North Savonia, 63°N, 27.°E	1979–2009	433.2	179.1	70–825	40.9	33.3	0–159	81.4	2
Finland, North Karelia, 63°N, 30°E	1977–2009	515.4	260.2	90–948	28.6	23.3	0–88	81.5	2
Finland, Kainuu, 64.5°N, 28°E	1977–2009	300.6	167.3	9–541	26.8	29.4	0–102	109.8	2
Finland, North Ostrobothnia, 65°N, 25°E	1980–2009	579.0	139.9	180–761	43.2	39.1	1–174	90.5	2
Finland, Tornio-Kemi, 65.9°N, 24.1°E	1977–2009	244.1	143.8	18–603	30.8	32.1	0–144	104.3	2
Finland, Lappi, 67°N, 26°E	1981–2009	162.8	100.9	0–357	7.4	8.1	0–33	109.9	2
Finland, all	1977–2009	11598.2	6026.3	1579–18750	623.8	470.6	74–2259	75.4	2
Sweden, Umeå, 64°N, 20°E	1980–2003	–	–	200–500	106.4	98.1	4–201	92.2	3
Sweden, Jämtland, 63.5°N, 15°E	1977–81	–	–	–	42.2	47.1	1–119	111.5	4
Sweden, Gästrikland, 60.5°N, 17°E	1978–84	299.4	152.3	77–437	26.1	13.0	14–48	49.7	5
Sweden, Värmland, 59.5°N, 13°E	1972–81	–	–	–	151.1	100.6	9–294	66.6	6
Norway, Kirkesdalen, 69°N, 18.8°E	1985–98	20.0	–	20	5.4	3.8	0–11	70.5	7
Norway, SE Norway, 61°N, 11°E	1974–8	–	–	–	2.8	3.0	1–8	108.3	8

Location	Period								Source
Norway, SE Norway, Hedmark, 61°N, 11.5°E	1971–83	–	–	–	11.6	14.5	0–40	124.9	9
Belgium, 50°N, 4°E	1985–2009	400.0	–	400	26.4	32.5	0–140	123.4	10
Germany, Baden-Württemberg, 48°N, 9°E	1963–8	–	–	–	5.7	2.9	2–10	52.0	11
Germany, Thüringen, 51°N, 11°E	1972–7	–	–	–	9.0	5.4	2–17	60.0	12
Germany, Saaletal, 51°N, 12°E	1975–8	–	–	–	13.3	6.1	8–22	45.9	13
Germany, Niedersachsen, Kaufunger Wald, 51.3°N, 9.6°E	1965–70	–	–	–	4.7	3.4	2–11	72.6	14
Germany, Hof, 50°N, 11°E	1985–96	38.5	23.2	1–69	6.6	5.1	1–17	78.1	15
Germany, München, Höhenkirchen, 48°N, 11.7°E	1985–96	28.6	8.4	5–56	1.8	2.7	0–8	152.4	15
Germany, Niedersachsen, West Harz, 51.8°N, 10.3°E	1977–2006	200.0	0.0	200	27.5	21.6	3–88	78.5	16
Czech Republic, Ore Mountains, 50.6°N, 13°E	2000–9	118.1	20.7	99–164	17.3	5.5	10–25	31.7	16
Switzerland, East Jura Mountains, 47°N, 6.5°E	1985–2008	194.3	56.2	138–246	22.7	12.6	8–57	55.3	17
Switzerland, West Jura Mountains, 46.5°N, 6°E	1986–2009	221.4	79.2	80–336	12.8	7.6	3–31	59.3	18
Italy, Cansiglio, 46°N, 12°E	1987–93	150.7	49.6	55–180	11.6	11.8	0–28	101.9	19
USA, Alaska, 64°N, 147°W	1981–2005	71.8	33.2	4–107	5.7	4.3	0–17	75.5	20
Canada, Yukon Territory, Kluane, 61°N, 138°W	1990–6	–	–	–	3.9	4.6	0–11	120.1	21

Sources: 1. Korpimäki (1987e), Korpimäki et al. (2009) and unpublished data, 2. Forsman et al. (1980), Jokinen et al. (1982, 1983), Haapa a and Saurola (1986a, 1986b, 1987, 1989), Haapala et al. (1990, 1991, 1992, 1993, 1994, 1995, 1996a, 1996b, 1997), Taivalmäki et al. (1998, 2001), Hannula et al. (2002), Björklund et al. (2003, 2009), Björklund and Saurola (2004), Honkala and Saurola (2006, 2007, 2008) and Honkala et al. (2010), 3. Hörnfeldt et al. (2005), 4. Holmberg (1982) and unpublished data, 5. Östlund (1984), 6. Carlsson U. (1983), 7. Strann et al. (2002), 8. Solheim (1983), 9. Sonerud (1985a), 10. Sorbi (1993, 1995) and unpublished data, 11. König (1969), 12. Ritter et al. (1978), 13. Rudat et al. (1979), 14. Scheller (1972), 15. Meyer H. et al. (1998), 16. Zárybnická (2009b) and unpublished data, 17. Ravussin et al. (2008), 18. Henrioux (2010), 19. Mezzavilla and Lombardo (1997) and Mezzavilla et al. (1994), 20. T. Swem, unpublished data, 21. Doyle and Smith J. (2001).

1984, Meyer H. et al. 1998, Meyer W. 2003, 2010) and in eastern Belgium (Scheuren 1970, Kämpfer-Lauenstein and Lederer 1992, Sorbi 1993, 1995, 2003).

A study site of boreal owls is located in the Ore Mountains in the Czech Republic (730–960 m asl), close to the border with Saxony. The habitat is characterised by a mosaic of open areas and forest fragments. Wood reeds *Calamagrostis villosa* and solitary trees (mainly European beech *Fagus sylvatica*) mostly cover clear-cuts and other open areas. Prickly spruce *Picea pungens*, birch *Betula* spp., European mountain ash *Sorbus aucuparia*, and European larch *Larix decidua* cover the forested areas. Within the study site (70 km^2), 120 nest-boxes for boreal owls were placed from 1999 onwards and breeding density and success, diet composition and parental behaviour have been recorded (Zárybnická 2008, 2009b, Zárybnická et al. 2009). There also are other study sites in the Czech Republic and Slovakia, where food samples from boreal owls have been collected (Kloubec and Vacik 1990, Pykal and Kloubec 1994, Kloubec and Pacenovsky 1996).

Two long-term study areas of boreal owls are situated in the Jura Mountains, in the western part of Switzerland, at 800–1400 m elevation (Ravussin et al. 1993, 1994, 2001a, 2001b, Ravussin 2004, Henrioux 2010). Approximately 70–130 natural cavities and 70–120 nest-boxes suitable for boreal owls have been inspected annually during 1985–2010 by Pierre-Alain Ravussin and co-workers in the Jura Mountains in the area (150 km^2) surrounding Sainte-Croix (Groupe Ornithologigue de Baulmes et Environs, www.chouette-gobe.ch) (Fig. 3.17). In addition, approx. 80–250 natural cavities and 20–80 nest-boxes have been inspected annually during 1986–2010 by Pierre Henrioux and his co-workers in a 126 km^2 area of the western Jura Mountains (Fig. 3.18). Reproductive success has been recorded. In addition, most fledglings, many female parents and some male parents have also been trapped and ringed at the nests in both

Fig. 3.17.   The long-term study area of boreal owls near Sainte-Croix in the Jura Mountains, Switzerland at 800–1400 m elevation. The owls mainly occupy mature forests dominated by spruce and beech. Photo: Pierre-Alain Ravussin.

Fig. 3.18.   The long-term study area of Pierre Henrioux and his co-workers at 800–1400 m elevation in the western Jura Mountains, Switzerland, near the border with France. Photo: Pierre Henrioux.

study areas (Ravussin et al. 2001a, 2010, Henrioux 2010). In the context of a nationwide monitoring project of owls and diurnal raptors in Germany, 13 sample areas of boreal owls covering a total of 3070 km^2 have been monitored during 1988–95 (Mammen 1997). The southernmost study sites of boreal owls in Europe are located in the north Italian Alps (Sperti et al. 1991, Mezzavilla et al. 1994, Mezzavilla and Lombardo 1997) and in the Pyrenees of northern Spain (Castro A. et al. 2008, Lopez et al. 2010).

The study sites of boreal owls in Montana, Idaho and northern Wyoming, USA, are in pristine boreal coniferous forests at high elevations (1580–2400 m), where the snow depth reaches 50–150 cm in winter and snow melts only in May (Hayward et al. 1992, 1993, Hayward and Hayward 1993). Therefore, the living conditions for owls are harsh, and breeding densities and clutch sizes are very low in comparison with the data from Europe (see Hayward and Hayward 1993). Thus, only 16 nests of boreal owls were found during a 5-year intensive research period in the main study area in Idaho (Hayward and Hayward 1993). The working conditions for the researchers were also severe, especially in winter and early spring in remote wilderness areas with plenty of snow. Considering these environmental restrictions, Dr. Gregory D. Hayward and his co-workers (1993) have done some good work. We still wonder, though, whether it would have been possible to find better study sites with higher owl densities so that the great effort put into this work could have produced more data and more convincing results; for instance on breeding parameters and survival.

The study sites of boreal owls in Alaska, USA are located near Fairbanks in the interior of the state (Whitman 2008, 2009, 2010) and near Anchorage on the west coast. Ted Swem and his co-workers (unpublished data) erected and maintained a network of 25–107 nest-boxes during 1984–2005. The annual number of nests of boreal owls varied

Fig. 3.19.   The study area of boreal owls in the forest near Fairbanks, interior of Alaska, USA, where snow covers the landscape in late spring (photo taken on 15 April by Jackson S. Whitman). (For colour version, see colour plate.)

from 1 to 13 and that of saw-whet owls from 1 to 18. Jackson S. Whitman's studies were conducted during 2003–7 in the boreal forests of the interior of Alaska (near Fairbanks) where all box routes were between 64.6°N and 65.5°N, and between 146.2°W and 148.7° W, at elevations from 110 to 690 m asl (Fig. 3.19). The number of nest-boxes was approx. 120 and a total of 111 nesting attempts were recorded during 2003–7 (Whitman 2010). Overstorey vegetation was highly variable. White spruce *Picea glauca*, often mixed with paper birch *Betula papyrifera*, eastern larch *Larix laricina* or balsam poplar *Populus balsamifera*, dominated lower elevations. Those at mid-slopes were highly variable, composed of monotypic stands of quaking aspen *Populus tremuloides*, paper birch, or black spruce *Picea mariana*, or some combination thereof. At higher elevations and on poorly drained soils, overstorey vegetation was dominated by black spruce. Shrub layers were composed largely of willow (usually *Salix alaxensis* or *S. bebbiana*), green alder *Alnus crispa* and blueberry *Vaccinium uliginosum*. Interior Alaska is typified by continental weather patterns, with generally mild summers and cold winters (Whitman

2009). In addition, numbers of breeding territories and diet of boreal owls were studied as a part of the boreal forest ecosystem project in Kluane, Yukon, Canada (Doyle and Smith J. 2001).

Other, relatively short-term study areas of boreal owls in Europe and North America are listed in Table 3.4.

4 Habitat use, roosts and nest-sites

Source: Drawn by Marke Raatikainen.

4.1.   Habitat use: foraging and breeding in safe patches

Animals need to decide which habitats to forage in and how long to spend in each habitat patch in heterogeneous landscapes, because this has consequences for individual fitness. Individual responses to heterogeneity include adjustment of total area of activity (home range size) or adjustment of time spent exploiting various habitats within their home range. Conventionally, spatial responses of individuals to variations in habitat character-istics have been estimated using home range size and habitat selection functions that are based on proportionate use of habitats in relation to their availability (Aebischer et al. 1993, Sutherland 1996). However, only rarely has the abundance of the main foods of avian predators been estimated in main habitat types when studying habitat selection. In addition, the spatial variation in the intra-guild predation risk for foraging predators has not often been assessed, although the concept of the 'landscape of fear' is becoming more important when studying the habitat use of foraging herbivorous animals (Brown J. and

Kotler 2007). Fear imposed by larger predators may induce smaller predators to hunt in safer habitat patches that, in turn, offer less abundant food resources.

Loss of, and alterations to, habitats are crucial human-induced factors affecting landscape heterogeneity and are the largest worldwide hazard to biodiversity and viability of populations (Fahrig 2003). In boreal environments, a long history of agricultural activity and forestry has profoundly altered the landscape composition, configuration and structure (Hansson 1992). In southern and central Finland, for example, it has left behind as little as 1% of pristine boreal forest. There is plenty of evidence that intensive forestry has detrimental impacts on the diversity and abundance of animal species at the community level (see reviews in Schmiegelow and Mönkkönen 2002, Fahrig 2003, Prugh et al. 2008). However, very little is known about how changes in the landscape affect home range size, habitat use, parental behaviour, body condition and, ultimately, reproductive success and survival of individual birds (e.g. Zanette 2000). This is the case especially for very mobile birds of prey occupying extensive home ranges (but see Redpath 1995, Laaksonen et al. 2004, Hinam and St. Clair 2008).

The boreal owl is a forest-dwelling predator species that, outside the breeding season, can be seen in very variable habitats ranging from farmyards to small remote islands, for example, those of the Baltic Sea in northern Europe and of the Great Lakes in North America. In general, however, the boreal owl is reluctant to pass through wide open areas, such as large agricultural fields, clear-cuttings and areas with open water. The boreal owl occupies Eurasian and North American coniferous forests. It breeds mainly in dense spruce *Picea* forests, but also occurs in mixed forests of pine *Pinus*, birch *Betula* and aspen *Populus*, and even in pure pine-dominated forests. In northern Europe, north of the coniferous forest belt, it can sometimes breed in pure birch forests (Norberg 1964). Availability of suitable nest-holes is an important determinant of breeding habitat quality, along with the availability of food.

4.1.1.   Habitat use in Europe

In western Finland, boreal owls were found to favour spruce-dominated forests as their breeding habitat: during 1966–79, about 15% of the boxes situated in spruce forests were occupied (total number checked 637), but only 8% in pine forests (254). The occupancy rate of boxes less than 200 m from agricultural fields was 15%, whereas it was only 6% in boxes >200 m from agricultural fields (Korpimäki 1981). Therefore, boreal owls appeared to avoid large uniform forest areas. The edges and small woods close to peatland bogs, the shores of lakes and ponds, as well as the yards of farmhouses close to forests were sometimes used as breeding habitat. In south-eastern Norway, boreal owls bred in forests composed of a mixture of spruce and pine, but in the nest-box area at about 150 m asl with a high proportion of agricultural land, the occupancy of boxes was higher than in areas at 500 m elevation dominated by peatland bogs and at 850 m elevation dominated by subalpine pine forest (Solheim 1983).

In our study area, occupancy of breeding territories and reproductive success were higher in areas with high proportions of spruce forest and agricultural land as well as low proportions of pine forest and peatland bog (Korpimäki 1988d). In addition, territories with a high proportion of spruce forest and agricultural land maintained higher densities of main foods (voles) in good vole years and higher densities and numbers of important alternative prey, such as small birds, in poor vole years (Hakkarainen et al. 1997a). Lifetime reproductive success of boreal owls also increased with a higher proportion of old-growth coniferous forest in the territory (Laaksonen et al. 2004), which underlines the importance of mature coniferous forests as the stronghold of boreal owls. In addition, nests were also found at the edges of agricultural land and lake shores, in small woods (<0.3 ha) in the middle of open agricultural land and peatland bog, and sometimes even in the yards of farms and family houses (Korpimäki 1981).

We radio-tracked 9 breeding boreal owl males in 2005 (an increase year of the 3-year vole population cycle) and 15 males in 2009 (a decrease year; see Fig. 1.1). It was evident that males delivering prey to their families avoided hunting in large open areas including agricultural fields and clear-cut areas, whereas they appeared to prefer spruce-dominated forests in both study years (Fig. 4.1). In addition, hunting males tended to prefer mixed forest stands consisting of spruce and pine in similar proportions in 2009 and, unexpectedly, pine-dominated forests in 2005. The avoidance of large open areas by hunting males was also evident in the nightly hunting bouts mapped in the four home ranges of radio-tracked males in 2009 (Fig. 4.2). In these maps, two panels (a and b) represent home ranges with a high proportion of open areas, and two panels (c and d) show home ranges with a high proportion of forest cover in the home range.

The more detailed ranking of habitat compositions of home range versus surrounding landscape was done in 2009 for 15 males in our study area. The expected result was that forest was preferred over all open habitat types. Within the forested landscape, older stands were the most preferred forest age, and mixed stands the most preferred forest type, with young and pine-dominated forests ranked last. Within the home range of male owls, forest was again preferred over all other open habitats, and old forest over the younger age classes. In addition, spruce forest was now the most preferred type, with pine forest again among the two least preferred in the ranking of forest types (Santangeli et al. 2012).

According to snap-trap captures made in four main habitat types of 9 males radio-tracked in 2005 and in four main habitat types of 22 home ranges including all the home ranges of 15 radio-tracked males in 2009, the abundance of small mammalian prey (voles, mice and shrews) varied greatly between nest-sites and also between the four habitat types (individuals trapped per 100 trap-nights, mean ± s.d.):

Year	Spruce forest	Pine forest	Clear-cut	Agricultural field
2005	4.0 ± 0.0	2.1 ± 1.6	2.8 ± 2.2	4.0 ± 3.5
2009	1.4 ± 1.4	0.9 ± 0.8	0.4 ± 0.9	1.1 ± 1.2

Spruce forest stands and agricultural fields appeared to support higher small mammal numbers than the average both in 2005 and in 2009, while lower than average food abundance was recorded in pine forests and clear-cut areas in both 2005 and 2009. In 2009, with a larger sample size, the abundance of small mammal prey was significantly lower in clear-cut areas than in pine and spruce forests and agricultural fields, but no other

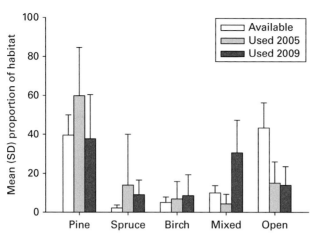

Fig. 4.1.   Mean (s.d.) proportions (%) of pine forests, spruce forests, birch forests, mixed forests with similar proportions of pine and spruce, and open areas (agricultural land and clear-cut areas) available and used by 9 male boreal owls in 2005 and 15 males in 2009 in the vicinity of their nests.

obvious differences were found in small mammal numbers between the other habitats, probably due to the large variation in prey abundance between nest-sites. In terms of species composition, bank voles were the most abundant small mammal prey in pine and spruce forests, and *Microtus* voles were the most abundant in agricultural fields. In the data from 2009, differences in habitat-specific biomass of small mammal prey were consistent with differences in numbers, but with higher biomass levels in agricultural field habitats due to the larger-sized *Microtus* voles found there (Santangeli et al. 2012). Therefore, the preference in hunting habitats of males radio-tracked in 2005 and 2009 (Fig. 4.1) was mostly consistent with the abundance of small mammal prey in four main habitats, although the avoidance of agricultural fields was evident despite the higher density and biomass of the *Microtus* voles living there. Main breeding habitats of boreal owls in central Europe are coniferous, mainly spruce-dominated, forests at high elevations on mountain slopes (e.g. König 1969, Schwerdtfeger 1984, Kloubec and Vacik 1990, Kloubec and Pacenovsky 1996, Zárybnická et al. 2009). Boreal owls can also breed in forests at lower elevations in many locations in Germany (Kuhk 1949, Schelper 1989, Meyer H. et al. 1998; summary in Mammen 1997), in the Bourgogne (Burgundy), Haut-Vecors and Jura regions of France (Joveniaux and Durand 1987), in eastern Belgium (Scheuren 1970, Kämpfer-Lauenstein and Lederer 1992, Sorbi 1995), and even in the State Forest of Gieten, in the eastern part of the Netherlands (Boerma et al. 1987). In general, they seem to prefer old-forest stands, where the availability of natural cavities is higher than in young forest stands (Rudat et al. 1979, Joveniaux and Durand 1987, Jörlitschka 1988).

Further south in Europe, for example in the Jura Mountains of Switzerland (Ravussin et al. 1994, 2001b) and in the Alps of northern Italy (Mezzavilla et al. 1994, Mezzavilla and Lombardo 1997), boreal owls occupy mature forests dominated by spruce and beech *Fagus sylvatica*. In the northern Dinaric Alps, central Slovenia, playback surveys revealed that boreal owls occupied higher altitudes (700–940 m asl), tawny owls were found at lower altitudes (320–850 m asl), while Ural owls occurred over the widest range

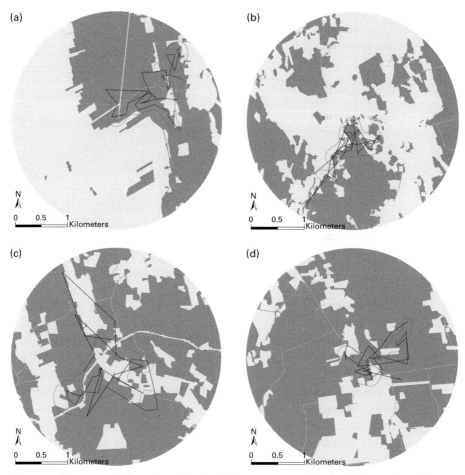

Fig. 4.2.   Coverage of open terrain (agricultural land and clear-cut areas, yellow) and forests (green) in four home ranges of male owls in 2009 in the 2-km radius around the nest-box (brown square). Brown line is the 95% kernel home range. Blue, dark red and violet lines represent the routes of male owls in radio-tracking and hunting bouts lasting for at least 2 hours on three different nights at the end of the nestling to post-fledging periods. Panels (a) and (b) represent home ranges where the proportion of open terrain is large (nest-boxes 278 and 351, respectively) and panels (c) and (d) represent home ranges where the proportion of forests is relatively large (nest-boxes 819 and 313, respectively). (For colour version, see colour plate.)

of altitudinal distribution (Vrezec 2003). In the Pyrenees, on the borders of France and Spain, the main breeding habitat has been characterised as mountain pine forests at 1200–2250 m elevation, with average annual temperatures below 6°C (Dejaifve et al. 1990, Prodon et al. 1990, Castro A. et al. 2008). More detailed analyses of habitat use during the breeding season indicated that breeding boreal owls in the Pyrenees were associated with mature but relatively open forests, where natural cavities, saplings, tree stumps and dead wood on the ground offered perches for hunting (Marine and Dalmau 2000). Modelling the spatial distribution of boreal owls to predict the distribution in the Iberian Peninsula revealed that the species is restricted to high altitudes in the Pyrenees, where the number of frost days is highest (Castro A. et al. 2008). These authors also suggested that the local population in Spain may be a sink population to which the owls immigrate from local populations in France and other central European countries.

4.1.2.   Habitat use in North America

Boreal owls in Canada and Alaska breed in forests dominated by black and white spruce *Picea mariana* and *P. glauca*, aspen *Populus tremuloides*, poplar *P. balsamea*, white birch *Betula papyrifera* and balsam fir *Abies balsamea*. In Alaska, breeding habitat was found to be highly variable, including a mixture of white spruce, white birch and poplar in riparian areas, aspen and mixed white spruce or birch in mid-elevations, and stands of black spruce at upper elevations. Elevation ranged from 200 to 650 m (Whitman 2008). Nests were found in boreal forests with balsam fir and white birch normally dominating at elevations ranging from 0 to 500 m in Nova Scotia, Canada (Lauff 2009). In north-eastern Minnesota, USA, singing male boreal owls were recorded as occupying older, upland mixed (aspen–conifer) forest stands and seemed to avoid relatively open brush–regenerative forest stands (Hayward and Hayward 1993, Lane et al. 2001). Further south, in the Rocky Mountains, Blue Mountains and Cascades, they occupy subalpine forests dominated by fir *Abies lasiocarpa* and Engelmann spruce *Picea engelmannii* (Hayward and Hayward 1993). Most breeding sites are above 1580 m asl in Idaho and Montana, USA, whereas in Colorado and northern New Mexico most locations exceed 3050 m asl (Hayward and Hayward 1993, Stahlecker and Duncan 1996).

The most detailed study on the habitat use of boreal owls was probably made in the Rocky Mountains of Montana, Idaho and northern Wyoming, USA (Hayward et al. 1993). The results appeared to indicate that boreal owls are more usually habitat special-ists in North America than conspecifics, as they are in Europe. The vast majority (88%) of hooting observations of boreal owls were located in subalpine fir habitats, and most (76%) were in mature forest stands. No owls were detected below 1290 m elevation, and 75% of the hooting locations were above 1580 m elevation. Owls bred in mixed conifer (39% of nest-sites), spruce–fir (18%), Douglas-fir *Pseudotsuga menziesii* (21%) and aspen (21%) stands, although aspen stands accounted for only <1% of the forest vegetation. The owls largely avoided breeding in the most common vegetation type – lodgepole pine (*Pinus contorta*) forest. The most characteristic structural features of nesting and hooting sites were a high density of large trees (>38 cm dbh – diameter at breast height), an open understorey and a multi-layered canopy (Hayward et al. 1993).

In the Rocky Mountains of Colorado, USA, Palmer (1986) and Ryder et al. (1987) studied habitat selection of 36 radio-tagged boreal owls during 1983–4. In general, boreal owls occupied higher-elevation forests (2700–3200 m asl) than saw-whet owls, and their densities were highest above 3000 m. Boreal owls occupied mainly mature spruce–fir forests interspersed with subalpine meadows. Owls avoided wide uniform stands of pines, most of which were very dense. Boreal owls also avoided stands of quaking aspens, which were the habitat type preferred by saw-whet owls in the same area. In addition, boreal owls inhabited sites with somewhat larger trees than sites inhabited by saw-whet owls. The ground in boreal owl territories was often covered with *Vaccinium* and *Arcina* forbs, and fewer grasses and sedges covered their territories than those of saw-whet owls. Moreover, preferred boreal owl territories had higher densities of red-backed voles *Myodes gapperi* and lower densities of deer mice *Peromyscus maniculatus*. Red-backed voles were the main food of boreal owls in the area, whereas deer mice were the main food of saw-whets (Palmer 1986, Ryder et al. 1987).

In the Rocky Mountains of Montana, Idaho and north Wyoming, roosting habitats also differed between winter and summer (Hayward et al. 1993). Winter roost sites were chosen in relation to their availability, whereas summer roosts were in dense shaded forests with greater canopy cover and tree density. Therefore, during the summer, owls appeared to select sites with a cool microclimate for roosting. Each of the 12 boreal owls radio-tracked both in winter and summer used summer roost sites with a higher average elevation; the difference in mean seasonal roost elevation being 186 m. The most productive foraging habitat was found in old spruce–fir stands, where prey population densities were also higher. The owls seemed to have conflicting preferences when choosing their nesting, foraging and roosting habitats, and therefore they used large home ranges (mean 1451 ha in winter and 1182 ha in summer) (Hayward et al. 1993).

In Colorado, all roosts were on steep slopes in fairly dense coniferous forests dominated by Engelmann spruce trees (Palmer 1986, Ryder et al. 1987). There were no apparent seasonal changes in roost sites and habitats. The conifer trees used for roosting appeared to offer better cover – and thus protection against enemies – above the owl, but less cover below, facilitating easier attacks on prey towards the ground surface. The mean height of roost trees was approx. 14 m with an average breast-height diameter of 34 cm (Palmer 1986, Ryder et al. 1987). In Ontario, boreal owls preferably roosted on trees with the foliage on the lower part restricted to the outer half of the branch, as quite often occurs in the balsam fir (Bondrup-Nielsen 1984).

In Montana, Idaho and north Wyoming, summer roost sites were found to be in cool microhabitats where the canopy cover and tree density were higher than those of random sites, whereas forest structure at winter roost sites did not obviously differ from random sites (Hayward et al. 1993). Roosts both in summer and winter were usually in the dominant tree species (66% and 70% of roosts, respectively). Roosts in summer were never in cavities: all were in conifer trees, with 64% of canopy cover. Owls roosted at an average height of 6 m in 25 cm dbh trees in summer, and at a height of 7 m in 27 cm dbh trees in winter. Around 75% of winter roosts occurred in the lower half of the tree (25% in the lowest quarter) in winter and 50% of roosts were in the lower third of the tree in summer (Hayward et al. 1993).

In Colorado, all winter and summer roosts were located in conifers at an average height of 4.7 m (range 1.4–14 m, $n = 174$) in 33 cm dbh trees and 22 cm from the trunk (Ryder et al. 1987). In north-east Minnesota, boreal owls typically preferred to roost in lowland areas characterised by thick coniferous forest stands, while upland mixed-type forests were largely avoided (Lane et al. 1997). Black spruce *Picea abina* was used as the roost tree in 82% of cases, followed by balsam fir *Abies balsamea* (9%) and northern white cedar *Thuja occidentalis* (4% of 115 roost sites; Lane et al. 1997). In Ontario, boreal owls preferably roosted in conifer trees (70%), but aspen and birch were also quite often used (30%). Roost height averaged 6 m and roosting owls averaged 37 cm from the trunk (Bondrup-Nielsen 1984).

4.1.3.   Home range size, food abundance and predation risk

In western Finland, the nocturnal home range size, estimated using the minimum convex polygon (MCP) method, averaged 151 ha (range 54–268 ha) for nine males providing food for their families and radio-tracked for 2–4 nights each in 2005 (M. Kukkonen and E. Korpimäki, unpublished data; two additional males fixed with transmitters in 2005 succumbed before the radio-tracking sessions, see Chapter 10.4). The corresponding home range size using the MCP-method averaged 225 ha (range 73–499 ha) for 15 males radio-tracked in 2009. For all 15 males radio-tracked in spring 2009, the data included a minimum of 32 fixes per male (mean ± s.d. 47 ± 9). This appeared to be sufficient to estimate 95% kernel home range size, because incremental analysis indicated an average saturation threshold of 31 fixes, consistent with that of 30 fixes suggested by Seaman et al. (1999). Home ranges produced by the 95% kernel estimator varied considerably in size, ranging from 41 to 293 ha (mean 114 ha). Home range size decreased with the increasing cover of spruce forest around the nest-site landscape (Santangeli et al. 2012). Spruce-dominated forest previously emerged among the less preferred forest type by hunting males. Only for six (out of nine) of the males radio-tracked in 2005, did we have enough fixes to calculate 95% kernel estimates of home range size (mean 59 ha), which was considerably less than the corresponding size in 2009 (114 ha). This was consistent with the data from small mammal abundance, which appeared to be higher in owl home ranges in 2005 than in 2009 (see Chapter 4.1.1). Individual mean roosting heights of nine males in 2005 varied from 1.0 to 5.3 m, with an overall mean of 2.5 m (s.d. = 1.0 m).

The nocturnal MCP home range size of four breeding boreal owl males hunting for their families averaged 187 ha (range 94–226 ha) and diurnal roosting sites were mostly within the nocturnal hunting area in south-eastern Norway (Sonerud et al. 1986, Jacobsen and Sonerud 1987). Individual mean roosting heights of 11 males varied from 2.6 to 7.5 m, with an overall mean of 4.3 m. The corresponding figures were from 2.7 to 6.5 m for five females, with an overall mean of 4.4 m and thus there was no apparent intersexual difference (Bye et al. 1992).

The maximum distance of a straight line recorded between the nests and a hunting male during different nights varied from 1030 to 1310 m, with an average of 1198 m in Norway (Sonerud et al. 1986). In our study area, the distance of night fixes to the nest ranged from 20 to 2880 m with an overall mean of 644 m (s.d. = 469 m) in 2009 (Santangeli et al. 2012).

Both Finnish boreal owl males (Figs. 4.1, 4.2) and Norwegian males (Sonerud et al. 1986, Jacobsen and Sonerud 1987) appeared to avoid hunting in the middle of clear-cut areas and agricultural land. In our study area, hunting males – as expected – preferred those forest habitats that offered abundant small mammal prey in terms of density and biomass. However, the avoidance of agricultural fields was evident despite the high density and biomass of *Microtus* voles there. The most likely reason is the probability of falling victim to larger birds of prey, including goshawks, Ural owls and eagle owls. The risk imposed by larger birds of prey is probably more frequent in the middle of wide agricultural fields and clear-cut areas than in dense forests. This avian predation risk probably creates a strong selection against boreal owls hunting in the middle of open areas far from forest refuges. In Norway, boreal owl males seemed to prefer mature forests with a quite open ground vegetation layer and edge habitats between agricultural land and clear-cut areas. The densities of voles (*Microtus* and *Myodes* spp.) tended to be higher in open habitats than in forests, but the higher ground vegetation layer in open country apparently induced poorer availability of voles for hunting males in open areas than in forests (Sonerud et al. 1986, Jacobsen and Sonerud 1987). This cannot be the main reason for the avoidance of open country in our study area, however, because most males were radio-tracked in spring and early summer when the ground vegetation layer of open areas had not yet grown.

In summary, home ranges in the breeding season appeared to be much more restricted in northern Europe, mostly 150–230 ha (mean MCP 187 ha in Norway (Sonerud et al. 1986, Jacobsen and Sonerud 1987), 151 ha in 2005 and 225 ha in 2009 in Finland (Santangeli et al. 2012)) than in North America, where it was mostly 1100–1300 ha (mean in Idaho, 1182 ha (Hayward et al. 1993) and in Minnesota, 1202 ha (Lane et al. 1997)). In comparison, 14 saw-whet owl males breeding in the transition zone between boreal forest and aspen parkland in Alberta, Canada had an average home range size of only 89 ha (range 12–137 ha; Hinam and St. Clair 2008). One should be aware that some differences in the number of fixes (e.g. Seaman et al. 1999) and in study methods to analyse home range size (e.g. Kenward 2001) could partly explain these intercontinental differences. Despite this, these large intercontinental differences indicate that boreal owls in North America breed on less productive habitats and thus suffer more from food scarcity, and they should therefore be expected to have considerably wider hunting areas than conspecifics breeding in northern Europe. This interpretation is also supported by our result that the home ranges of hunting males were on average 33% larger in 2009, with a lower abundance of small mammal foods, than in 2005, with more abundant small mammal prey on home ranges. In addition, our few observations of two males that we attempted to radio-track in the poor vole years of 1984 and 1987 indicated that during these 'lean' periods, hunting males often travelled up to 2–3 km from their nests. However, the assumption that the home range sizes of boreal owls are considerably expanded when food is scarce still needs additional supporting data from radio-tracked males in different areas of Europe and North America. Breeding barn owls radio-tracked in Scotland caught more of their prey closer to the nest in years of high than low vole abundance (median 500 m vs. 750 m, respectively) (Taylor I. 1994). In fragmented woodlands of the UK, the home range size of tawny owls increased with decreasing wood size, averaging 134 ha in woods of <4 ha in size, 73 ha in woods of 4–10 ha and 26 ha in woods of >10 ha in size. For male tawny owls, wood isolation and wood size accounted for 80% of the variation in home range size (Redpath 1995).

4.2.   Roosting behaviour: avoidance of enemies

To our knowledge, the only detailed study on the roosting behaviour of boreal owls has been done by Hayward et al. (1993) in the Rocky Mountains of Idaho, USA. During the daytime, boreal owls spent most of the time in roosts, with their eyes closed. On 16 occasions when the owls were watched for ≥2-hour periods (in total 46 hours), they perched quietly with eyes closed 77% of the time (Hayward et al. 1993). Periods of sleep did not often last more than 40 minutes and were interrupted by brief periods (2–5 minutes) of preening and surveying the surroundings. The owls preened parts of their plumage and feet, stretched their wings, and sometimes preened their entire plumage. Extended preening periods of the entire plumage lasted for 20–30 minutes and always preceded leaving the roost to initiate foraging at night. The owls spent 6% of the daytime observation periods in preening, 10% of the time actively looking around, 4% of the time eating, 1% of the time hunting during daylight, and 1% of the time in gular fluttering (a cooling behaviour), but only on warmer days (Hayward et al. 1993).

Boreal owls in Idaho retrieved stored prey and ate a proportion of it on 63 occasions, and prey cached near the roosting owls was also observed on 20 other occasions (Hayward et al. 1993). Roosting with a prey item was never recorded in our study area. In Idaho, stored prey was usually cached in the fork of a tree. The owls only very rarely foraged for an extended period during daylight hours: on only one of 16 days with a ≥2-hour observation period, the owl was observed to forage for 15 minutes (Hayward et al. 1993). In addition, the owls attacked prey from the roost tree on seven occasions. They seldom consumed stored prey whole: more than 50% of the cached prey recorded in roosts was headless and the front half of the prey was consumed in pieces. On four occasions, the owl extracted the intestines from the abdominal cavity of a small mammal and then consumed the rear half of the prey. The owls tended to possess stored prey more often in summer than in winter (cached prey at 17% vs. 4% of roosts, respectively). The consumption of stored prey happened mostly in the afternoon, both in summer and winter, and was preceded by casting of a pellet (Hayward et al. 1993). On average, boreal owls cast 1.2 pellets per day (Korpimäki 1981).

Roosting sites of boreal owls are widely dispersed throughout their home ranges, and new roost sites appear to be chosen each day. Daytime roosts often represent the end of a nocturnal foraging trip. Linear distances between successive daytime roosts of seven females and seven males averaged 934 m in summer and 1540 m in winter (range 0–6935 m). The owls usually roost alone and far from the nest and their mate. Five breeding male owls roosted at a distance of >1000 m from their nests in 85% of the cases (mean 1729 m, range 200–5600 m) (Hayward et al. 1993). Distances of daytime roost sites to the nests of males in our study area averaged 763 m (range 206–1004 m) during 2005 and 920 m (range 148–2710 m) during 2009. Distances between successive roost sites of males averaged 508 m (range 3–2262 m) in our study area (Santangeli et al. 2012).

Of radio-marked members of five mated pairs in Idaho, females and males were observed to roost within 150 m of each other on only 6% of occasions (Hayward et al. 1993). Members of one pair were found to roost together four times: on 14 March in adjacent trees 6.5 km from a future nest-site, on 15 March 30 m apart from each other 200 m from the nest, on 18 March together 2.6 km from the nest without being together on the previous day, and on 27 March 30 m apart from each other 3 km from the nest. These observations suggest that the members of this pair travelled together during some nocturnal movements up to as far as 6.5 km from the nest (Hayward et al. 1993).

In summary, most roost sites of boreal owls were 5–6 m above ground in North America, whereas they tended to be only 2–5 m above ground in northern Europe.

We suggest that boreal owls select daytime roost sites so as to avoid mobbing by small birds (review in Caro 2005) and the risk of being killed by larger birds of prey (reviewed by Sergio and Hiraldo 2008). A cool microclimate at the roost site could also be important in more southern populations, while a warm microclimate at the roost site is probably important in winter in northern populations. During the breeding season, males roost far from their nests to avoid attracting mobbing birds and predators to the vicinity of their nests and do not defend their broods against enemies during the daytime. In contrast, males defend their offspring against enemies in the darkness of night (Hakkarainen and Korpimäki 1994c).

4.3.   Nest-sites and nest-hole shift: the need for space, insulation and safety

4.3.1.   Natural cavities and nest-boxes

Boreal owls, like most other cavity-nesting species, do not build their own nest, apart from scraping a shallow dent in the soft material at the bottom of the cavity. In Eurasian coniferous forests, the boreal owl is mainly dependent on cavities excavated by black woodpeckers (Fig. 4.3). Further south in Europe, it can sometimes also accept cavities hollowed out by green woodpeckers *Picus viridis*, but even there boreal owls are mainly secondary users of black woodpecker cavities. For example, in the Swiss Jura Mountains (>850 m asl), a total of 161 natural cavities suitable for breeding boreal owls were found by systematic searches over 15 years in a 850 km^2 area (about 0.2 cavities/km^2) (Fig. 4.4): 160 of these were located between 1040 and 1390 m asl. The vast majority (98.1%) of the cavities had been excavated by black woodpeckers in large beech trees either in timber forests or in small islands of trees (Ravussin et al. 1994). Cavities hollowed by great spotted woodpeckers (*Dendrocopos major*), which are much more common than those of larger woodpeckers in northern Europe, are too small to be entered by boreal owls. Only if larger woodpeckers or pine martens happen to enlarge an entrance hole, can boreal owls enter cavities hollowed by great spotted woodpeckers, but even then they are usually too small for breeding. In North America, the pileated woodpecker *Dryocopus pileatus* and common flicker *Colaptes auratus* are the main cavity-makers for boreal owls (Hayward and Hayward 1993).

   Unlike, for example, the hawk owl and the Ural owl, which quite often breed in stick-nests or even on the ground, the boreal owl is a strictly hole-nesting species. We are aware of only one observation in Finland where a boreal owl nest has been found between two spruce trunks, and three nests in buildings (Korpimäki 1981, Lehtoranta 1981). Elsewhere, the owl has bred in nests made by hooded crows *Corvus corone*, magpies *Pica pica*, and red squirrels *Sciurus vulgaris*, in a rock hole, in a barn, under a plate on the turfed roof of a building, and in the attic of a cottage (summary in Glutz von Blotzheim and Bauer 1980, Korpimäki 1981, see also Jacobsen and Sonerud 1993). These are, however, very rare exceptions, and probably only occur under conditions of extreme scarcity of nest-holes.

Fig. 4.3.   In Finland, most natural cavities excavated by black woodpeckers and later used by boreal owls are in aspen trees. Photo: Benjam Pöntinen. (For colour version, see colour plate.)

Fig. 4.4.   In the Jura Mountains of Switzerland, the vast majority of cavities used by boreal owls were excavated by black woodpeckers in large beech trees in timber forests. Photo: Pierre Henrioux.

In North America, relatively few nests of boreal owls have been found in natural cavities. Eighteen (out of 19) nest-cavities of boreal owls were excavated by pileated woodpeckers and one probably by common flickers in the Rocky Mountains of Idaho (Hayward et al. 1993), whereas in Colorado, common flickers hollowed out most of the cavities used by owls (Ryder et al. 1987). In Idaho, most nest-cavities were in ponderosa pine (53%), followed by aspen (37%), Douglas-fir (5%) and Engelmann spruce (5%) (Hayward et al. 1993). Ten nests were found in snags and the rest were on living trees in Idaho (Hayward et al. 1993), whereas all nests were in conifer snags in Colorado (Ryder et al. 1987). Entrance holes of cavities averaged 102 mm high (range 64–150 mm) and 95 mm (56–148 mm) wide (Hayward et al. 1993). The corresponding dimensions were 6–17 cm and 6–7 cm in Ontario, Canada (Bondrup-Nielsen 1984). The cavities ranged from 7 to 50 cm deep inside and from 15 to 26 cm (mean 19 cm) inside diameter, with the average tree diameter at the cavity 41 cm (26–61 cm). The owls selected relatively high cavities, averaging 12.7 m above ground (range 6–25 m) in Idaho (Hayward et al. 1993) and range 11–17 m in Ontario (Bondrup-Nielsen 1984). Height of the cavity averaged 51% of the tree height (Hayward et al. 1993). In trees or snags with multiple cavities, owls usually occupied one of the uppermost cavities, which suggested a preference for high nest-sites. The forest in the immediate vicinity of nest-trees appeared to have an open structure with the density of 2.5–23 cm dbh trees within a 0.01-ha plot around the nest averaging only 398 and the density of trees larger than 23.1 cm dbh averaging 212. The former density of trees was three times lower than the average for winter roosts in the same area (Hayward et al. 1993).

4.3.2.   Many users of cavities and nest-boxes

The concept of keystone species suggests that, in many ecosystems, certain species have important effects on several other species, often far beyond what might be expected from a simple consideration of their density or biomass (review in Simberloff 1998). Therefore, a keystone species has important functions that might be influential on the function and structure of the ecosystem. If a local population of a keystone species declines, many other species will decline along with them. The black woodpecker is probably an important keystone species in Eurasian coniferous forests, because it is the only species excavating tree-holes large enough for bigger hole-nesting species. Therefore, the cavities made by black woodpeckers have many secondary users in European forests apart from boreal owls. The common goldeneye *Bucephala clangula*, the stock dove *Columba oenas*, and the red squirrel are the most common cavity-users in northern Europe, and the stock dove and the nuthatch *Sitta europaea* in central Europe (Table 4.1). In addition, the jackdaw *Corvus monedula* frequently uses cavities made by black woodpeckers close to agricultural farmland. In the Kauhava region, western Finland, where stock doves and jackdaws have only distributed recently, the keenest competitors of boreal owls for nest-holes are the common goldeneye and the red squirrel. It is noteworthy that the worst predators of boreal owls, the pine marten *Martes martes* in northern Europe and the stone marten *M. foina* in central Europe, also use the same cavities for roosting and reproduction.

Table 4.1. Utilisation of natural cavities (CA) made by black woodpeckers in the Kauhava region (western Finland), in farmland (FA) and forest (FO) areas of Uppland (central Sweden), in Småland (southern Sweden), in Saaletal (Germany) (HI and LO = high-quality and low-quality cavities, respectively), and in Thüringen, (Germany) (DT = deciduous trees, CT = coniferous trees). Utilisation of nest-boxes (NB) of boreal owl size in the Kauhava region is given for comparison.

	Kauhava		Uppland			Saaletaal		Thüringen	
	CA	NB	FA	FO	Småland	HI	LO	DT	CT
Total number of hole years	576	677	176	151	61				
Unoccupied	488	461	80	120	25				
Water-filled	0	0	0	1	7				
Boreal owl	49	104	3	2	2	13	7	301	109
Tawny owl	0	0	0	0	1	2	1	36	4
Stock dove	0	0	15	11	0	42	16	741	2
Common goldeneye	6	20	8	7	2	0	0	0	0
Black woodpecker	14	0	1	5	2	25	3	450	26
Grey-headed woodpecker	0	0	0	0	0	0	0	4	0
Green woodpecker	0	0	0	0	0	0	0	1	0
Great spotted woodpecker	1	0	0	0	0	0	0	3	0
Wryneck	0	0	1	0	0	0	0	0	0
Starling	0	1	0	0	9	6	1	69	4
Tits (*Parus* spp.)	1	0	3	1	4	1	3	16	6
Eurasian nuthatch	0	0	1	0	0	33	3	250	24
Jackdaw	3	0	72	0	8	0	0	0	0
Red squirrel	11	75	1	1	0	5	8	21	5
Siberian flying squirrel	2	12	0	0	0	0	0	0	0
Pine marten	1	0	0	3	0	0	0	0	0
Stone marten	0	0	0	0	0	1	2	0	0
Marten spp.	0	0	0	0	0	0	0	7	4
Edible dormouse	0	0	0	0	0	0	0	6	0
Bats (Chiroptera)	0	0	0	0	1	0	0	0	0
Bee (*Apis* spp.)	0	0	0	1	0	0	0	13	4

Sources: Kauhava region, pooled data from 1966–2008 (Korpimäki 1984 and unpublished data); Uppland and Småland, pooled data from 1986–7, farmland was defined as nest-holes <200 m from agricultural fields covering at least 10 ha (Johnsson et al. 1993); Saaletaal, data from 1978 (Rudat et al. 1979); Thüringen, data from 21 years (Meyer W. 2003).

Detailed studies in Sweden and Germany on the quality and users of natural cavities showed that many holes were of poor quality due to shallow depth, narrow entrance, and water flowing down into the hole on rainy days (Rudat et al. 1979, Johnsson et al. 1993, Meyer W. 2003). In addition, red squirrels in northern European coniferous forests are still quite abundant, although decreasing (Selonen et al. 2010). They collect much nesting material, including sticks, moss, lichen and grass, in the cavities and nest-boxes. By filling the cavities and boxes up to the entrance hole, red squirrels usually make these holes unsuitable for breeding boreal owls. In contrast, Siberian flying squirrels *Pteromys volans*, which in Finland also occupy natural cavities made by black woodpeckers (Table 4.1), do not usually use sticks as their nest material. Therefore, flying squirrels do not substantially reduce the accessibility of natural cavities to other secondary hole-nesters. In central Europe, nuthatches can block up the entrance holes of cavities of black woodpeckers and large nest-boxes with mud to make the entrance holes inaccessible to larger secondary users. Moreover, there are several records, at least in Germany, in which this mud-walling has even happened during the breeding season of boreal owls, so that a female boreal owl with eggs or young has been imprisoned inside her nest-cavity (König 1968a). Although the male owl continued to feed its family through the small entrance hole, the female and young would have died inside their 'jail' if nest-inspectors had not been able to come in time to break the mud wall (König and Weick 2008). Common goldeneyes only use cavities where the diameter of the entrance hole is >75 mm, while stock doves and jackdaws can use entrance holes with a smaller diameter (Johnsson et al. 1993). The smallest entrance hole diameter through which a boreal owl female has forced herself was recorded to be 5.7 cm (Schelper 1972, Korpimäki 1981), but they can accept boxes and holes with an entrance diameter of 18 cm.

4.3.3.　Nest-box experiments

Four main types of nest-boxes to provide nest-sites for owls and other medium-sized and large cavity-nesters have been used in Finland. The starling/pygmy owl type has an internal diameter of 13–15 cm and diameter of entrance hole of 4.5–5.5 cm. The corresponding diameters for the goldeneye/boreal owl type are 17–21 cm and 8–10 cm, for the tawny owl type 22–24 cm and 12–13 cm, and for the Ural owl type 26–35 cm and 15–18 cm. These four types of nest-boxes were available in the field during 1965–75 in South Ostrobothnia and Suomenselkä, western Finland, and were checked annually (Korpimäki 1976). Boreal owls clearly preferred to breed in the nest-boxes suitable for their own type, but nests were quite often also found in the tawny owl box type, and sometimes even in the Ural owl box type and the pygmy owl box type (Table 4.2). In the latter case, the starling boxes had unusually large entrance diameters (5.7–6.7 cm), which a female boreal owl was able to get through. In these areas, tawny owls are quite rare (Korpimäki 1987b) and only occupy their boxes very infrequently, thus leaving these boxes as safe breeding sites for boreal owls. The other most common users of boreal owl boxes have been common goldeneyes, great tits and starlings, but it should be noted that the starling population in Finland has decreased dramatically from the early 1980s onwards (Korpimäki 1978, Solonen et al. 1991), and therefore very few starlings breed in owl boxes nowadays. All black woodpeckers have been found breeding in cavities excavated by themselves, but great spotted woodpeckers can also sometimes breed in pygmy owl type boxes, where starlings and great tits have been the most common users. Pygmy owls do not usually accept starling boxes as their breeding sites, but in the late 1980s, bird-ringers in Tampere, central Finland found that pygmy owls willingly accept boxes that have thick (at least 5 cm) front walls with an entrance diameter of 4.5 cm. This discovery has proved very useful, and from the late 1990s onwards hundreds of pygmy owls have been breeding in boxes in Finland (for example, >500 in 2002, 2006 and 2008, >900 in 2009; Haapala et al. 1996a, 1997, Björklund et al. 2003, Honkala and Saurola 2006, 2007, 2008). Great tits, boreal owls and common goldeneyes were the most frequent users of tawny owl

Table 4.2. Utilisation of different-sized natural cavities and nest-boxes in South Ostrobothnia and Suomenselkä, western Finland (data from 1965–75, Korpimäki 1976).

Species	Pygmy owl size	Boreal owl size	Tawny owl size	Ural owl size
Boreal owl	2	161	22	3
Hawk owl	0	1	0	0
Tawny owl	0	1	4	0
Ural owl	0	2	4	67
Pygmy owl	6	0	0	0
Stock dove	0	1	0	0
Goosander	0	5	2	3
Common goldeneye	0	236	14	11
Black woodpecker	0	18	0	0
Great spotted woodpecker	13	0	0	0
Three-toed woodpecker	1	0	0	0
Wryneck	1	1	0	0
Swift	3	0	0	0
Jackdaw	0	3	1	0
Starling	27	47	14	1
White wagtail	0	3	2	0
Common redstart	3	20	3	0
Pied flycatcher	5	3	2	0
Eurasian treecreeper	0	1	0	0
Great tit	27	173	23	1
Red squirrel	2	62	9	0
Siberian flying squirrel	4	28	3	0
Bee (*Apis* spp.)	0	2	0	0
Total	94	768	103	86

type boxes. Large Ural owls clearly preferred their own box type, which was also quite often occupied by common goldeneyes.

In our study area, the boreal owl population nested mainly in nest-boxes (90% of 287 nests found in 1966–82; Korpimäki 1984). The vast majority of boreal owls also breed in nest-boxes in many other local populations of northern Europe (Lagerström 1980, Sonerud 1985a, Hörnfeldt et al. 1990), central Europe (Schelper 1972, Schwerdtfeger 2008) and southern Europe (Mezzavilla and Lombardo 1997, Lopez et al. 2010). In our study area, the users of the nest-boxes of boreal owls are the same as for natural cavities, but black woodpeckers only breed in their own cavities and do not accept nest-boxes (Table 4.1). The occupancy rate of nest-boxes by breeding boreal owls was markedly higher than that of black woodpecker cavities (15.0% of boxes vs. 8.5% of cavities occupied by boreal owls; data from Table 4.1). This preference for nest-boxes over natural cavities was found despite the fact that the natural cavities of black woodpeckers in our study area are usually 7–10 m above ground (mean 7.3 m in southern and central Sweden; Johnsson et al. 1993), whereas our boreal owl nest-boxes are usually only 4–6 m above ground. Many secondary users of natural cavities prefer to breed in cavities high up (Johnsson et al. 1993), and this is also true for boreal owls in North America (Hayward et al. 1993). Also, in northern Italy, boreal owls preferred nest-boxes over natural cavities

(Mezzavilla et al. 1994). In the Jura Mountains, Switzerland, breeding boreal owls have gradually shifted to breed in nest-boxes, although the availability of natural cavities has remained quite constant (Ravussin et al. 2001b). This shift has been mainly because nest-boxes give better protection against nest-robbers such as martens.

As far as we know, only five nest-box experiments on boreal owls have been carried out in North America. In the Chamberlain study area of Idaho, a total of 45 wooden nest-boxes constructed from 2-cm-thick lumber were erected in 1984 in three main vegetation types (15 boxes in mixed conifer, 11 in Douglas-fir, and 19 in lodgepole pine) within a 9-km^2 plot (Hayward et al. 1993). Nest-boxes measured 44 cm high, 25 cm wide, 18 cm deep, had a 9 cm-diameter entrance hole, and were hung 4–15 m above ground. During 1985–8, courting or nesting was recorded only at three nest-boxes, although radio-marked boreal owls frequently foraged and roosted near boxes. One nest of boreal owls was in the nest-boxes of mixed conifer stands in 1987 and 1988, and one nest in a box in a Douglas-fir forest in 1987 (Hayward et al. 1993).

A larger-scale experiment was carried out in Payette National Forest, central Idaho, where 283 boxes were hung in 1987 and an additional 167 boxes were erected in 1989 (Hayward et al. 1992). Nest-boxes were constructed from rough-cut 3-cm-thick pine and fir. Inside-box dimensions were as follows: bottom 20×20 cm, front height 46 cm, back height 51 cm and entrance hole diameter 9 cm. Boxes were spaced at 0.5-km intervals along haul roads in three separate drainage areas at elevations of 1520–2140 m. An almost 2-m layer of snow accumulates in most of the areas and half of the ground is not snow-free until after 1 June in most years. During the three years, the occupancy rate of boreal owls averaged only 4.2% (3.1%, 3.7% and 5.8% in 1988, 1989, 1990, respectively; Hayward et al. 1992). The authors concluded that the use of nest-boxes to monitor demographic responses of boreal owls to habitat change was too costly, in particular when including the costs of labour (US$10/hour for the crew leader and US$6/hour for crew; Hayward et al. 1992). Possible reasons for the low occupancy rates of boreal owl nest-boxes remained open in these two experiments, but may include the abundance of suitable natural cavities excavated by pileated woodpeckers and common flickers in these pristine forests. In addition, the very deep snow layer covering the ground for most of the breeding season of boreal owls might have prevented access to small mammals, which might have resulted in a scarcity of food during the egg-laying period.

A total of 105 nest-boxes of boreal owls were erected in boreal forests of the southern Yukon, Canada, and inspected during the five years commencing in 1984. Only 1% of the nest-boxes were occupied by boreal owls although they are fairly common in the forested areas of Yukon (Mossop 1997). Some boxes were also used by American kestrels *Falco sparverius*, buffleheads *Bucephala albeola*, Barrow's goldeneyes *B. icelandica*, northern flickers and American red squirrels *Tamiasciurus hudsonicus*. The author concluded that apparently sufficient numbers of adequate natural cavities existed in the study area. Nest-box experiments were much more successful in Alaska, USA, where 91 nest-boxes were erected in 2004, and 17 additional boxes in 2005, in the interior of Alaska (near Fairbanks) (Fig. 4.5). A total of 342 nest-box years were available during 2005–7, and 111 nests of boreal owls were found in these boxes during 2005–7 (Whitman 2008, 2010). In addition, a total of 1781 nest-box years yielded 142 nests of boreal owls and

Fig. 4.5.   Nest-boxes used for boreal owls near Fairbanks, Alaska, USA. Photo: Jackson S. Whitman.

174 nests of saw-whet owls during 1984–2005 near Anchorage, Alaska (T. Swem and co-workers, unpublished data).

Selective logging of large trees resulted in few large-diameter trees suitable for owl nest-sites at Lianhuashan Mountain, Gansu, China. Therefore, an owl nest-box project was initiated in 2002–3, when 67 boxes suitable for boreal owls were placed in coniferous trees at 4–6 m above ground. Four nests of boreal owls were found in the boxes in 2003, and during 2003–7 between four and seven nest-boxes have been occupied each year by boreal owls (Fang et al. 2009).

Four main types of nest-boxes have been available for boreal owls in our study area. Most boxes (79%) were made from darkened board and had square bottoms. Three size classes correspond to the boreal owl, tawny owl and Ural owl box types described above. The fourth type was made from short logs with an internal diameter of 17–20 cm and the diameter of the entrance hole was 8–10 cm (Fig. 4.6). The bottom area of this box was thus smaller than that of the square-bottomed board boxes, but the diameter of the entrance hole was the same as that of the boreal owl box type. The numbers of these four nest-box types available in our study area and the percentage used by boreal owls were as follows (pooled data from 1979–85; Korpimäki 1987e): boreal owl size class (board boxes) 760 box years and 9.6%, boreal owl size class (log boxes) 677 box years and 6.8%, tawny owl size class (board boxes) 134 box years and 11.9%, and Ural owl size class (board boxes) 165 box years and 3.0%. Of the four nest-box types available, boreal and tawny owl type board boxes were preferred to the Ural owl type, while the log boxes seemed to be occupied less frequently than the two smaller size-classes of board boxes. This suggests that the bottom area of the boxes is an important criterion in nest-site selection in our study area, because the log boxes had smaller bottom areas, although

Fig. 4.6.   Most boxes (79%) for boreal owls in our study area were made from board and had square bottoms. In addition, some boxes were also made from short logs with an internal diameter of 17–20 cm (log-box on the left). Some of the board boxes have been in the field for more than 20 years and their entrance holes have been enlarged by woodpeckers. The senior author (left) and Rauno Varjonen (right) are shown in the background. Photo: Marke Raatikainen.

bark-covered log-boxes are better camouflaged on the tree trunk than board boxes and thus provide better protection against enemies.

In the Swiss Jura Mountains, various types of nest-boxes were also installed to investigate their security against nest predators, mainly pine martens. It appears that the type of nest-box was important to nest-site selection and breeding success. In particular, the nest-boxes made from PVC drainage pipes were preferred to all other types that were also protected by metal sheets against martens and other mammalian nest predators (Chapter 15.3 in this volume; see also figures 6 and 8 in Ravussin et al. 2001b).

4.3.4.   Larger clutches in nest-boxes?

Boreal owls in our study area produced significantly larger clutches in nest-boxes than in natural cavities (mean ± s.d. = 5.80 ± 1.25, $n = 217$ vs. 4.96 ± 1.35, $n = 28$, t-test, $p < 0.001$; pooled data from 1966–82, Korpimäki 1984). The same tendency also appeared, although not significantly so, for the number of fledglings produced (2.88 ± 2.17, $n = 231$ vs. 2.47 ± 1.85, $n = 30$). There was no obvious difference in the median date of laying the first egg (4 April in nest-boxes vs. 3 April in natural cavities) or in the proportion of eggs from which chicks hatched (86.7% vs. 79.8%, respectively). The marked difference in the clutch size was not due to a preference for natural nest-sites, because the proportions of pairs breeding in natural cavities was only 6.2% in good vole years and 6.8% in poor vole years, when the breeding density was low and possible interspecific competition for natural cavities was supposedly relaxed.

In Thüringen, Germany, the number of fledglings produced by boreal owls averaged 3.35 (n = 34) in nest-boxes and 4.18 (n = 22) in natural cavities (Ritter et al. 1978). This difference was mainly due to a higher percentage of eggs hatched in natural nesting places and was not significant. Ritter et al. (1978) suggested that the good hatching success in natural holes was caused by better insulation against cold spells, because the walls of natural cavities were thicker than those of their nest-boxes. The relative air humidity can also be higher in natural cavities than in nest-boxes. The larger data set from 21 years, collected in the same province of Germany, included 387 nests in natural cavities and 58 nests in nest-boxes. Boreal owls breeding in natural cavities reared on average 2.5 fledglings per brood, but significantly fewer in nest-boxes (1.3). The lower productivity in nest-boxes was mainly due to the fact that pine and stone martens predated 59% of owl nests in nest-boxes but only 24% in natural cavities (Meyer W. 2003). The reverse appeared to be true in the Pyrenees of Spain, where the incidence of predation (mainly by pine martens) was higher in natural cavities than in nest-boxes (50% vs. 15%, respectively) (Lopez et al. 2010).

The results from the Finnish and German study areas appeared to be inconsistent. There are several possible reasons that could explain this difference. Old holes of black woodpeckers may have several entrance holes, which reduce the insulation of the cavity, so that the thermoregulation costs of the female owl probably increase. Boreal owls may initiate egg-laying and incubation very early (in late February to early March) when ambient temperature can be as low as −15 to −20°C. This probably makes good insulation of the nest-cavity very important in northern conditions. This interpretation was also supported by the fact that in good vole years, in particular, with early initiation of egg-laying and incubation of boreal owl nests, nesting was most successful in thick-walled boxes with a small-diameter (8 cm) entrance hole (Korpimäki 1985a). During rainy periods, water may flow into the cavity, forming ice or a pool of water at the bottom. Rudat et al. (1979) found that 20% of owl nests in natural cavities in Germany failed due to water collecting at the bottom, but this has not been observed in our study area. Instead, red squirrels and Siberian flying squirrels can fill up the cavities with nest material or food stores. This probably makes the cavities too shallow and thus energetically unfavourable for boreal owls.

We suggest that the main reason for the larger clutch sizes of boreal owls in nest-boxes rather than in natural cavities is that the bottom area of the cavities excavated by black woodpeckers is smaller (mean hole diameter 20.1 cm; Rudat et al. 1979) than those of the most commonly used nest-boxes of boreal owls (bottom area 20 × 20 cm, entrance diameter 8 cm; Korpimäki 1985a). In addition, the diameter of the entrance hole of natural cavities is larger (vertical diameter 7.6 cm, horizontal diameter 10.6 cm; Johnsson et al. 1993) than those of boreal owl nest-boxes. By using the pooled data from different-sized nest-boxes in the field during 1966–82, we have shown that mean clutch size and the bottom area of the nest-box correlated positively in good vole years (r = 0.313, p < 0.01), with clutch size increasing on average by 0.005 eggs per cm^2 increase in bottom area. This means that the clutch size will increase by 0.56 eggs when the area of the square bottom of the nest-box increases from 289 cm^2 (=17 × 17 cm) to 400 cm^2 (=20 × 20 cm). This 'area effect' was not due to the fact that egg-laying started earlier in larger boxes than in small ones. However, in poor vole years this 'area effect' was not found in boreal owls, nor was it found for other vole-eating birds of prey, such as the Eurasian kestrel, in our study area (Valkama and Korpimäki 1999). This 'area effect' has been recorded for many small hole-nesting passerine birds including the great tit, willow tit *Parus montanus*, the marsh tit *P. palustris*, the pied flycatcher, but excluding the starling (summary in Korpimäki 1985a). The main reason for the bigger clutches of boreal owls in larger nest-boxes is probably that many eggs, fat females, and more prey items in the food stores at the bottom of the box need more space in good vole years. The mean number of prey items in the food stores of boreal owls was also larger in medium-sized and large boxes than in small boxes (6.1 vs. 3.5 prey items; Korpimäki 1985a).

In addition, there was a lower proportion of totally destroyed nests in boxes with a small entrance hole (diameter 5.5–8.0 cm) than in boxes with medium-sized (9–11 cm) and large entrance holes (12–18 cm) (Korpimäki 1985a). This was attributable to the fact that the larger-hole-nesting tawny and Ural owls are not able to enter boxes with a small entrance hole. These larger owls are enemies of boreal owls and can even kill them (Mikkola 1983, Hakkarainen and Korpimäki 1996). The main predator of boreal owls and their nests in northern Europe, the pine marten, mainly uses boxes with large entrance holes (average diameter >10 cm) as nesting and daytime roosting places (Nyholm 1970, Ahola and Terhivuo 1982) but apparently is somewhat reluctant to enter a small hole. Therefore, pine martens find boreal owls nests in boxes with a larger entrance hole more frequently than in those with a smaller one. The wall thickness of the box did not have any obvious effect on clutch size and number of fledglings produced, but there were significantly more unhatched eggs in boxes with thin walls (0.5–3.0 cm) than in boxes with thick walls (3.5–10 cm) (Korpimäki 1985a). This suggests that good insulation of the nest-box is important for the efficiency of incubation and therefore for hatching success.

4.3.5.   Occupancy rate of boxes decreases with box age

It has been shown that the occupancy rate of nest-boxes used by breeding boreal owls decreases with increasing age of the boxes. In our study area, the percentage use of suitable nest-boxes by boreal owls was >20% for boxes 1–4 years of age, whereas it was only 10–18% for boxes 5–10 years of age, and <10% for boxes 11–18 years of age (Korpimäki 1987e). Suitable nest-boxes were defined as those that were used by boreal owls for nesting at least once during the study period (1966–85). To find out whether this decreasing usage of nest-holes with increasing age of the hole is also true for natural cavities, we sampled 36 natural cavities that were inspected annually and were in the field for at least 10 years. As control nest-holes we used the closest nest-box, which was usually 1–2 km away from the natural cavity. Consistent with the earlier results (Korpimäki 1987e, see also Sonerud 1985a), the proportion of natural cavities and nest-boxes occupied by breeding boreal owls decreased significantly with increasing age of the nest-hole, but this decreasing trend appeared to be steeper for natural cavities than for nest-boxes (Fig. 4.7).

Boreal owls preferred to breed in new cavities and nest-boxes, which can be explained by at least two, not mutually exclusive, hypotheses. First, the *nest-hole quality hypothesis* posits that this decrease is due to the deteriorating quality of the cavity or box and predicts

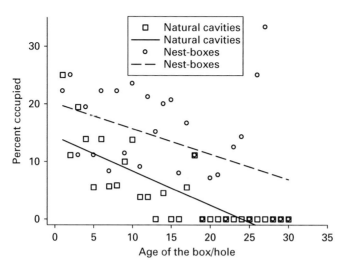

Fig. 4.7.   Percentage of natural cavities and nest-boxes occupied by breeding boreal owls plotted against the age of the nest-hole. Pooled data from 36 natural cavities that had been in the field for at least 10 years and their closest 36 nest-boxes (pooled data from 1966–2008) (Pearson correlation, $r = -0.78$, $p < 0.01$, $n = 30$ for natural cavities and $r = -0.42$, $p < 0.05$, $n = 30$ for nest-boxes) (data from Korpimäki 1987e and unpublished).

that the breeding frequency in nest-boxes should increase when an old box is replaced with a new box at the same place on the tree (Korpimäki 1987e, 1993a). Puddles of water, or old nest material placed there by squirrels, often fill cavities (see above). Because there are often many entrance holes in the cavities, and old cavities and boxes sometimes crack in the field, the insulation of old nest-holes may be poorer than that of new ones. Boreal owls may select nest-holes with good insulation, as they start to lay early under adverse weather conditions (Korpimäki 1987f, Hörnfeldt et al. 1990).

Second, the *predation risk hypothesis* suggests that the decrease in the breeding frequency in nest-holes with age is a response to the mammalian predators, in particular martens (*Martes* spp.), having a long-term memory of the holes on their home ranges (Sonerud 1985a, 1989). Because the predation risk may be higher for old holes than for new ones (Sonerud 1985a), there should be selection for breeding in new cavities. This hypothesis predicts that the use of old nest-boxes should increase if they were relocated.

We selected 162 nest-boxes from the middle of our study area with the oldest boxes in autumn 1986. These boxes were divided into four categories at random:

1. 28 boxes were replaced by new boxes at the same place on the tree
2. 28 boxes were transferred 40–100 m and re-erected within the same wood at the original height (5–6 m above ground) on the same tree species (spruce)
3. 27 boxes were both renewed and relocated within the same wood
4. 79 old boxes were left in the same place on the tree and served as controls.

All the new boxes were of the same type as the old boxes (small or medium-sized board boxes or log boxes; see Chapter 4.3.3). After being transferred, none of the boxes could be seen from where they had been situated earlier, but the visibility of the box was similar to its earlier location. The boxes of the four categories were checked in late March to April and again in late May to early June during 1987–91. Pellets and other prey remains accumulated at the bottom of the boxes were removed after the breeding season to study diet composition. This might have reduced the load of detrimental ectoparasites (Møller 1989a) and increased the use of the boxes but did not bias the comparisons between different categories. The breeding frequency in the boxes was estimated as the number of breeding attempts (at least one egg laid) per year.

The mean breeding frequency (s.d., number of boxes) of experimental nest-boxes during 1987–91 was as follows (Korpimäki 1993a): control boxes 0.242 (0.175, 79), renewed boxes 0.211 (0.158, 28), relocated boxes 0.186 (0.159, 28), and renewed and relocated boxes 0.147 (0.183, 27). The breeding frequency for control and renewed boxes tended to be higher than that for relocated, and renewed and relocated boxes, but the between-treatment difference was not significant (Kruskal–Wallis test, $H_{3,158} = 5.49$, two-tailed $p = 0.14$). Because the renewal and/or the relocation of the boxes did not have obvious effects on the breeding frequency in the boxes, the outcome of the nest-box experiments did not support the nest-hole quality and the predation risk hypotheses. It may be that the variance in the nest-box quality in our study area was too small to essentially influence the breeding dispersal of boreal owls. In addition, nest-boxes were cleaned after the breeding season by removing old prey remains and pellets, which offer good reproductive and living conditions for the detrimental ectoparasites of owls. Because the parasite loads of the old boxes were reduced, the selection advantage in promoting breeding dispersal may have been reduced.

In general, the results from nest-hole shifts of parent boreal owls in successive breeding seasons (i.e. breeding dispersal) in our study area lend support to the *food depletion hypothesis*, which proposes that a breeding owl pair may reduce vole density in the vicinity of the nest-box. This food depletion, in turn, increases nest-hole shifts and thus the occupancy rate of new nest-boxes (Korpimäki 1993a, Chapter 9.3). Food depletion may be particularly important in explaining nest-hole shifts of boreal owls in areas with pronounced temporal between-year variations in main food abundance (for example in northern Europe). Spatial variation in nest-hole quality and predation risk, in turn, might be more important in areas where environmental conditions are more stable between years.

In summary, a nest-box with a square bottom at least 20 × 20 cm, entrance hole diameter 7.5–8.0 cm, and wall thickness at least 3 cm appeared to be optimal for breeding boreal owls in northern European conditions with low ambient temperatures during the early phases of the breeding season. The provision of nest-boxes to offer high-quality nest-sites for boreal owls has been very successful in all parts of Europe, provided that the network of nest-boxes is annually managed and that new boxes are set up at least every fifth year. In addition, the provision of nest-boxes has also been successful in China and Alaska. Therefore, we would like to suggest that these kinds of nest-boxes should also be erected on a large scale in coniferous forests managed for modern forestry in Canada and the USA. This would be important in order to finally find out whether boreal owls really accept nest-boxes to the same degree that conspecifics in Eurasia do, or whether mainly natural nest-sites should be provided for the persistence of boreal owl populations in North American boreal forests south of Alaska.

5 Interactions with prey animals

Source: Drawn by Marke Raatikainen.

5.1.   Foraging behaviour: locating and capturing hidden prey at night

Food abundance appears to be the primary factor governing the life-history, demography and behaviour of vertebrate predators, including boreal owls. Boreal owls are mostly nocturnal when hunting (Fig. 5.1), even during the breeding season in the far north, where midsummer nights are short and light (Klaus et al. 1975, Korpimäki 1981, Mikkola 1983). On the other hand, nights in winter are very long, and therefore their senses should be fully adapted to hunt in full darkness even in dense coniferous forests. Boreal owls mainly use auditory location of prey, in which they are really skilful, because the asymmetrical location of their ear-openings on different sides of the skull highly improves the accuracy of auditory location (Norberg 1968, 1978). It is very difficult to locate prey making rustling sounds when concealed in dense vegetation or below a snow layer. The main advantage of asymmetrical ears is that the horizontal and vertical directions of a sound source can be determined simultaneously, with the same accuracy in both planes. Even more importantly, asymmetrical ears seem to be indispensible for the localisation and capture of hidden prey that moves, because the precise location of a mobile prey can be continuously tracked (Norberg 1987). With symmetrical ears, horizontal and vertical directions have to be determined one at a time, with a tilting of the head in between (Norberg 1987).

Vision is apparently less important in locating prey concealed in dense vegetation or below a snow layer, although the eyes of nocturnal birds are highly adapted to viewing in darkness, with a retina dominated by rods (Martin G. 1990). Like other owls, boreal owls have binocular vision, looking at a target with both eyes to ascertain its accurate location. Binocular vision becomes more effective the farther apart the eyes are situated. Smaller owl species, including boreal owls, also have a flatter skull, which increases the wide spacing of the eyes still further. Owls can improve their three-dimensional vision by

Fig. 5.1.   Boreal owls spend a majority of time searching for prey from low perches: usual perch height ranges from 1 to 3 metres. Photo: Benjam Pöntinen.

bobbing their heads so that both eyes see the object under scrutiny from many different angles. The bobbing or weaving movements of the head make boreal owls look very comical when they carefully assess what they are seeing before taking action.

Boreal owls also use vision in hunting, in particular during light midsummer nights and in snowy areas in the north where snow cover markedly increases the light levels at night. Small rodents mark their territories with urine and faeces which are visible in ultraviolet (UV) wavelengths of light (Desjardins et al. 1973, Viitala et al. 1995). We have shown experimentally that wild Eurasian kestrels can see vole scent marks in UV light and use them as cues when searching for vole patches for hunting purposes in the field (Viitala et al. 1995). UV vision might be advantageous for owls in the wild, because females could use it to find a suitable breeding area, and males might use it to search for food patches on their home ranges. We performed a laboratory experiment with 14 young and 14 older (<6 and >6 months old, respectively) boreal owls. These owls had a choice between four adjacent arenas: an arena with vole urine and faeces in UV light, an arena with vole urine and faeces in visible light, a clean arena in UV light, and a clean arena in visible light. Neither young nor older owls scanned more arenas with vole urine and faeces in UV and visible light than clean arenas (Koivula et al. 1997), although UV light is also present in the night sky. Therefore, our results suggest that boreal owls do not rely on visual detection of scent marks when hunting, at least under laboratory conditions. This result is consistent with the results from a previous study on the eye structure of nocturnal tawny owls, where no UV-receptors were found (Bowmaker and Martin G. 1978). It was also unlikely that we failed to detect UV-sensitivity in boreal owls as a consequence of laboratory artefacts, because we used the same experimental design and room where diurnal Eurasian kestrels had shown a preference for scanning on the arena with vole urine and faeces in UV light (Viitala et al. 1995). The experimental situation was not unnatural, as wild boreal owls sometimes hunt inside small buildings in both winter and summer (Korpimäki 1981). The prey selection of boreal owls has also been studied in an aviary, where they willingly hunt voles (Koivunen et al. 1998a, 1998b).

Dr. R. Åke Norberg (1970) was the first to study and photograph the hunting behaviour of boreal owls. The study was done during light summer nights in northern Sweden. Boreal owls mainly use the 'travel-sit-and-wait' mode when hunting, in which they move through a forest following an irregular or zig-zag route. They spend the majority of time searching for prey from low perches (perch height: mean 1.7 m, range 0.5–8 m in northern Sweden, Norberg 1970; mean 4 m, range 0.5–9 m in Idaho, USA, Hayward et al. 1993) (Fig. 5.1). Owls flew an average of 25 ± 8 m between hunting perches, the median distance between perches was 17 m, and >90% of recorded flights were estimated to be <40 m (Hayward et al. 1993). Individual mean hunting perch height was 3.1 m (range 1.7–8.7 m) in the pooled data from south-eastern Norway and south-eastern Sweden, whereas daytime roost sites were markedly higher on trees: mean 4.3 m for both males and females (Bye et al. 1992). Perches from which a successful and an unsuccessful attack were launched appeared not to have obvious differences in height: 2.6 m and 2.4 m, respectively (Bye et al. 1992). Boreal owls strike the prey from a mean distance of about 5 m (Norberg 1970), and attack distance varied from 2.2 to 12.6 m, with an overall mean of 5.6 m (Bye et al. 1992). Before striking, the owl lowers its head, makes

wing beats during the first part of the strike flight, and at the end of the strike it glides
for approx. 1 m with its whole body directed towards the prey item. The owl draws its
head back some 20 cm from the prey, extends its legs and stretches its feet forwards
with talons fully spread so that both feet cover an area of 4 × 6.5 cm. The owl closes its
eyes just before impact, probably to protect them when plunging into the dense ground
vegetation. After a strike, the owl spreads its tail and wings to take support from the ground.
The prey is usually killed by bites in the head or back of the neck (see detailed pictures in
Norberg 1970).

   During perch-hunting, individual mean giving-up time was 134 seconds (range
45–232 s), while individual detection time was substantially longer (mean 273 seconds,
range 184–396 s). Most (66%) recorded perches were left after only 2 minutes (Bye et al.
1992), or 78% of perches were occupied <5 minutes, 64% for <3 minutes, and 27% for
<1 minute (Hayward et al. 1993). The owls are able to use the travel-sit-and-wait hunting
mode even in dense coniferous forests, and they seemed to prefer dense patches of forest
(Norberg 1970). Rustling sounds of prey moving among vegetation, dry grass and leaves,
or under the snow, are the clue that probably first directs the attention of boreal owls to the
prey. There is also evidence that boreal owls can locate roosting birds and nests of
incubating or brooding birds by using this hunting mode (Norberg 1970, Hayward et al.
1993). This was also confirmed by our camera recording at the nests. In a poor vole year,
the male delivered, for example, first a female chaffinch and then its 3–4 chicks at
intervals of a few minutes (Korpimäki 1981). Boreal owls are well adapted to hunt in
dense forests because of their small size and skilful flight in dense vegetation. Because of
the use of low perches and short strike distances, the owl can scan very effectively and
make good use of auditory clues. The search area – and thus the capture rate – of a
predator relying mostly on auditory clues will decline quadratically with increasing perch
height. If prey is hidden in ground cover that is at least twice the height of prey, a predator
simultaneously using vision and hearing will maximise its capture rate by perching as
close to the ground as possible, because the auditory search area increases more rapidly
with declining height than the visual search area declines (Rice 1983, Bye et al. 1992). It
is really fascinating that, in the course of evolution, boreal owls seem to have obtained
'knowledge' of these basic principles of physics!

5.2.   Do onomies restrict hunting to the hours of darkness?

Boreal owls usually initiate hunting after sunset, because male owls usually deliver their
first prey to the nest 30–60 minutes after sunset in our study area (Korpimäki 1981) and
120 minutes after sunset in the western Czech Republic (Zárybnická 2009a; see also
Klaus et al. 1975). The highest peak in nest visits occurred between 22:00 and 23:00
hours and a lower peak was recorded between 01:00 and 02:00 hours in our study area. In
the course of the breeding season, with shorter nights these two peaks merged and there
was only one activity peak between 23:00 hours and midnight (Korpimäki 1981).

   Direct observations at the nests in the course of a 3-year vole cycle with varying main
food abundance also revealed that the frequency of prey deliveries in the increase and

decrease phases of the vole cycle was much higher before midnight than after midnight, whereas in the low phase the opposite was true (Hakkarainen and Korpimäki 1994a). In Germany, with longer nights the first peak in nest visit frequency was between 20:00 and 22:00 hours and the second lower peak between 02:00 and 04:00 hours (Klaus et al. 1975). A similar bimodality was also recorded in prey delivery rates of male owls in the Czech Republic (Zárybnická 2009a). Males normally stop active hunting before sunrise, as last prey deliveries occur 30–40 minutes (in Finland) or 30–60 minutes (Czech Republic) before sunrise (Korpimäki 1981, Zárybnická 2009a). Only very occasionally have prey deliveries at the nests been recorded in the morning, during daylight hours (Korpimäki 1981), even when the main food is scarce, when egg-laying starts late, and when the hunting period at night lasts <4 hours in June. This leads to the peculiar situation that males spend >20 hours per day in roosts, and the family is therefore left without food deliveries for >80% of the day. In years of food abundance, breeding is initiated 1 month earlier, which means that in April–May the night length in northern Europe is much longer, leaving 5–8 hours of darkness per night for foraging by boreal owl males. This reduces the roosting time of males to between 67% and <80% of the day.

An important question is whether nocturnal boreal owls could be able to provide food for their offspring during daylight to buffer against the short northern nights. We studied prey delivery rates of male owls in our study area with short nights (night length, mean ± s.d. 4:57 ± 1:21 h) and in the western Czech Republic with considerably longer nights (8:03 ± 0:23 h). Finnish males delivered a higher number of prey items to their nests in a 24-hour period than Czech males did (7.8 ± 2.9 vs. 5.6 ± 1.4 prey items/nest per day). Also, at night, Finnish males delivered more prey items to their nests than did Czech males (7.0 ± 2.2 vs. 5.2 ± 1.3 prey items/nest per day), whereas no such a difference was recorded during the daytime (0.8 ± 1.2 vs. 0.3 ± 0.1 prey items/nest per day). Thus, Finnish male owls did not increase their prey delivery rates during the daytime despite the short nights. At the Czech site, wood mice (*Apodemus* spp.) and voles of the genera *Microtus* and *Myodes* were the main foods of boreal owls, whereas in our study site shrews were the main foods, along with *Microtus* and *Myodes* voles. As a result, the mean prey weight of Finnish owls was substantially smaller than that of Czech owls (19.3 ± 1.6 g vs. 24.0 ± 0.4 g; Zárybnická et al. 2009). There was no obvious inter-areal difference in the number of fledglings produced, but the number of owlets hatched per nest (6.1 ± 0.7 vs. 4.4 ± 1.0) and nestling mortality (2.0 ± 1.5 vs. 0.8 ± 1.1 dead nestlings/ nest) was higher in Finland than in the Czech Republic. These results suggest that Finnish males were not able to deliver sufficient biomass of food for their young during the short nights and also appeared to be reluctant to extend their hunting period and prey delivery rates to the nest during daylight hours in the morning.

We assume that the reason for no compensation of food scarcity by delivering prey during daylight hours might be the higher predation risk due to the presence of large diurnal (mainly the goshawk) and crepuscular (mainly Ural and eagle owls) birds of prey at the Finnish site. Breeding densities of larger birds of prey were instead very low at the study site in the Czech Republic. If the male boreal owls were to hunt during the day, because of lack of food or because of the briefness of the night, their risk of being predated would be greater than at night. During the twilight hours and during the day, diurnal and crepuscular birds of prey can easily find boreal owls and they are also the target of mobbing by other non-raptorial birds (review in Caro 2005), which makes it easy for their predators to locate them. The hunting mode of boreal owls tends to make them particularly vulnerable to larger predators, because owls perching close to the ground and making short flights between perches close to the ground are particularly visible to other avian predators hunting from higher perches or flying above the canopy. In addition, boreal owls are probably also vulnerable to ground-searching mammalian predators, such as stoats *Mustela erminea*, weasels *M. nivalis*, pine martens and red foxes, which also hunt in high-density patches of voles.

5.3.   Do food stores buffer against temporary food shortages?

Food storage during the breeding season has been recorded in owls, diurnal raptors and shrikes (*Lanius* spp.) (Källander and Smith H. 1990). Because owls lack a crop in which to store food, prey storage may have a higher adaptive value for owls than for diurnal raptors. Prey storing has been reported both in captive and in wild owls, including wild boreal, hawk, short-eared, tawny, Ural, great grey, snowy and barn owls (review in Korpimäki 1987d; see also Roulin 2004). In addition, pygmy owls store food in tree-holes in late autumn to be consumed during the snowy season (Solheim 1984, Halonen et al. 2007, Suhonen et al. 2007), possibly to improve over-winter survival.

The following four, not mutually exclusive, hypotheses have been put forward to explain the evolution of food storing in birds and mammals.

1. Predators living in cold regions must store food to buffer themselves against temporary food shortages, usually resulting from periods of adverse weather conditions (the *insurance hypothesis*; Korpimäki 1987d).
2. The *large prey hypothesis* states that young birds have difficulties in dismembering and eating large prey items (Korpimäki 1987d). Therefore, they consume the smallest prey items first and dismember large prey items once the benefits of eating them are worth the handling and processing costs.
3. In nocturnal species such as boreal and barn owls, males deliver food only at night, while brooding females and young owlets attempt to spread the consumption of meals over 24 hours. Therefore, the *feeding time hypothesis* states that a function of food stores is to allow the offspring to eat at a time when they are hungry and when parents cannot hunt and deliver prey (Roulin 2004).
4. Prey storing may secure food resources when several co-existing predator species have similar prey preferences (the *food security hypothesis*) (Oksanen T. 1983).

During our visits to boreal owl nests, we identified and counted prey items in the food stores, and most stored prey items were also sexed and body mass measured. We counted a total of 12 554 prey items in the food stores during 4249 visits in 1973–2009 (Fig. 5.2). The total weight of prey animals in food stores was 248.6 kg during these 37 years.

The largest food stores (up to 35 prey items) were found during the egg-laying and hatching periods, but there were regularly food stores in boxes prior to the egg-laying, incubation and

Table 5.1. (a) The number of prey items per visit (n = number of visits) in the different phases of the breeding cycle (pooled data from 1973–85) (Korpimäki 1987d). (b) The storage period (n = number of observations) of the most numerous prey species (pooled data from 1975–80) (Korpimäki 1987d).

(a) Phase	Mean ± s.d	Range	n
Prior to egg-laying	1.5 ± 2.1	0–10	69
Egg-laying period	4.7 ± 5.2	0–32	107
Incubation	2.7 ± 4.4	0–35	318
Hatching	4.6 ± 5.7	0–35	175
First three weeks of nestling period	2.4 ± 4.9	0–32	283
Last week of nestling period	0.1 ± 1.0	0–12	232

(b) Prey species	Mean ± s.d.	Range	n	Mean weight (g)
Common shrew	1.4 ± 0.7	1–3	44	7.5
Harvest mouse	1.0 ± 0.0	1	6	8.0
Willow warbler	1.7 ± 1.2	1–3	3	9.5
Yellowhammer (nestling)	2.2 ± 1.1	1–3	5	10.0
Bank vole	1.3 ± 0.5	1–3	61	16.5
Thrush spp. (nestling)	1.3 ± 0.6	1–3	15	20.0
Chaffinch	1.9 ± 2.0	1–9	18	21.5
Sibling vole	1.7 ± 0.8	1–3	12	23.5
Field vole	2.0 ± 1.1	1–4	18	25.0
Yellowhammer	1.3 ± 0.6	1–2	3	32.0
Redwing	2.0 ± 1.0	1–3	4	58.0

Fig. 5.2.   Female owls pile up the prey items at the edges of the nest-box. In the case of big food stores, the bottom of the box is full of *Microtus* and bank voles. Photo: Jorma Nurmi. (For colour version, see colour plate.)

early nestling periods as well (Table 5.1a, Fig. 5.2). The finding of the highest numbers of prey items in the food stores of boreal owls during the egg-laying and hatching periods was also recorded in Germany (Schwerdtfeger 1988). The probable reason is that the energy needs of the female are highest prior to and during egg-laying. Moreover, during hatching the small owlets do not tolerate long fasting periods and food stores help the brooding females and hatchlings to survive temporary food shortages (Korpimäki 1987d). In the early breeding season, the risk of experiencing adverse weather conditions is higher than later in the season. In contrast, during the late nestling and fledging periods when the chicks are >3 weeks old and the female is no longer brooding them, prey items were very infrequently recorded in nest-boxes. Larger young consume much more food and parent owls are usually not able to feed them so much that food stores would accumulate. In addition, large nestlings do not suffer from temporary food shortages to the same extent that small young do.

The yearly mean number of prey items per visit was highest in 1974, 1977, 1989, 1991 and 2003 (yearly mean range 3.7–5.5 prey items per visit), whereas it was lowest in 1973, 1993, 1997 and 1998 (range 0.1–1.1 prey items per visit; Fig. 5.3). The yearly mean prey weight stored in nest-boxes per visit was largest in 1974, 1977, 1989 and 2003 (yearly mean range 85–118 g per visit), whereas it was smallest in 1973, 1978, 1993, 1998 and 2001 (range 2–16 g per visit; Fig. 5.3). Both in terms of prey number and prey weight, the food stores per nest visit during 1973–2009 were larger in years of *Microtus* vole abundance than in those of vole scarcity (Fig. 5.3). Because the yearly mean clutch size was also positively correlated with the spring trap index of *Microtus* voles during 1973–2009 ($r = 0.727$, $n = 37$, $p < 0.01$), we controlled for this close relationship in the next partial correlation analyses. There was a close positive correlation, though not quite statistically significant, between the yearly mean number of prey items per visit in the food stores and the yearly mean clutch size during 1973–2009 (Fig. 5.4; $r = 0.328$, $n = 34$, $p = 0.05$), whereas for the yearly mean weight of food stores per visit and yearly mean clutch size the correlation turned out to be significant (Fig. 5.4; $r = 0.351$, $n = 34$, $p = 0.04$). Controlling for the *Microtus* vole abundance, there was no obvious relationships between the yearly mean number of prey items in the food stores and the yearly mean number of hatched or fledged young raised by owls. The same also appeared to be true for the yearly mean prey weight of food stores (data not shown). In Germany, there was a positive correlation between the yearly mean number of prey items in food stores and the yearly mean clutch size during 1979–87 (Schwerdtfeger 1988), but in this area the probable effect of between-year variations in small rodent abundance was not controlled for.

Female owls carefully piled the prey items at the edges of the nest box and, in the case of bigger food stores, the bottom of the box surrounding females and eggs was totally covered by several layers of prey items (Fig. 5.2). The storage time of different prey species varied from 1 to 9 days, was usually 1–2 days, and tended to increase with the body mass of different prey species (Table 5.1b). Because half of the boreal owl females start to lay before 4 April (Korpimäki 1987f), ambient temperatures are usually low in March to early May. Therefore, very few of the stored prey items became rotten, because the older prey items were usually consumed first. Boreal owls are also able to thaw prey, and need approx. 20 minutes to thaw a mouse at −16°C initial temperature (Bondrup-Nielsen 1977).

During 1976–85, about 50% of stored prey was partially eaten (at least the head of the prey animal was consumed). Camera recordings showed that 94% of the prey items were

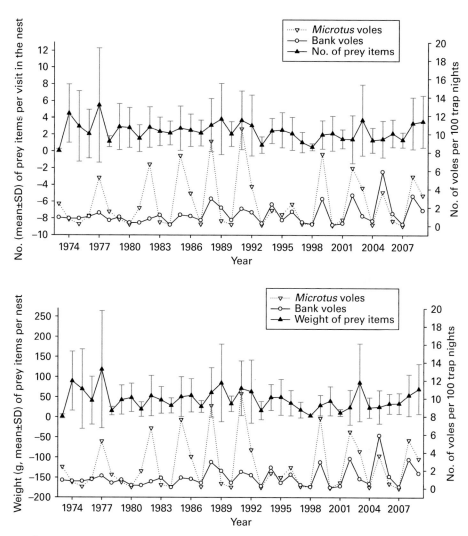

Fig. 5.3.   Yearly mean number per visit per nest (s.d.) (upper panel) and yearly mean pooled weight per visit per nest (s.d.) (lower panel) of prey items stored by boreal owls during the egg-laying, incubation, hatching and early nestling periods in 1973–2009. The yearly spring trap index values for *Microtus* and bank voles during 1973–2009 shown in Fig. 1.1 were also plotted in both panels. Both yearly mean number and yearly mean pooled weight of prey items per visit per nest in the food stores were positively correlated with the spring trap index for *Microtus* voles ($r = 0.35$, $p = 0.03$ and $r = 0.33$, $p = 0.045$, respectively), but not with the spring trap index of bank voles ($r = 0.16$, $p = 0.35$ and $r = 0.15$, $p = 0.38$, respectively).

not already decapitated when the male delivered them to the nest-hole (Korpimäki 1981). Therefore, the female must have eaten, or fed to its young, the front parts of the prey animals. The proportion of whole prey animals in the stores decreased with increasing weight of prey species. For example, 80–100% of prey species weighing <15 g (common shrew, lesser shrew, water shrew, harvest mouse, nestlings of robin and chaffinch) were

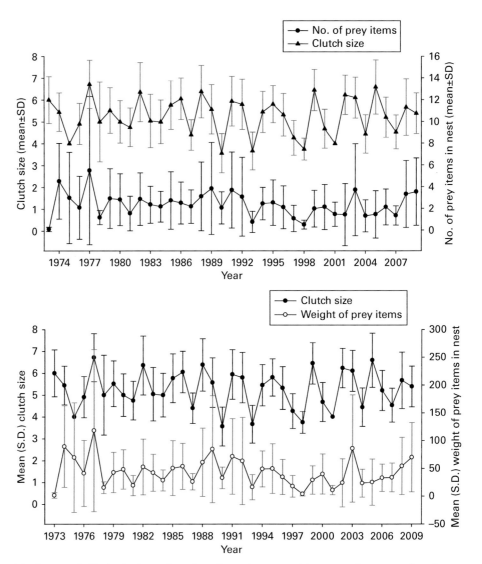

Fig. 5.4.   Yearly mean (s.d.) clutch size and number of prey items per visit per nest (upper panel) and yearly mean (s.d.) pooled weight of prey items per visit per nest (lower panel) stored by boreal owls during the egg-laying, incubation, hatching and early nestling periods in 1973–2009.

stored as whole specimens, while the corresponding proportion was only 30–70% for prey species weighing >15–30 g (bank vole, field vole, sibling vole, house mouse, great tit, chaffinch and nestlings of redwing and song thrush), and only 0–25% for prey species weighing >30 g (water vole, song thrush, redwing and yellowhammer). Therefore, there was a tendency for the largest prey animals to be eaten gradually and the smallest in one sitting. However, this cannot be considered as evidence for the 'large prey hypothesis', which states that young birds have difficulty in dismembering and eating large prey items, because the incubating and brooding females dismember stored prey items and

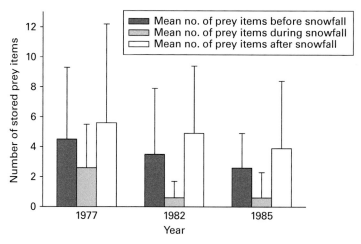

Fig. 5.5. Mean number (s.d.) of prey animals stored in the nests of boreal owls before, during and after a snowfall period in 1977, 1982 and 1985. The snowfall period started when snow fell for a duration of at least 3 days and with fallen snow at least 10 cm deep. The data were collected during the following periods: 2–11 April 1977 before snowfall, 12–21 April 1977 during snowfall, 22 April–1 May 1977 after snowfall; 28 March–6 April 1982 before snowfall, 7–16 April 1982 during snowfall, 17–27 April 1982 after snowfall; 23 March–1 April 1985 before snowfall, 2–11 April 1985 during snowfall, 12–24 April 1985 after snowfall (Korpimäki 1987d).

eat them themselves or divide them between the chicks when food stores are piled in the nests.

Because boreal owls start to lay eggs early in the year, in late February, March and April, adverse weather conditions, such as snowfalls, were usual, at least during the 1960s to the late 1980s. During 1973–85 such periods with a duration of at least 3 days and with fallen snow at least 10 cm deep occurred four times in the breeding season (1973, 1977, 1982 and 1985). In 1977, 1982 and 1985, the number of stored prey animals was significantly higher both before and after the snowfall period than during it (Fig. 5.5). Unfortunately, the data from 1973 were too scanty and did not allow us to study the effect of snowfall in that spring. The finding that long snowfall periods increase the risk of nest failure is further supported by the fact that, in 1985, 6 out of 47 females deserted their clutches at that time, probably because of food shortages.

In addition to boreal owls, the pygmy owl, the long-eared owl, the short-eared owl, the hawk owl, the Ural owl, the Eurasian kestrel and the common buzzard use voles of the genera *Microtus* and *Myodes* as their main foods in our study area (Korpimäki 1981, 1985d; also see Chapter 14.3). Boreal owls mainly hunt and roost within 1 km of the nest (Chapter 4.1.3). Therefore, when these owls and diurnal raptors bred <1 km from the nests of boreal owls, these birds of prey were assumed to hunt at least partly in the same area, and these boreal owl pairs were classed as 'neighbours', whereas other pairs were classed as 'non-neighbours'. Non-neighbouring boreal owl nests seemed to contain larger food stores than those that were 'neighbours', particularly in 1977 and 1979, while in 1982 and 1985 there was no obvious difference in the size of food stores between these two categories (Fig. 5.6). This result was obtained despite the fact that the data were

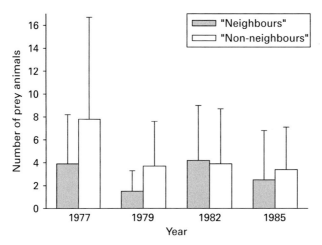

Fig. 5.6.   Mean number (s.d.) of prey animals stored in the nests of neighbouring and non-neighbouring boreal owl pairs during the egg-laying, incubation and hatching periods of 1977, 1979, 1982 and 1985. When the long-eared owl, the short-eared owl, the hawk owl, the Ural owl, the Eurasian kestrel or the common buzzard bred <1 km from the nest of a boreal owl, these boreal owl pairs were classed as neighbours, while other pairs were classed as non-neighbours (Korpimäki 1987d).

mainly collected in good vole years, when it was likely that the densest vole patches were those near which boreal owls bred close to other vole-eating birds of prey, and poorer vole patches were those near which only 'non-neighbours' bred. Because the hypothesis of Oksanen T. (1983) predicts that neighbouring boreal owl pairs should have larger prey stores than 'non-neighbours' so as to secure food resources when several co-existing predator species have similar prey preferences, our results were inconsistent with the prediction of this hypothesis. The tendency for the neighbouring boreal owl pairs to have smaller prey stores than 'non-neighbours' may be caused by interspecific competition for food.

To conclude, probably the most important function of the food stores of boreal owls is that they provide a buffer for incubating and brooding females and small chicks against temporary food shortages that quite often and unpredictably occur due to the adverse weather conditions of the northern boreal climate. Long cold spells and snowfall periods regularly occur in the boreal forest zone in spring and they considerably increase the energy needs of owl families and, more importantly, also considerably reduce the hunting success of males. Cold spells probably decrease the mobility of prey animals and snowfalls cover small mammals below the protective snow layer, where they are much less available to hunting male owls. In addition, with ongoing climate change, the probability of heavy rainfall in spring is predicted to increase, at least in northern Europe (Lehikoinen et al. 2009, 2011), which will probably also decrease the hunting success of boreal owl males. Because of their small body size, boreal owls cannot accumulate internal fat reserves to the same extent as larger birds of prey. Internal food stores may also reduce mobility and thus hunting success of the owls. Probably those boreal owl males who are most efficient at collecting food stores in order to buffer their female and/or owlets against low prey delivery rates due to temporary poor hunting success will be most successful in the population. In addition, the food stores also probably buffer boreal owl chicks against daily 16–20-hour non-delivery periods of prey by males due to the northern short hunting nights, and thus allow females to feed their chicks at any time of the day. This gives some support to the 'feeding time hypothesis' of Roulin (2004), but nevertheless we suggest that the longer-term insurance provided by food stores is more important at northern latitudes, because they are likely to decrease the probability of total nest failures of boreal owls. Our results that the yearly mean number and weight of prey items in the food stores was closely correlated with the yearly mean clutch size (Fig. 5.4), even after controlling for the between-year variation in vole abundance, give relatively strong support for the 'insurance hypothesis'.

5.4.   Dietary diversity in Europe and North America

The life of predators, including their growth, age of first reproduction, clutch and brood sizes, reproductive success, juvenile and adult survival, and mobility (dispersal) are largely governed by their food supply. Therefore, quantifying dietary requirements, both in term of quality (prey species) and quantity (how much prey) has long been, and probably continues to be, one of the first steps in studying the life-history traits, demography and behaviour of predators. A widespread traditional assumption is that many owls and diurnal raptors are highly opportunistic in the prey they eat (see, e.g., Jaksic and Braker 1983, Steenhof and Kochert 1985, Marti et al. 1993a, Korpimäki and Marti 1995; but see Korpimäki 1987c, 1987d). If this dietary opportunism holds true over large geographical areas, examination of patterns in the diets of birds of prey should be largely explained by the distribution patterns of their prey animals. Owls and diurnal raptors are convenient species for the examination of animal community patterns involving predator–prey interactions. Most of the prey of birds of prey, in particular of owls, can be relatively easily identified to the species level, which allows accurate estimates of the kind, number and size of prey taken (Marti 1987, 1987, Marti et al. 2007). Also, the prey chosen by larger birds of prey are relatively seldom influenced by the need of the predator to minimise the risk of being killed (Korpimäki and Marti 1995). In smaller birds of prey including boreal owls, however, the intra-guild predation risk imposed by larger birds of prey is probably an important factor affecting hunting habitat choice and activity times, along with prey density (see Chapters 4.1.3 and 5.2).

In literature surveys, we found a total of 26 European studies and unpublished data sets (Tables 5.2, 5.3) but only 3 North American studies (Table 5.4) on the diet composition of boreal owls in the breeding season. We selected the data sets meeting the following criteria: minimum of 100 identified prey items per study area, vertebrate prey identified at least to genus, invertebrate prey identified at least to order, geographical area of data collection well defined, and season(s) and year(s) of data collection stated. This gave breeding season diet data from 27 different sites in Europe, ranging from northern Italy to the Kola Peninsula in Russia and to Lapland in Sweden, as well as from eastern Belgium to the Czech Republic, Slovakia and eastern Finland (Table 5.2).

Table 5.2. Diet composition (proportion of prey by number) of boreal owls in the breeding season according to the analyses of pellets and prey remains in Europe.

Prey species or group	RK	SL	FNO	FNS	FCO	FSO	FSÄ	SUV	FKH	FNK	SG	NS	SV	GH	GT	BR	BS	GSi	GJ	GNW	C	CK	CJ	GSt	SJ	IC
Common mole *Talpa europaea*	–	–	–	–	–	–	–	–	–	–	–	–	–	–	–	–	–	–	–	–	0.0	–	–	–	0.0	–
Common shrew *Sorex araneus*	2.8	18.4	18.7	10.9	26.6	23.1	25.1	16.9	16.1	–	11.0	7.6	–	2.1	22.9	29.8	–	21.3	0.5	18.9	19.3	17.3	14.5	14.2	19.0	34.9
Masked shrew *S. caecutiens*	1.8	–	1.6	0.9	0.5	–	1.2	–	–	–	–	–	–	–	–	–	–	–	–	–	–	–	–	–	–	–
Pygmy shrew *S. minutes*	–	0.4	1.0	0.6	0.7	0.6	0.9	1.2	0.8	–	2.0	–	–	0.7	1.7	1.2	2.8	1.9	0.5	4.4	2.4	2.8	2.8	–	0.2	2.4
Taiga shrew *S. isodon*	–	–	–	1.0	–	–	–	–	–	–	–	–	–	–	–	–	–	–	–	–	–	–	–	–	–	0.2
Alpine shrew *S. alpinus*	–	–	–	–	–	–	–	–	–	–	–	–	–	–	–	–	–	–	–	–	0.2	–	0.5	0.3	–	–
Least shrew *S. minutissimus*	–	–	0.1	0.0	–	–	–	–	0.1	–	–	–	–	–	–	–	–	–	–	–	–	–	–	–	–	–
Shrew spp. *Sorex spp.*	0.9	–	0.8	4.3	0.9	5.2	1.2	–	0.2	4.9	–	5.1	38.6	–	1.7	–	17.3	–	–	13.4	–	–	–	13.8	–	–
Water shrew *Neomys fodiens*	–	–	0.2	0.5	1.1	0.4	1.2	0.6	0.2	–	–	–	–	–	–	–	0.1	0.1	–	0.2	0.7	0.2	0.1	–	–	–
Miller's water shrew *N. anomalus*	–	–	–	–	–	–	–	–	–	–	–	–	–	–	–	–	0.0	0.1	–	–	0.0	0.1	0.0	–	–	–
Greater white-toothed shrew *Crocidura russula*	–	–	–	–	–	–	–	–	–	–	–	–	–	–	–	–	0.0	0.3	–	0.1	–	–	–	–	–	–
Lesser white-toothed shrew *C. suaveolens*	–	–	–	–	–	–	–	–	–	–	–	–	–	–	–	–	–	–	–	–	0.0	–	–	–	–	–
Bicoloured shrew *C. leucodon*	–	–	–	–	–	–	–	–	–	–	–	–	–	–	–	–	–	–	–	–	–	0.0	–	–	–	–
Shrew total	5.5	18.8	22.5	18.2	29.8	29.2	29.6	18.7	17.3	4.9	13.0	12.7	38.6	2.8	26.3	31.1	20.2	23.7	1.0	37.1	22.7	20.4	18.0	–	19.2	37.5
European whiskered bat *Myotis mystacinus*	–	–	–	–	–	–	–	–	–	–	–	–	–	–	–	–	–	–	–	–	–	–	0.0	–	–	–
Natterer's bat *M. nattereri*	–	–	–	–	–	–	–	–	–	–	–	–	–	–	–	–	–	–	–	–	0.0	–	0.0	–	–	–
Daubenton's bat *M. daubentoni*	–	–	–	–	–	–	–	–	–	–	–	–	–	–	–	–	–	–	–	–	–	–	0.0	–	–	–
Bechstein's bat *M. bechsteini*	–	–	–	–	–	–	–	–	–	–	–	–	–	–	–	–	–	–	–	–	0.0	–	–	–	–	–
Brown long-eared bat *Plecotus auritus*	–	–	–	–	–	–	–	–	–	–	–	–	–	–	–	–	–	–	0.5	–	0.0	0.0	–	–	–	–
Northern bat *Eptesicus nilssoni*	–	–	–	–	–	–	0.1	–	0.2	–	–	–	–	–	–	–	–	–	–	–	–	–	–	–	–	–
Least weasel *Mustela nivalis*	–	–	–	–	–	0.0	–	–	–	–	–	–	–	–	–	–	–	–	–	–	–	–	–	–	–	–
Eurasian red squirrel *Sciurus vulgaris*	–	–	0.1	–	–	0.0	–	–	–	–	–	–	–	–	–	–	–	–	–	–	0.0	–	–	–	–	–

Species																									
Siberian flying squirrel *Pteromys volans*	–	–	–	–	–	–	–	–	0.3	–	–	–	–	–	–	–	–	0.1	–	–	–	–	–	–	–
Forest dormouse *Dryomys nitedula*	–	–	–	–	–	–	–	–	–	–	–	–	–	–	–	–	–	0.1	–	–	–	–	–	–	–
Edible dormouse *Myoxus glis*	–	–	–	–	–	–	–	–	–	–	–	–	–	–	–	–	–	–	–	–	–	–	0.0	–	–
Hazel dormouse *Muscardinus avellanarius*	–	–	–	–	–	–	–	5.3	–	–	1.2	0.9	4.3	4.1	–	3.3	5.0	–	1.5	1.4	1.2	2.6	–	–	5.9
Garden dormouse *Eliomys quercinus*	–	–	–	–	–	–	–	–	–	–	–	–	–	–	–	0.4	–	–	–	–	–	0.2	–	–	–
Northern birch mouse *Sicista betulina*	–	–	–	–	0.1	–	–	–	–	–	–	–	–	–	–	–	2.1	–	–	–	–	–	–	–	–
European pine vole *Microtus subterraneus*	–	–	–	–	–	–	–	–	–	–	–	0.3	–	–	–	–	2.5	–	–	0.1	1.4	2.8	1.4	–	3.1
Norway lemming *Lemmus lemmus*	–	4.1	–	–	–	–	–	1.5	–	–	–	–	–	–	–	–	–	–	–	–	–	–	–	–	–
Wood lemming *Myopus schisticolor*	–	–	0.0	–	–	0.1	–	–	–	4.8	–	–	–	–	–	–	–	–	–	–	–	–	–	–	–
Bank vole *Myodes glareolus*	19.3	14.9	28.0	45.9	24.5	42.7	20.6	47.6	15.4	24.9	22.0	7.6	18.0	7.2	16.4	16.6	13.8	7.8	1.1	16.6	19.1	33.6	–	–	18.8
Grey-sided vole *M. rufocanus*	2.8	28.4	–	–	–	–	–	11.7	–	–	–	–	–	–	–	–	–	–	–	–	–	–	–	–	–
Myodes spp.	–	0.8	40.3	–	–	–	–	–	–	–	–	–	–	–	–	–	0.2	–	–	–	–	–	–	–	–
European water vole *Arvicola terrestris*	–	0.2	0.5	–	0.5	0.3	0.6	0.8	–	1.0	–	0.1	–	–	–	0.2	0.2	–	–	0.1	–	0.5	–	–	–
Common vole *Microtus arvalis*	22.0	31.5	33.6	15.7	11.2	7.0	38.3	9.9	51.2	31.4	23.4	27.3	15.4	46.6	12.4	11.8	31.6	36.1	41.4	–	–	–	5.9	4.7	7.9
Field vole *M. agrestis*	–	–	–	–	10.0	1.7	–	6.5	–	–	–	–	–	–	–	–	–	–	–	3.3	12.1	4.1	2.2	20.0	2.6
Sibling vole *M. rossiaemeridionalis*	2.8	–	–	–	–	–	–	0.5	–	–	–	–	–	–	–	–	–	–	–	–	3.6	11.4	–	–	–
Root (tundra) vole *M. oeconomus*	–	–	–	–	12.6	7.7	–	14.5	–	–	–	–	2.5	–	–	–	–	6.6	9.4	–	6.2	16.3	–	–	–
Microtus spp.	–	–	–	–	–	–	–	–	–	–	–	–	–	–	–	–	–	–	–	–	–	–	–	–	0.2
European snow vole *Chionomys nivalis*	–	–	–	–	–	–	–	–	–	–	0.1	–	–	–	–	–	–	–	–	–	–	–	–	–	0.2
Vole spp.	0.2	3.3	–	–	–	–	–	15.7	–	0.1	3.3	2.6	11.9	–	14.2	–	–	–	–	–	–	2.2	–	–	4.5
Microtus spp. total	31.5	40.3	33.6	15.7	33.8	16.4	38.3	24.4	57.7	31.4	23.9	27.3	15.7	46.6	11.8	11.8	31.6	39.0	24.8	42.2	41.4	27.7	3.6	–	0.2
Voles total	46.8	75.9	68.5	74.4	61.6	58.8	59.3	62.8	72.0	79.4	77.2	74.0	49.5	33.5	72.0	28.9	49.3	51.8	49.1	21.2	47.7	60.2	68.9	37.6	26.1
Brown rat *Rattus norvegicus*	–	–	0.0	–	0.1	–	–	–	–	–	–	–	–	–	0.1	–	–	–	–	–	–	–	–	–	–
Harvest mouse *Micromys minutes*	–	–	0.5	0.8	1.3	2.4	–	1.7	–	–	–	–	–	–	–	0.1	0.0	–	–	–	–	0.0	–	–	–
House mouse *Mus musculus*	–	1.7	0.3	0.8	0.9	1.4	–	2.4	–	4.4	0.9	0.0	–	–	0.0	–	0.0	–	0.1	–	0.0	0.0	–	–	–
Mus/*Micromys* spp.	–	–	–	–	–	–	–	–	–	0.8	–	–	–	–	–	–	–	–	–	–	–	–	–	3.6	6.8
Striped field mouse *Apodemus agrarius*	–	–	–	–	–	–	–	–	–	–	–	–	–	–	–	–	–	–	2.1	–	–	–	–	–	0.2

Table 5.2. (cont.)

Prey species or group	RK	SL	FNO	FNS	FCO	FSO	FSÄ	SUV	FKH	FNK	SG	NS	SV	GH	GT	BR	BS	GSi	GJ	GNW	C	CK	CJ	GSt	SJ	IC
Yellow-necked field mouse	–	–	–	–	–	–	1.1	1.3	–	–	–	–	–	2.8	–	0.6	0.5	8.5	–	–	–	–	14.2	–	–	5.5
A. flavicollis																										
Long-tailed field mouse	–	–	–	–	–	–	–	10.5	–	–	–	2.0	–	6.4	–	–	0.4	8.1	67.9	–	–	–	1.2	–	–	6.2
A. sylvaticus																										
Apodemus spp.	–	–	–	–	–	–	–	–	–	–	0.1	–	–	–	0.8	13.7	23.6	5.3	–	27.5	10.9	26.8	–	–	37.1	13.1
Mouse spp.	–	–	–	–	–	–	–	–	–	–	–	1.0	2.6	–	–	14.3	–	–	–	–	–	–	–	–	–	–
Murids total	–	–	1.7	0.8	1.6	2.2	5.0	16.2	4.2	0.8	0.1	3.0	2.6	9.2	0.8	28.6	24.5	21.9	69.9	27.6	10.9	26.8	15.5	6.8	37.1	24.9
Cricetidae	21.1	–	–	–	–	–	–	–	–	–	–	–	–	–	–	–	–	–	–	–	–	–	–	–	–	–
European hare	–	–	–	–	–	–	–	–	–	–	–	–	–	–	–	–	–	–	–	–	–	–	–	–	–	–
Lepus europaeus																										
European rabbit	–	–	–	–	–	–	–	–	–	–	–	–	–	–	–	–	–	–	–	–	–	–	–	–	–	–
Oryctolagus cuniculus																					0.0					
Mammals total	73.4	98.8	92.9	93.5	93.0	90.3	94.1	97.8	93.9	79.7	97.3	94.4	96.0	34.0	99.2	94.4	95.3	99.0	96.9	97.3	95.2	96.5	96.5	93.8	98.2	97.6
Birds total	19.3	1.2	7.1	6.5	7.0	9.0	5.9	2.2	6.1	20.3	2.7	5.1	4.0	66.0	0.8	5.6	4.7	0.9	3.1	2.7	4.5	3.5	3.4	5.5	1.8	2.4
Frogs total	–	–	–	–	–	–	–	–	–	–	–	–	–	–	–	–	–	–	–	–	–	–	–	–	–	–
Insects total	7.3	–	0.1	–	–	0.7	–	–	–	–	–	0.5	–	–	–	–	–	0.1	–	–	0.3	–	0.1	0.6	0.0	–
Total	100	100	100	100	100	100	100	100	100	100	100	100	100	100	100	100	100	100	100	100	100.0	100.0	100.0	100.0	100.0	100.0
No. of prey items	109	489	2869	690	1216	27759	1130	1281	1155	123	1538	197	495	141	118	161	2698	1644	193	968	5324	7288	9719	325	6834	421
Study period	1981, 84, 91	1964	80–84	76–84	60–66	73–09	60–67	53–62	60–63	80	68–74	48–49	58–62	75–77	71	68, 79–80	89–96	69–81	72–74	55–63	83–89	00–09	92–96	30–44	91–09	89–90
Period years	3	1	5	9	7	37	8	10	4	1	7	2	5	3	1	3	8	13	3	9	7	10	5	15	19	2
Source	1	2	3	4	5	6	7	8	9	10	11	12	13	14	15	16–17	18	19	20	21	22	23	24	25	26	27

Abbreviations of localities and sources: RK = North-west Russia, Gulf of Kandalaksha area (Boiko and Shutova 2007), SL = Sweden, Lapland (Lindhe 1966), FNO = Finland, North Savonia (Jäderholm 1987), FCO = Finland, Central Ostrobothnia (Sulkava P. and Sulkava S. 1971), FSO = Finland, South Ostrobothnia (Kauhava region) (Korpimäki 1988b and unpublished), FSA = South Ostrobothnia (Ähtäri region) (Sulkava P. and Sulkava S. 1971), SUV = Sweden, Uppland and Värmland (Fredga 1964), FKH = Finland, Kanta-Häme (Sulkava P. and Sulkava S. 1971), FNK = Finland, North Karelia (Kuhlman and Koskela 1980), SG = Sweden, Gävleborg (Ahlbom 1976), NS = South Norway (Hagen 1952), SV = Sweden, Västergötland (Norberg 1964), GH = Germany, Harz Mountains (Zang and Kunze 1978), GT = Germany, Thüringen (Ritter and Zienert 1972), BR = Belgium (Scheuren 1970), BS = Belgium (Libois and Gailly 1984), BS = Belgium (Sorbi 1993, 1995 and unpublished), GSi = Germany, Siegerland (Bülow and Franz 1982), GJ = Germany, Jena (Klaus et al. 1975), GNW = Germany, Nordrhein-Westfalen (Gasow 1968), C = Czech Republic (Kloubec and Vacík 1990), CK = Czech Republic, Ore Mountains in 1999–2008 (Zárybnická 2008, 2009b and unpublished), CJ = Czech Republic, Jizerské hory and Giant Mountains (Pokorný 2000), GSt = Germany, Stuttgart (Uttendörfer 1952), SJ = Switzerland, West Jura Mountains (Henrioux 2010), IC = Italy, Cansiglio (Mezzavilla et al. 1994).

Table 5.3. Total numbers of various bird species found in the diets of boreal owls based on prey items in the food stores and on analyses of pellets and prey remains in nest-boxes in northern Europe and in central and southern Europe. Pooled data from 26 studies summarised in Table 5.2.

Species		Weight (g)	Northern Europe	Central and Southern Europe
Hazel grouse (y)[a]	*Bonasa bonasia*	30	1	–
Boreal owl (n)[a]	*Aegolius funereus*	80	77	57
Common swift	*Apus apus*	37	9	–
Eurasian wryneck	*Jynx torquilla*	37	1	2
Woodpecker species	*Picus* spp.	149	–	2
Great spotted woodpecker	*Dendrocopos major*	88	1	3
Lesser spotted woodpecker	*Dendrocopos minor*	50	1	2
Eurasian skylark	*Alauda arvensis*	37	–	2
Barn swallow	*Hirundo rustica*	19	36	1
Tree pipit	*Anthus trivialis*	22	17	3
Meadow pipit	*Anthus pratensis*	18	4	6
Pipit species	*Anthus* spp.	20	–	2
Grey wagtail	*Motacilla cinerea*	21	–	3
White wagtail	*Motacilla alba*	21	5	1
Winter wren	*Troglodytes troglodytes*	10	–	4
Dunnock	*Prunella modularis*	18	–	19
European robin	*Erithacus rubecula*	17	8	52
European robin (n)		10	4	–
Common redstart	*Phoenicurus phoenicurus*	16	2	5
Redstart species	*Phoenicurus* spp.	16	–	4
Whinchat	*Saxicola rubetra*	17	4	3
Ring ouzel	*T. torquatus*	94	–	4
Common blackbird	*T. merula*	90	–	24
Fieldfare	*T. pilaris*	106	55	1
Song thrush	*T. philomelos*	67	45	62
Redwing	*T. iliacus*	58	146	1
Mistle thrush	*T. viscivorus*	115	3	1
Thrush species	*Turdus* spp.	87	28	34
Thrush species (n)		20	17	–
Common grasshopper warbler	*Locustella naevia*	13	–	3
Common whitethroat	*Sylvia communis*	15	–	1
Garden warbler	*S. borin*	20	–	1
Eurasian blackcap	*S. atricapilla*	18	–	12
Sylviid warbler species	*Sylvia* spp.	20	–	5
Wood warbler	*Phylloscopus sibilatrix*	9	–	1
Common chiffchaff	*P. collybita*	8	1	2
Willow warbler	*P. trochilus*	10	10	4
Leaf-warbler species	*Phylloscopus* spp.	9	118	16
Goldcrest	*Regulus regulus*	6	6	19
Kinglet species	*Regulus* spp.	6	–	6
Spotted flycatcher	*Muscicapa striata*	16	2	–
European pied flycatcher	*Ficedula hypoleuca*	13	14	–
Long-tailed tit	*Aegithalos caudatus*	8	–	1
Marsh tit	*Parus palustris*	11	–	2
Willow tit	*P. montanus*	11	33	3
Siberian tit	*P. cinctus*	12	1	–

Table 5.3. (cont.)

Species		Weight (g)	Northern Europe	Central and Southern Europe
Crested tit	*P. cristatus*	12	14	3
Coal tit	*P. ater*	9	–	14
Eurasian blue tit	*P. caeruleus*	11	–	2
Great tit	*P. major*	20	38	21
Tit species	*Parus spp.*	14	42	18
Eurasian nuthatch	*Sitta europaea*	20	–	1
European treecreeper	*Certhia familiaris*	9	1	2
Short-toed treecreeper	*C. brachydactyla*	9	–	–
Treecreeper species	*Certhia* spp.	9	–	1
Red-backed shrike	*Lanius collurio*	30	–	8
Eurasian jay (n)	*Garrulus glandarius*	20	1	–
Common starling	*Sturnus vulgaris*	74	3	1
House sparrow	*Passer domesticus*	32	2	2
Eurasian tree sparrow	*P. montanus*	23	–	3
Sparrow species	*Passer* spp.	26	–	2
Common chaffinch	*Fringilla coelebs*	22	807	215
Common chaffinch (n)		5	4	–
Brambling	*Fringilla montifringilla*	22	10	22
Finch species	*Fringilla* spp.	22	–	58
Fringillidae	*Fringillidae*	22	–	1
European serin	*Serinus serinus*	12	–	5
European greenfinch	*Carduelis chloris*	29	2	10
European goldfinch	*C. carduelis*	16	–	2
Eurasian siskin	*C. spinus*	13	9	15
Common linnet	*C. cannabina*	18	–	3
Common redpoll	*C. flammea*	14	15	–
Carduelis species	*Carduelis* spp.	14	1	1
Red crossbill	*Loxia curvirostra*	41	–	2
Parrot crossbill	*Loxia pytyopsittacus*	56	1	–
Crossbill species	*Loxia* spp.	41	1	–
Eurasian bullfinch	*Pyrrhula pyrrhula*	34	45	16
Hawfinch	*Coccothraustes coccothraustes*	55	–	1
Yellowhammer	*Emberiza citrinella*	32	51	22
Yellowhammer (n)		10	5	–
Rustic bunting	*Emberiza rustica*	19	1	–
Common reed bunting	*Emberiza schoeniclus*	18	1	–
Emberiza species	*Emberiza* spp.	23	–	1
Passerine species	Passeriformes	15	117	139
Goldcrest size		6	–	–
Willow tit size		11	–	–
Great tit size		20	353	–
Chaffinch size		22	488	–
Yellowhammer size		32	–	–
Thrush size		77	120	9
Bird species	Aves spp.	22	–	276
Birds total			2781	1250

[a] n = nestling, y = juvenile individual.

Table 5.4. Diet composition (proportion of prey number) of boreal owls at Chamberlain Basin, Idaho, USA in winter (Wi) and summer (Su; 31 nests) during 1981 and 1984–8 (Hayward et al. 1993), in the interior of Alaska, USA during the breeding seasons 2004–6 (Whitman 2009), and at Kluane, Yukon, Canada in the breeding season in 1994 (7 nests) and 1995 (1 nest) (Doyle and Smith J. 2001).

Prey group	Idaho		Alaska			Yukon	
	Wi	Su	2004	2005	2006	1994	1995
Mammals							
Sorex spp.	12	11	2	3	4	2	73
Myodes gapperi	49	31	–	–	–	–	–
Myodes rutilus	–	–	86	70	39	12	5
Thomosys talpoides	7	11	–	–	–	–	–
Microtus spp.	11	8	3	2	–	–	–
M. pennsylvanicus	–	–	5	20	4	16	5
M. oeconomus	–	–	–	3	34	32	–
M. miurus	–	–	–	–	<1	9	–
M. richardsoni	<1	<1	–	–	–	–	–
M. xanthognathus	–	–	–	<1	2	–	–
Synaptomys borealis	–	–	–	<1	10	–	–
Lemmus trimucronatus	–	–	–	<1	2	–	–
Phenacomys intermedius	2	4	–	–	–	13	0
Peromyscus maniculatus	7	5	–	–	–	10	0
Glaucomus sabrinus	5	<1	–	–	–	–	–
Tamias amoenus	1	2	–	–	–	–	–
Zapus princeps	–	2	–	–	–	–	–
Z. hudsonius	–	–	–	<1	<1	1	1
Neotoma cinerea	–	<1	–	–	–	–	–
Ochotona princeps	–	<1	–	–	–	–	–
Tamiasciurus hudsonicus	–	–	–	–	<1	–	–
Lepus americanus	–	–	–	–	<1	1	9
Spermophilus parryii	–	–	–	–	–	1	–
Mustela spp.	<1	–	–	–	–	–	–
Birds	5	5	3	5	6	5	9
Insects	1	18	–	–	–	–	–
No. of prey items	242	672	58	503	1321	217	22

Small mammals were the main foods of boreal owls in the breeding season, comprising on average $92.1 \pm 12.9\%$ of prey number and $90.5 \pm 13.2\%$ of prey weight in Europe (Table 5.2) and $88.6 \pm 10.1\%$ of prey number in North America (Table 5.4). The proportion of birds remained generally low, averaging $7.5 \pm 12.6\%$ of prey number and $9.5 \pm 13.2\%$ of prey weight in Europe and $5.3 \pm 0.4\%$ of prey number in North America. These figures should be interpreted with caution, however, because the diet data were mainly collected in good small rodent years when the number of owl nests, and thus food samples collected at their nests, are usually considerably higher than in poor rodent years. This probably underestimates the usage of prey animals other than small rodents. Two additional prey classes were recorded in the diet, namely amphibians (frogs, *Rana* spp.) and insects (mainly coleopterans and hymenopterans), but only in very low numbers.

A total of 47 mammal species have been recorded in the breeding-season diet of boreal owls in Europe. Terrestrial small rodent species, particularly voles of the genera *Microtus* and *Myodes*, were the primary prey of boreal owls in both Europe and North America (Tables 5.2, 5.4). *Microtus* voles mainly occupying open country including agricultural fields and grass-lands averaged 27.8 ± 16.4% of prey items and 35.1 ± 19.8% of prey weight in Europe, and forest-dwelling *Myodes* voles an additional 23.6 ± 14.0% of prey items and 22.4 ± 14.3% of prey weight (Table 5.2). Among *Microtus* voles, the field vole *M. agrestis* was the most frequent prey species in most parts of Europe (Table 5.2; see also Schwerdtfeger 1988, Pokorny et al. 2003). However, in our study area the sibling vole *M. rossiaemeridionalis* was almost as numerous a prey species, and in some areas of Germany and the Czech Republic its close relative, the common vole *M. arvalis*, was the most numerous vole species recorded in the diet. Among *Myodes* voles, the bank vole *M. glareolus* was the most frequent species (Fig. 5.7) with the exception of Swedish Lapland, where the grey-sided vole *M. rufocanus* was the most numerous species. In most parts of Europe, shrews (Soricidae) served as the most important alternative prey, but in the Sumava Mountains, south-western Czech Republic, they appeared to be the main prey (Pokorny et al. 2003). In some areas of temperate Europe, particularly in Germany, the Czech Republic, Slovakia, Belgium, Switzerland and Italy, mice of the genus *Apodemus* also appeared as an important alternative prey group. The common shrew *Sorex araneus* was the most abundant shrew species in the diet (Fig. 5.8), and the wood mouse *Apodemus sylvaticus* was the most frequent mouse species in most areas (Fig. 5.9), while the yellow-necked field mouse *A. flavicollis* emerged as the most numerous mouse species in some areas. Capturing aerial bats is very infrequent, probably because of scarcity of bats in the north or at high altitudes, where boreal owls usually live.

A total of at least 66 bird species were included in the breeding season diets of boreal owls in Europe (Table 5.3). Small forest-dwelling passerine birds were the most numer-ous avian prey. The chaffinch *Fringilla coelebs* emerged as the most frequent avian prey species in Europe (Fig. 5.10), followed by the redwing *Turdus iliacus*, the song thrush *T. philomelos*, the great tit *Parus major*, the yellowhammer *Emberiza citrinella*, the bullfinch *Pyrrhula pyrrhula*, and the fieldfare *T. pilaris*. Boreal owls also often consume the smallest nestlings of their own broods that have succumbed in the nest, which

Fig. 5.7.   Boreal owl male about to deliver a decapitated bank vole to the nest. Photo: Benjam Pöntinen. (For colour version, see colour plate.)

Fig. 5.8.   Male delivers a common shrew to the nest. Photo: Benjam Pöntinen. (For colour version, see colour plate.)

Fig. 5.9.   Boreal owl male about to deliver a wood mouse to the nest in Belgium. Photo: Serge Sorbi. (For colour version, see colour plate.)

explains their regular inclusion in the diet. Fifteen avian species (the dunnock *Prunella modularis*, the robin *Erithacus rubecula*, the blackbird *Turdus merula*, the song thrush, the blackcap *Sylvia atricapilla*, the goldcrest *Regulus regulus*, the coal tit *Parus ater*, the blue tit *P. caeruleus*, the great tit, the chaffinch, the brambling *Fringilla montifringilla*, the greenfinch *Carduelis chloris*, the siskin *C. spinus*, the yellowhammer, and the bullfinch) totalled >10 items in the diets in central and southern Europe (a total of 1250 bird prey items; Table 5.3). In the Harz Mountains of Germany, great tits, robins, chaffinches, goldcrests, coal tits, song thrushes, blackbirds, dunnocks and bramblings were the most frequent avian prey found in the food stores of boreal owls (Schwerdtfeger 1988). In northern Europe, 15 avian species also totalled >10 among the 2781 bird prey

Fig. 5.10.   Male owl delivers a male chaffinch to the nest. Photo: Benjam Pöntinen. (For colour version, see colour plate.)

Fig. 5.11.   Male delivers a young water vole to the nest. Photo: Benjam Pöntinen. (For colour version, see colour plate.)

items recorded in the diet during the breeding season (the barn swallow *Hirundo rustica*, the tree pipit *Anthus trivialis*, the fieldfare, the song thrush, the redwing, the willow warbler *Phylloscopus trochilus*, the pied flycatcher *Ficedula hypoleuca*, the willow tit *P. montanus*, the crested tit *P. cristatus*, the great tit, the chaffinch, the brambling, the redpoll *Carduelis flammea*, the yellowhammer, and the bullfinch; Table 5.3). A pygmy owl has been recorded in the winter diet of boreal owls in our study area.

The mean prey weight of boreal owls in Europe was 20.0 ± 3.4 g. The average weight of mammal prey was markedly heavier than avian prey (19.1 ± 2.1 g vs. 31.6 ± 19.2 g, respectively). The largest mammalian prey were the red squirrel, the European hare *Lepus timidus*, the rabbit *Oryctolagus cuniculus*, the water vole *Arvicola terrestris* (Fig. 5.11) and the Siberian flying squirrel (Table 5.2). Probably the individuals of these large prey mammal species that were killed and transported to the nest were not full-grown, because the body mass of adult specimens greatly exceeds the body mass of male boreal owls. The smallest mammalian prey included the least shrew *Sorex minutissimus* and the pygmy shrew *S. minutus*. The largest avian prey species were the mistle thrush *Turdus viscivorus* and the fieldfare, and the smallest was the goldcrest. Because adult mistle thrushes and fieldfares have been recorded in the food stores of boreal owls, male owls have probably killed these large specimens, but they normally decapitate these large prey items prior to transportation to the nest.

In our study area, there exist long-term data on the diet composition of boreal owls both in winter, during the egg-laying, incubation and early nestling periods (prey items identified in food stores), and during the late nestling period (prey items identified in the prey detritus on the bottom of the nest-boxes) (Table 5.5). The diet of boreal owls consisted of most of the common small mammal and bird species over-wintering and breeding in the area, including a total of 16 mammal and 39 bird species. As elsewhere in Europe, small mammals formed the most important prey group in winter (76% of prey number and 69% of prey weight), in the early stage (96% and 95%) and in the late stage (90% and 86%) of the breeding season. Birds accounted for approx. one-quarter of prey number and nearly one-third of prey weight in winter, but their proportions decreased to 3–5% in the early breeding season and then increased to 9–14% in the late breeding season. Quite a small number of food samples from 3 years in Belarus included mainly small mammals (82% of prey number and 88% of prey weight) in the breeding season and somewhat less (67% and 83%, respectively) outside the breeding season. There were only minor between-season differences in the proportions of small birds (4% and 6% vs. 5% and 6%, respectively), whereas the proportion of insects (mainly beetles) was relatively high in the breeding season diet (13% and 2%) but no insects were found in the diet outside the breeding season (Sidorovich et al. 2003).

The proportion of small mammals was considerably higher in the food stores than in the prey detritus of the nests (Table 5.5). *Microtus* voles (i.e. field and sibling voles) were the main prey in both winter and spring, but their proportion of both prey number and prey weight was considerably higher in the early breeding season (in the food stores of owl nests) than in winter or in the late breeding season (in prey remains accumulated at the bottom of the boxes). Of the two *Microtus* vole species, sibling voles were more frequent than field voles in the winter diet, whereas field voles were taken more often in the early breeding season, and the importance of these two species was almost equal in the late breeding season. Bank voles were the most important alternative prey and their occurrence in the diet followed the same seasonal trend as for *Microtus* voles. Shrews, in particular common shrews, were the second most important alternative prey, along with small birds. The consumption of shrews peaked in the late breeding season and was considerably lower in winter and in the early breeding season. Murids, including the harvest mouse *Micromys minutus* (Fig. 5.12), house mouse *Mus musculus* and young Norway rats *Rattus norvegicus*, were regularly taken in low numbers without any obvious differences between winter and spring. Resident great, crested and willow tits,

Table 5.5. Diet composition of boreal owls by number and by weight in winter (data from Korpimäki 1981 and unpublished), during the egg-laying, incubation, hatching and early nestling periods (as identified in the food stores), and in the late nestling period (as identified in the prey detritus accumulated on the bottom of the boxes; data from Korpimäki 1988b and unpublished).

Prey species or group	Weight (g)	Source	Winter pellets 1970/1–1988/9			Storage 1973–2009			Nest remains 1973–2008		
			n	Number %	Weight %	n	Number %	Weight %	n	Number %	Weight %
Common shrew *Sorex araneus*	7.5	1	114	11.0	4.1	1146	9.1	3.4	6400	23.1	9.8
Pygmy shrew *S. minutes*	3.5	1	28	2.7	0.5	39	0.3	0.1	157	0.6	0.1
Masked shrew *S. caecutiens*	5.5	2				2	0.0	0.0			
Taiga shrew *S. isodon*	10.5	2				1	0.0	0.0			
Water shrew *Neomys fodiens*	14.5	1	6	0.6	0.4	35	0.3	0.2	99	0.4	0.3
Shrew species *Sorex* spp.	7	1	5	0.5	0.2	13	0.1	0.0	1453	5.2	2.1
Shrews total			153	14.7	5.2	1236	9.7	3.7	8109	29.2	12.3
Eurasian red squirrel *Sciurus vulgaris*	295	2							1	0.0	0.1
Least weasel *Mustela nivalis*	42					1	0.0	0.0	2	0.0	0.0
European hare *Lepus europaeus*	193					1	0.0	0.1			
Hares total			0	0.0	0.0	1	0.0	0.1	0	0.0	0.0
Wood lemming *Myopus schisticolor*	27.5	2				1	0.0	0.0			
Bank vole *Myodes glareolus*	16.5	1	212	20.4	16.8	4679	37.0	30.8	6814	24.5	23.0
European water vole (y)[a] *Arvicola terrestris*	65	1	2	0.2	0.7	9	0.1	0.2	143	0.5	2.3
Field vole *Microtus agrestis*	25	1	91	8.8	10.9	4379	34.6	43.7	3105	11.2	15.9

Prey	Weight (g)	n	%	%	n	%	%	n	%	%
Sibling vole	23.5[1]	120	11.6	13.5	1444	11.4	13.5	2764	10.0	13.3
M. rossiaemeridionalis										
Microtus spp.	24.5[1]	171	16.5	20.1	29	0.2	0.3	3502	12.6	17.5
Vole spp.	21.7	4	0.4	0.5	122	1.0	1.1			
Microtus spp. total		382	36.8	44.6	5852	46.3	57.5	9371	33.8	46.7
Voles total		600	57.8	61.9	10 662	84.3	89.6	16 328	58.8	72.0
Brown rat	66	3	0.3	1.0	4	0.0	0.1	23	0.1	0.3
Rattus norvegicus										
Eurasian harvest mouse	8[1]	27	2.6	1.0	174	1.4	0.6	365	1.3	0.6
Micromys minutes										
House mouse	15[1]	7	0.7	0.5	94	0.7	0.6	236	0.9	0.7
Mus musculus										
Murids total		37	3.6	2.5	272	2.2	1.2	624	2.2	1.6
Mammals total		790	76.1	69.5	12 173	96.2	94.7	25 064	90.3	86.0
Birds total		245	23.6	30.5	476	3.8	5.3	2508	9.0	14.0
Coleoptera	0.2[4]	3	0.3	0.0				117	0.4	0.0
Hymenoptera	0.2[4]							70	0.3	0.0
Insects total	0.2[4]	3	0.3	0.0		0.0	0.0	187	0.7	0.0
Total		1038	100.0	100.0	12 649	100.0	100.0	27 759	100.0	100.0
Diet width index[b]		7.6			3.6			6.9		
No. of samples or nests		133			1525			392		

[a] y = juvenile individual.
[b] Diet width calculated using Levins' (1968) formula.

Table 5.6. Bird prey of boreal owls according to their body mass in winter (based on pellet analyses during 1970/1 to 1988/9), in the early breeding season (prey items found stored in nest-boxes during 1973–2009) and in the late breeding season (prey items identified from prey detritus in nest-boxes during 1973–92) in the Kauhava region of western Finland.

| Weight class (g)[a] | Winter | | Breeding season | | | |
| | | | Early | | Late | |
	No.	%	No.	%	No.	%
10–19	49	20.0	130	27.3	578	30.7
20–29	58	23.7	231	48.5	1006	53.4
30–39	78	31.8	50	10.5	66	3.5
40–49	13	5.3	19	4.0	2	0.1
50–59	14	5.7	25	5.3	124	6.6
60–120	33	13.5	21	4.4	108	5.7
Total	245	100.0	476	100.0	1884	100.0

[a] Body mass of the various bird species recorded in the diet in Table 5.3.

Fig. 5.12.   Male provides a harvest mouse to the nest. Photo: Benjam Pöntinen. (For colour version, see colour plate.)

treecreepers *Certhia familiaris*, goldcrests, redpolls, red crossbills *Loxia curvirostra*, bullfinches and yellowhammers were the most common avian prey in winter. Migratory chaffinches, song thrushes, redwings and fieldfares emerged as the most frequent avian prey in the breeding season. Because resident great, crested and willow tits were quite often taken in the breeding season, the prey birds were markedly smaller than in winter (Table 5.6). The diet of boreal owls also included two small predators of the vole-eating guild, the least weasel *Mustela nivalis* and the pygmy owl. Most of the changes in diet composition from winter to the late breeding season were due to seasonal variation in the availability of most important prey groups rather than to different methods of analysing diet composition (Korpimäki 1986e, 1988b, see also Chapter 5.7).

Small mammals accounted for 79% of prey number and >95% of prey weight of boreal owls in Idaho, USA (Table 5.4). Their primary foods were southern red-backed voles *Myodes gapperi*, which accounted for 36% of prey number and 37% of prey weight (pooled data from summer and winter), while red-backed voles accounted for 54% of 72 prey items and *Microtus* voles comprised another 25% of the diet in Colorado, USA (Palmer 1986). Northern pocket gophers *Thomodys talpoides* (10% and 26%, respectively), shrews *Sorex* spp., *Microtus* spp. voles, deer mice *Peromyscus maniculatus* and small birds served as the most important alternative prey in Idaho. In winter, northern flying squirrels *Glaucomys sabrinus* (mean body mass 140 g) were frequently (11 specimens) captured by female owls representing 45% of prey weight, but only one specimen was captured by males (Hayward et al. 1993). The diet of boreal owls appeared to include all but one of the mammal species smaller than 50 g known to occupy the study area (Hayward et al. 1993). Bird prey included seven passerine and one woodpecker species. The proportion of insects was unusually high, with crickets as the dominant group in summer, although their importance may have been underestimated by pellet analyses (Hayward et al. 1993). There were also considerable inter-individual differences in the diets of male owls, which included unusual numbers of less common prey taken by particular males. The diet of a male consisted of 10% of birds by prey number in summer, and this same male caught 54% of the yellow-pine chipmunks *Tamias amoenus* killed by males in summer. Two other males took many western jumping mice *Zapus princeps* and crickets. Despite these dissimilarities, red-backed voles were the most numerous prey for all males in both winter and summer (Hayward et al. 1993).

At least 11 species of small mammals and 14 species of birds were identified in the food stores of boreal owls during three years in Alaska (Whitman 2009). Tundra (or northern) red-backed voles *Myodes rutilus* were the primary food of boreal owls, accounting for 49% of prey number and 46% of prey weight (Table 5.4). Their proportion in the diet was markedly higher in 2004 and 2005 than in 2006, consistent with the yearly trap indices of northern red-backed voles (13.6 individuals trapped per 100 trap-nights in 2004, 29.4 in 2005, and 3.1 in 2006). The root vole (also known as the tundra vole) *Microtus oeconomus* was the second main prey, accounting for 24% of prey number and 27% of prey weight. Its proportion in the diet was higher in 2006 than in 2004 and 2005. The food stores also quite often included northern bog lemmings *Synaptomus borealis*, meadow voles *Microtus pennsylvanicus*, brown lemmings *Lemmus trimucronatus* and shrews. Avian prey accounted for 6% of food stores by both number and weight. Dark-eyed juncos *Junco hyemalis*, yellow-rumped warblers *Dendroica coronata*, grey jays *Perisoreus canadensis* and black-capped chickadees *Poecile atricapillus* were the most frequent avian prey.

In Yukon, Canada, three *Microtus* vole species (the root vole (tundra vole), the meadow vole and the Alaska vole *M. miurus*) were the main foods in summer, and heather voles *Phenacomys intermedius*, tundra (northern) red-backed voles, deer mice and shrews were the most important alternative prey (Table 5.4). In the Kluane project, boreal owls were studied as a member of a predator assemblage subsisting mainly of snow-shoe hares *Lepus americanus*, and boreal owls also killed many young hares in the early increase phase of the 10-year hare cycle, when vole populations crashed between

1994 and 1995 (Doyle and Smith J. 2001). Boreal owls captured forest-dwelling heather voles and deer mice much more often than any other co-existing avian predator, including hawk owls and American kestrels *Falco sparverius*. As expected, this showed that boreal owls hunted more in closed forest than other co-existing avian predators.

In summary, voles of the genera *Microtus* and *Myodes* emerged as the main (staple) food of boreal owls in both Europe and North America (Tables 5.2, 5.4). Shrews of the genus *Sorex* and small passerine birds were the most important alternative prey in northern Europe, whereas mice of the genus *Apodemus* were an important alternative prey in central and southern Europe, particularly in terms of prey biomass. In North America, the lower densities of *Microtus* and *Myodes* voles and higher species diversity of small to medium-sized mammals led to the situation where boreal owls used many mammal species, including deer mice, shrews, pocket gophers, flying squirrels and lemmings, as their alternative prey, along with small passerine birds, during scarcity periods of their staple prey. In any case, there is an urgent need to conduct more studies on the diet composition of boreal owls and its between-year variation in North America and Asia. In particular, it would be very interesting to compare the dietary diversity of co-existing boreal and saw-whet owls in the USA and Canada.

5.5.   Regional trends in dietary diversity

Earlier analyses of the diet data from the breeding season of European owl and diurnal raptor species showed that the number of prey species and the food niche breadth of mammal-eating birds of prey generally decreased from south to north and from west to east in Europe. The opposite trend was found for the mean prey mass of mammal-eating birds of prey (Korpimäki and Marti 1995). Mammal-eating birds of prey included boreal, short-eared, long-eared, barn, great grey and hawk owls, and Eurasian kestrels. We reanalysed the large data set from Table 5.2 and found that the dietary diversity (the diet width index) of boreal owls tended to decrease with longitude and latitude, but these relationships almost reached significance only for the west-to-east trend (Fig. 5.13). For barn owls, a simpler niche measure – the number of prey species making up to 80% of the diet by number of items – was used (Taylor I. 1994). This food niche estimate closely positively correlated with the diet width index of boreal owls (Spearman rank correlation, $r_s = 0.62$, $n = 26$, $p = 0.001$) and was also significantly negatively related to longitude ($r_s = -0.48$, $n = 26$, $p = 0.01$) and almost significantly to latitude ($r_s = -0.37$, $n = 26$, $p = 0.06$). In most south-western studies on boreal owl diets, some 5–17 prey species made up to 80% of the diet by number of items, whereas in the north-eastern studies, the corresponding figure was only 3–6 prey species. In contrast, we did not find any obvious trend in the mean prey weight of boreal owls in relation to latitude or longitude within Europe.

The majority of European owls subsist mainly on small mammals as their main food (Mikkola 1983, Marti et al. 1993a), and populations of small mammals become multi-annually more cyclic northwards and eastwards in Europe (Hansson and Henttonen 1988, Hanski et al. 1993, Korpimäki et al. 2005b). Climate is also more continental,

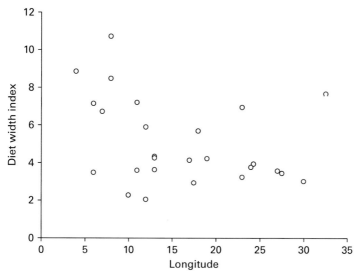

Fig. 5.13.   Diet width indices of boreal owls plotted against the longitude in Europe (data from Table 5.2; Spearman correlation, $r_s = -0.34$, $n = 26$, $p = 0.09$). Diet width (diet diversity) was calculated using Levins' (1968) formula. In the calculations, the prey were identified to species level as far as possible, because supraspecific levels of prey identification consistently underestimate diet width (Greene and Jaksic 1983).

with long and cold winters and deeper snow layers in the east. During population peaks, owls concentrate their feeding on abundant small rodent supplies, but when rodent populations decline, owls skip breeding or disperse over long distances to search for areas of vole abundance (Korpimäki 1992a, 1992c). This response probably reduces their dietary diversity. In the south, abundant supplies of small mammals exist only locally and infrequently (Korpimäki et al. 2005b, Fargallo et al. 2009), and owls are forced to capture all the available small and medium-sized mammals; the number of available species is also larger in temperate than in boreal and arctic zones of Europe (Goszczynski 1977, Erlinge et al. 1983, Korpimäki 1992c, Jedrzejewski et al. 1996). The species diversity of breeding birds also increases towards the south-west in Europe (Järvinen O. 1979) and this probably creates a wider dietary diversity of birds there. These factors probably also explain the smaller dietary diversity of boreal owls in north-east than in south-west populations of Europe. Unfortunately there are too few diet studies on boreal owls in North America to analyse large-scale continent-wide trends in dietary diversity.

5.6.   Wide between-year variation in diet composition

Foraging theories attempt to explain the prey choice and spatio-temporal variation in the diet composition of predators (for reviews see, for example, Pyke et al. 1977, Pyke 1984, Stephens and Krebs 1986, Sih and Christensen 2001, Stephens et al. 2007). Conventional optimal foraging theory is based on the assumption that predators simply choose prey so as to maximise fitness, and that there is some 'currency' – in terms of energy, time or nutrients – that influences fitness (e.g. Pyke 1984, Stephens and Krebs 1986). Detailed energy- and time-budget studies on breeding long-eared owls in the Netherlands showed that they mostly attempted to maximise their energy intake (Wijnandts 1984) but very few food competitors and enemies of long-eared owls were present in this large open-country study area. However, wild small to medium-sized foragers are only rarely free to 'choose' their prey items on the basis of energy content only. Therefore, the success of optimal diet theory in predicting variations in diet composition of predators hunting mobile prey has remained low, mainly because variations between prey in encounter rate and capture success are often more important than variation in the active prey choice of predators in determining diet composition (Sih and Christensen 2001). In addition, vertebrates demonstrate complex avoidance behaviours of their enemies and do not passively wait to be captured and eaten (Caro 2005). The more realistic mechanistic approach to diet selection and its spatio-temporal variation tries to include the possible effects of ecological interactions, for example the anti-predatory behaviours of prey animals, the presence of food competitors, and predation risk, on the diet selection of predators (Sih 1993).

The main prey species of boreal owls in our study area are field and sibling voles (*Microtus* spp.), while bank voles are the most important alternative prey, and shrews (mainly the common shrew) and small birds another important alternative prey (Table 5.5). Cross-correlation analyses of long-term snap-trapping data in the same study area revealed that population density indices of *Microtus* voles and bank voles fluctuated in close synchrony (upper panel of Fig. 1.1). The usual cycle length was 3 years for both *Microtus* and bank voles (Korpimäki et al. 2005a). The density indices for common shrews also varied in weak synchrony with *Microtus* voles (lower panel in Fig. 1.1), while density indices for mice (mainly the harvest mouse) appeared to peak between 6 months and 1 year before those of *Microtus* voles (Fig. 1.1). This high-amplitude 3-year cyclic variation in population densities of main prey, which is synchronous with most important alternative mammalian prey, induces wide between-year variation in food conditions for boreal owls. 'Fat' and 'lean' periods follow each other in a predictable manner that also is strongly reflected in the prey number, and thus diet composition, of boreal owls.

During the 1-month nestling period, the consumption rates of young owls ranged from 22.0 to 29.7 g of food per chick per day (mean 24.2 ± 1.8 g) (Whitman 2009). The large inter-annual variation in brood size (Chapter 8.1) induces a wide variation in the number of prey items delivered to the nest, which can be found in the prey detritus layers accumulated on the bottom of the nest-boxes. This number of prey items per nest is also closely positively correlated with the number of fledglings produced ($r_s = 0.44$, $n = 208$, $p < 0.001$; pooled data from 1981–9, when prey detritus layers of most nests in the study population were analysed). The same is also true for the mean number of prey items stored in nest-boxes per visit and number of fledglings produced ($r_s = 0.22$, $n = 197$, $p = 0.002$; pooled data from 1981–9). The traditional way to study the diet composition of predators is to compare, for example, between-year variations in the proportions of different prey species or groups in the diet. We have used this approach to study the between-year variation in the diet composition of birds of prey (e.g. Korpimäki and Sulkava S. 1987, Korpimäki 1988b, 1992a). However, the proportions that describe different prey species or groups must add up to 100% over all prey types (the 100% sum constraint; see Aebischer et al. 1993 for a similar problem arising from the proportional use of different habitat types of animals). The consequence is that the preference of animals for one prey type will almost invariably lead to an apparent avoidance of other prey types. Therefore, to account for the wide variation in the number of prey items found per nest and to avoid this kind of dependency between various prey types, subsequent between-year comparisons of various prey types in the diet of boreal owls were done by using the number of prey types per nest as an observation unit.

The yearly mean number of *Microtus* voles per nest visit in the food stores varied from 0.00 to 3.66 (in 1977) with a mean of 0.72 (s.d. 0.93) during 1973–2009 in our study area (upper panel of Fig. 5.14). The corresponding figures for bank voles ranged from 0.00 to 2.52 (in 2008) with a mean of 0.91 (s.d. 0.54) (lower panel of Fig. 5.14), for shrews from 0.01 to 1.46 (in 2001) with a mean of 0.38 (s.d. 0.36) (upper panel of Fig. 5.14), and for birds from 0.00 to 1.94 (in 1975) with a mean of 0.18 (s.d. 0.36) (lower panel of Fig. 5.14). There was no obvious long-term trend in the mean numbers of *Microtus* voles, bank voles, shrews and small birds per visit in the food stores during 1973–2009 (Fig. 5.14). The yearly mean number of *Microtus* voles in the food stores significantly increased with increasing density indices of these voles in the field, whereas the number of water voles and small birds decreased with increasing density indices of *Microtus* voles (Table 5.7). The annual mean number of bank voles in the larders also increased

Table 5.7. (a) Spearman rank correlations between the spring trap indices of small mammals and the yearly mean number of the most important prey groups per nest visit in food stores of boreal owls during 1973–2009 (number of years = 37). (b) Spearman rank correlations between the spring trap indices of small mammals and the yearly mean number of the most important prey groups identified in the prey detritus accumulated at the bottom of the nest-boxes during 1973–94, 1996, 1999–2000, 2003–6 and 2008 (number of years = 30).

	Spring trap index of		
	Microtus voles	Bank voles	Common shrews
(a) Food stores			
Microtus voles	$r_s = 0.60, n = 37, p < 0.001$	$r_s = 0.56, n = 37, p < 0.001$	$r_s = 0.12, n = 37, p = 0.48$
Bank voles	$r_s = 0.51, n = 37, p = 0.001$	$r_s = 0.57, n = 37, p < 0.001$	$r_s = 0.15, n = 37, p = 0.37$
Shrews	$r_s = -0.25, n = 37, p = 0.14$	$r_s = -0.26, n = 37, p = 0.12$	$r_s = 0.06, n = 37, p = 0.69$
Water voles	$r_s = -0.39, n = 37, p = 0.02$	$r_s = -0.27, n = 37, p = 0.11$	$r_s = -0.24, n = 37, p = 0.15$
Mice	$r_s = 0.44, n = 37, p = 0.007$	$r_s = 0.46, n = 37, p = 0.004$	$r_s = 0.29, n = 37, p = 0.08$
Birds	$r_s = -0.46, n = 37, p = 0.004$	$r_s = -0.19, n = 37, p = 0.25$	$r_s = 0.10, n = 37, p = 0.55$
(b) Prey detritus			
Microtus voles	$r_s = 0.64, n = 30, p < 0.001$	$r_s = 0.51, n = 30, p = 0.004$	$r_s = 0.21, n = 30, p = 0.27$
Bank voles	$r_s = 0.42, n = 30, p = 0.02$	$r_s = 0.36, n = 30, p = 0.05$	$r_s = 0.03, n = 30, p = 0.86$
Shrews	$r_s = 0.01, n = 30, p = 0.98$	$r_s = -0.20, n = 30, p = 0.29$	$r_s = 0.18, n = 30, p = 0.34$
Water voles	$r_s = 0.02, n = 30, p = 0.90$	$r_s = 0.00, n = 30, p = 0.99$	$r_s = -0.19, n = 30, p = 0.32$
Mice	$r_s = 0.24, n = 30, p = 0.21$	$r_s = 0.12, n = 30, p = 0.54$	$r_s = -0.06, n = 30, p = 0.75$
Birds	$r_s = -0.22, n = 30, p = 0.25$	$r_s = -0.29, n = 30, p = 0.12$	$r_s = -0.20, n = 30, p = 0.30$

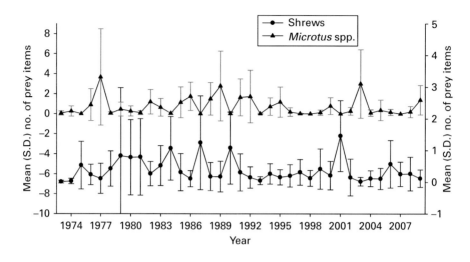

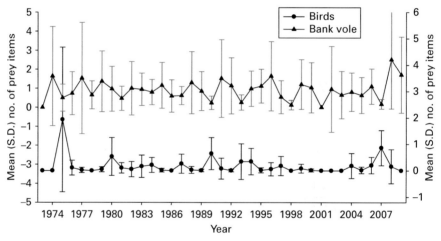

Fig. 5.14. The yearly mean (s.d.) number of *Microtus* voles (left y-axis) and shrews (right y-axis) per nest visit in the food stores of boreal owls during 1973–2009 (upper panel). The yearly mean (s.d.) number of bank voles (left y-axis) and birds (right y-axis) per nest visit in the food stores of boreal owls during 1973–2009 (lower panel). No obvious long-term trends were found for the yearly mean number of *Microtus* voles (Spearman correlation, $r_s = -0.07$, $n = 37$, $p = 0.70$), bank voles ($r_s = 0.07$, $n = 37$, $p = 0.69$), shrews ($r_s = 0.15$, $n = 37$, $p = 0.38$) and birds ($r_s = -0.02$, $n = 37$, $p = 0.89$)

with increasing density indices of these voles in the field. The yearly mean number of mice in the food stores remained small but was still significantly positively correlated with the spring trap indices of both *Microtus* and bank voles.

The yearly mean number of *Microtus* voles identified in the prey detritus accumulated at the bottom of the nest-box in the late nestling period varied from 0.8 to 73.7 (in 2003) and averaged 19.8 (s.d. 17.7) during 1973–94, 1996, 1999–2000, 2003–6 and 2008 (upper panel of Fig. 5.15). The corresponding figures for bank voles ranged from 1.0 to 49.1 (in 1974) with a mean of 17.7 (s.d. 13.0) (lower panel of Fig. 5.15), for shrews from 1.8 to 86.5 (in 1984) with a mean of 24.6 (s.d. 19.5) (upper panel of Fig. 5.15), and for birds 0.9

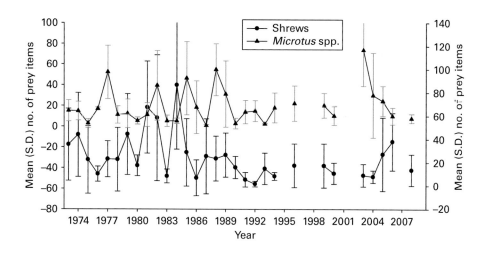

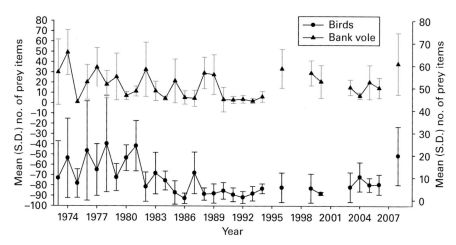

Fig. 5.15. Annual mean (s.d.) number of *Microtus* voles (left y-axis) and shrews (right y-axis) identified in the prey detritus accumulated at the bottom of the nest-box in the late nestling period during 1973–94, 1996, 1999–2000, 2003–6 and 2008 (upper panel). The same for bank voles (left y-axis) and small birds (right y-axis; lower panel). No obvious long-term trends were found for the yearly mean number of *Microtus* voles (Spearman correlation, $r_s = 0.11$, $n = 30$, $p = 0.58$) and bank voles ($r_s = -0.09$, $n = 30$, $p = 0.65$), whereas the annual mean number of shrews ($r_s = -0.38$, $n = 30$, $p = 0.04$) and birds ($r_s = -0.42$, $n = 30$, $p = 0.02$) decreased with year.

to 25.2 (in 1978) with a mean of 8.3 (s.d. 7.1) (lower panel of Fig. 5.15). No obvious long-term trends emerged in the yearly mean numbers of *Microtus* and bank voles in the prey detritus layers, while the annual mean number of shrews and small birds in the diet significantly decreased with year during 1973–94, 1996, 1999–2000, 2003–6 and 2008 (Fig. 5.15). The annual mean number of *Microtus* voles found in the prey detritus significantly increased with increasing density indices of these voles in the field (Fig. 5.16), whereas the numbers of other important prey groups were not associated with the density indices of *Microtus* voles (Table 5.7). The yearly mean number of bank

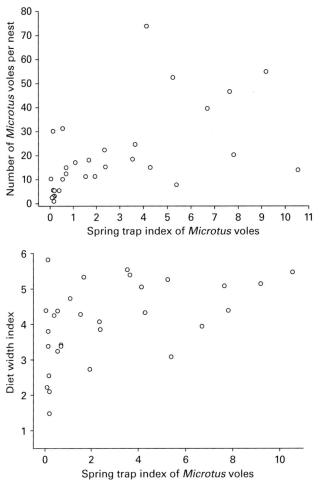

Fig. 5.16.   Annual mean number of *Microtus* voles found in the prey detritus accumulated at the bottom of the nest-box in the late nestling period plotted against the density indices of these voles in the field during 1973–94, 1996, 1999–2000, 2003–6 and 2008 (Spearman rank correlation, $r_s = 0.64$, $n = 30$, $p < 0.001$; upper panel). The same but for the yearly diet width index (Levins' formula) plotted against the density indices of *Microtus* voles in the field during 1973–94, 1996, 1999–2000, 2003–6 and 2008 ($r_s = 0.45$, $n = 30$, $p = 0.012$; lower panel) (data from Korpimäki 1988b and unpublished).

voles determined in the prey detritus also increased with increasing density indices of these voles in the field. The yearly mean number of shrews identified in the prey detritus was not obviously correlated with the spring trap indices of shrews in the field.

Conventional optimal diet theory (ODT) attempts to predict diets and dietary shifts of predators in response to changes in prey availability. ODT assumes that the preferences of predators are based on profitability of prey types and not on the relative densities of various prey types in the field (e.g. Pyke et al. 1977, Pyke 1984). Because the profit-abilities of the different prey types of boreal owls remained unknown and our food samples included prey items brought to nests, our results from between-year variation in the diet composition of boreal owls cannot be viewed as a test of ODT. However, it is possible to relate the between-year variation in diet composition to the predictions of conventional ODT (see summary of other examples of avian predators in Korpimäki 1992a).

Boreal owls took more *Microtus* voles with the increasing abundance of these voles in the field (Fig. 5.16, upper panel). This was consistent with the first prediction of conventional ODT that predators should feed on the most valuable prey types in years of food abundance (Pyke 1984), assuming that these voles were the most profitable prey (see Chapter 5.8). We found that the annual number of *Microtus* and bank voles in the food stores and prey detritus increased with increasing density indices of these voles in the field, but no corresponding relationships were found for shrews and mice (Table 5.7). This was in agreement with the second prediction of ODT that the inclusion of a food type in the diet does not depend on its abundance, but on the abundance of the preferred prey type, supposing that these voles were the preferred prey (see Chapter 5.8). Annual dietary diversity – as measured by Levins' (1968) formula – increased at high densities of *Microtus* voles and decreased at low densities of these voles in the field (Fig. 5.16, lower panel). In addition, the diet width also expanded at high densities of bank voles and shrank at low densities of these voles in the field ($r_s = 0.47$, $n = 30$, $p = 0.009$). These results are in apparent disagreement with the third prediction of ODT that dietary diversity (diet width) should expand when the abundance of the preferred prey decreases and shrink when it increases (Pyke 1984).

Our results thus show that boreal owls in our study area are only partially following the grand foraging theory. Results that were consistent with all three above-mentioned predictions of conventional ODT have previously been obtained for inter-annual variation in the dietary diversity of Eurasian kestrels, Ural owls and long-eared owls in Finland (Korpimäki 1986a, 1992a, Korpimäki and Sulkava S. 1987) and for three diurnal raptor species in Idaho, USA (Steenhof and Kochert 1988). In addition, the dietary diversity of Montagu's harriers increased with decreasing densities of common voles *Microtus arvalis* in France (Salamolard et al. 2000). The possible reason for the divergence of boreal owl diet variation from that of the other birds of prey studied so far is that boreal owls are more specialised on small mammal species that generally vary in temporal synchrony in our study area. When main prey, *Microtus* and bank voles, and the most important alternative prey, shrews, and even mice, increased and decreased nearly synchronously (Korpimäki et al. 2005a), there were very few alternative prey (mainly small birds) available in the field at low densities of main prey. The hunting mode of boreal owls is not well suited for hunting smaller birds (Chapter 4.1), and the presence of enemies in the daytime probably restricts their hunting to the hours of darkness (Chapter 4.1.3), when the availability of most small birds is low. In addition, small boreal owls may suffer from food and interference competition by larger birds of prey even at high densities of main prey (Korpimäki 1987c, 1987d). These factors probably induce decreasing dietary diversity of boreal owls at low densities of main prey.

The diet of breeding boreal owls in our study area mainly consists of *Microtus* and bank voles in the increase and decrease years of the 3-year vole cycle, whereas the numbers of shrews and small birds in the diet are larger than those of the voles in the low years of the vole cycle. The numbers of *Microtus* and bank voles in the diet also fluctuate in accord with the abundance of these voles in the field. Because *Microtus* (field and sibling) voles occupy agricultural fields, grasslands and grassy clear-cut areas, boreal owls must quite often hunt these voles at least on the edges of open areas, although they are mainly adapted to hunt in forests. The reason may be that *Microtus* voles are considerably larger (Table 5.5), and thus a more profitable prey, than bank voles and common shrews living mainly in woodland. In our study area, the bank vole is the most important alternative prey in the breeding season, because its yearly number in the diet correlated positively with its abundance in the field. Shrews, in particular common shrews, are more important alternative prey than small birds. The probable reasons are that they are more frequent than birds and the hunting technique of boreal owls is better suited to hunting shrews than mostly day-active small birds. In addition, shrews are much noisier than voles, particularly in spring (Heikura 1984, Korpimäki 1986e); therefore boreal owls, which mainly locate their prey by auditory clues, can successfully hunt shrews even in summer, when the dense vegetation cover usually protects voles against avian predators. In low vole years, some male boreal owls take plenty of small birds. For example, one male owl took a total of 34 barn swallows in the low vole year of 1980. These males are probably older, more experienced individuals, which have learnt to search for roosting birds and take adults and their young from the nests. Familiarity with their home ranges may improve their success when hunting birds. Harvest and house mice are a quite regular alternative prey, probably because they are fairly common and the hunting mode used for voles is also suitable in catching them. Boreal owls can also occasionally take very large prey (e.g. young water voles, brown rats, red squirrels, fieldfares, mistle thrushes) or very small prey (coleopterans and pygmy shrews) in comparison with their own body size. When the abundance of main prey, *Microtus* voles, decreased in the field, boreal owls seemed to shift to alternative prey in the following order: bank voles, shrews and birds.

5.7.   Seasonal variation in diet composition

In birds of prey subsisting on voles as their main foods at northern latitudes, breeding generally occurs relatively early in the spring, although vole numbers peak in autumn (Fig. 1.1), long after the nestling phase of predatory birds. The seasonal peak density of voles in autumn increases the over-winter survival probabilities of boreal owls (Hakkarainen et al. 2002), but egg production usually happens near the seasonal low phase of vole populations. However, the largest clutch and brood sizes are observed in the earliest breeding pairs rather than the latest, e.g. in the Eurasian kestrel (Cavé 1968, Korpimäki and Wiehn 1998), the long-eared owl (Wijnandts 1984) and the boreal owl (Korpimäki 1987f, Hörnfeldt and Eklund 1990, Korpimäki and Hakkarainen 1991). Boreal owls are the earliest breeders among the north European birds of prey. For example, half of the females started to lay eggs before 4 April in our study area during 1970–86 (Korpimäki 1987f; see also Chapter 8.1). During the egg-laying period, the mean long-term snow depth in our study area is approx. 35 cm (Solantie 1975, 2000), which reduces the availability of small mammals for hunting owls. Later in spring, usually in April and early May, the snow melts and new vegetation starts to grow, changing the hunting opportunities available to owls. This raises the question of whether changes in vole density and in snow and vegetation cover induce seasonal changes in the diet composition of boreal owls. If so, the next questions are whether owls select directly between different prey types on the basis of their encounter rates and energy content, or do they choose indirectly by selecting habitat patches with the highest profit and then hunting unselectively within these patches?

We analysed seasonal changes in the diet composition of boreal owls from winter to midsummer in the situations when vole populations were either increasing or decreasing during the first half of the year (Korpimäki 1986e). We used the data from 1973–85 because during this period there were still normal winters with regular snow cover in our study area, and we also collected plenty of food samples in the winter. In good vole years (i.e. in the increase and decrease phases of the vole population cycle), *Microtus* voles were the most frequent prey group in both winter and the breeding season, comprising 52% of prey items (Table 5.8). Their proportion in the diet peaked at the end of March, but after mid-April their number remained relatively constant. Bank voles were most often taken during the second half of April and during the first half of May. In the breeding season (16 March to 30 June), the proportion of shrews (the most numerous species being the common shrew) increased almost continuously, and in winter their proportion was also high. Birds were most often caught in winter and at the end of the nesting season (16–31 May and all of June) but there were few of them in the diet between mid-March and late April. In June, nestlings and young birds also played an important role in the diet. There was irregular variation in the small proportion of mice in the diet.

In poor vole years (i.e. in the decline and low phase of the vole population cycle), bank voles were the most important prey, followed by shrews, *Microtus* voles and birds (Table 5.8). *Microtus* voles were the main prey in winter and their proportion decreased dramatically towards the end of the breeding season. The number of bank voles in the diet peaked at the beginning of the breeding season (16 March to 15 April) and decreased markedly as the season proceeded. The reverse was evident for shrews, but the capturing of mice varied only irregularly. Adult birds were the second most common prey group by number in winter, and in May and June their proportion was also high. In particular at the end of May and in June, nestlings and young birds formed an important prey group, because they were the second most frequent prey after shrews.

The bird species in the diet of boreal owls showed great seasonal variation. In winter the most important prey birds were great, willow and crested tits, the bullfinch, the yellow-hammer and the red crossbill. These six species are common over-wintering birds in our study area. At the end of March and in April, prey birds were either wintering species (such as the bullfinch and the yellowhammer) or the earliest migratory species (the chaffinch). In May, migratory birds, particularly thrushes and chaffinches, emerged as the most common birds in the diet. In June, nestlings and juveniles of thrushes, chaffinches, yellowhammers and warblers *Phylloscopus* spp. formed an important prey group.

Table 5.8. Seasonal changes in the proportions (as percentages by prey number) of the most important prey groups in the diet of boreal owls during the first half of the year (a) in increase and peak years (pooled data from 1973, 1977, 1982 and 1985) and (b) in decline and low years (1974–6, 1978–81 and 1983–4) of vole abundance.

Prey group	Time period							Total
(a) Increase and peak years	1 Jan–15 Mar	16–31 Mar	1–15 Apr	16–30 Apr	1–15 May	16–31 May	1–30 Jun	
Shrews	17.3	2.4	4.6	15.9	10.5	15.8	20.0	11.8
Bank voles	17.3	11.2	15.5	34.0	34.1	27.2	20.0	27.7
Microtus spp.	34.7	84.0	74.0	46.6	49.0	45.1	48.0	52.4
Mice	6.7	2.4	5.5	2.8	5.6	–	–	4.1
Birds adults	24.0	–	0.5	0.7	0.7	12.0	8.0	3.9
nestlings and young	–	–	–	–	–	–	4.0	0.1
total	24.0	–	0.5	0.7	0.7	12.0	12.0	4.0
Diet width[a]	5.04	2.49	3.12	4.02	3.98	5.15	4.25	
No. of prey items	150	125	219	427	602	184	25	1732

Prey group	Time period						Total
(b) Decrease and low years	1 Jan–15 Mar	16 Mar–15 Apr	16–30 Apr	1–15 May	16–31 May	1–30 Jun	
Shrews	1.9	17.1	26.0	24.0	36.8	45.8	25.2
Bank voles	9.9	51.6	46.5	42.3	19.6	14.5	35.4
Microtus spp.	42.9	25.1	23.1	18.8	7.7	4.2	21.8
Mice	1.4	4.8	3.2	3.6	4.2	–	3.2
Birds adults	34.0	0.9	1.0	11.2	26.7	18.7	12.3
nestlings and young	–	–	–	–	4.9	16.3	1.9
total	34.0	0.9	1.0	11.2	31.6	34.9	14.2
Diet width[a]	8.18	3.05	3.24	3.95	5.20	4.13	
No. of prey items	254	351	624	421	285	166	2141

[a] Diet width calculated using Levins' (1968) formula.

The seasonal variation in diet composition from January to June in both increasing and decreasing conditions of main prey abundance shows that boreal owls respond flexibly to the availability of prey. The probable reasons for these marked changes in the diet composition are variations in snow and vegetation cover and in the abundance and behaviour of prey animals. Because changes in the food were nearly similar both in the good and poor vole years, these changes were not induced by the higher proportions of shrews and birds in late nests of poor vole years only (Korpimäki 1981). However, there is still one other factor that may affect these seasonal differences. It is possible that, at the beginning of the breeding season, the high proportion of *Microtus* voles in the diet is due to the fact that hunting mostly happens in agricultural fields and clear-cut areas. Later in the season, boreal owls in our study area may turn to hunting more in forests, which could cause the increased proportion of bank voles and birds in their diets. Because there are no data on changes in hunting habitat utilisation during the breeding period in our study area, the assessment of this possibility is difficult.

In south-eastern Norway, two radio-marked male boreal owls hunted more in the forest than expected from random use, and bank voles were the main prey, when the ground was snow-covered (Jacobsen and Sonerud 1993). When the ground was partially snow free, these males used clear-cut areas and forests in proportion to their availability, and *Microtus* spp. (field and root (tundra) voles *M. oeconomus*) were the most numerous prey group in their diet (Jacobsen and Sonerud 1993). A similar dietary shift from mainly bank voles in late winter to *Microtus* voles as the snow cover disappears in spring has also been documented for boreal owls in the vicinity of Umeå, northern Sweden (Hörnfeldt et al. 1990) and for hawk owls in south-eastern Norway (Nybo and Sonerud 1990). But in both of those study areas the snow cover is more than twice as deep as in our study area in late winter, which probably explains the reverse dietary shift in our study area (i.e. from *Microtus* voles in winter to bank voles in spring). In south-east Norway, snow-tracking showed that long small-mammal trails (>1 m) were most frequent in forests, short trails (<1 m) were equally frequent in clear-cuts and forests, and ventilation holes of voles were most frequent in the clear-cut areas (Jacobsen and Sonerud 1993). Therefore, voles occurring above the snow were likely to be more readily available to aerial predators in forests than in clear-cut areas, but when the snow melted, prey availability increased in clear-cuts relative to that in forest. Thus, boreal owls seemed to choose their small mammal prey indirectly by selecting the hunting habitat according to the potential profit (Jacobsen and Sonerud 1993).

In our study area, the depth of the snow layer generally increases from January to mid-March, after which the snow usually begins to melt (Solantie 1975, 2000). The deep snow cover protects small mammals against aerial predators (Korpimäki 1985e, 1987g, Sonerud 1986). This protective effect probably explains the lower proportion of *Microtus* voles and higher proportion of birds in the winter diet than in the breeding season, even in years of increasing vole abundance. Bank voles living mainly in forests move more above the snow than *Microtus* voles living in open areas (Hansson 1982), and they also often climb on trees in winter (Pulliainen and Keränen 1979, Viro and Sulkava S. 1985). The snow cover in forests is shallower than in open areas. These factors increase the availability and vulnerability of bank voles in comparison with *Microtus* voles, and probably cause the higher proportion of the former in the winter diet.

In late March and April, the snow melts most rapidly in open habitats. The vegetation of the previous summer has been flattened down on the ground and no new vegetation has yet grown. The melt-water forces *Microtus* voles to leave their holes and winter nests, and to come out into the open, where they are easy to catch. That is the most likely reason for the high proportion of *Microtus* voles in the diet at that time (Table 5.8). In May, the vegetation starts growing in the agricultural fields, decreasing the availability of small mammals in this habitat. At that time the snow has just melted in forests, forcing bank voles to come up out of the ground and increasing their risk of getting caught by owls. At the end of May and beginning of June, the vegetation in forests also becomes denser as the leaves of bushes and bilberries develop. This reduces the availability of bank voles in forests, and thus birds nesting above the ground and their young become the easiest prey. Moreover, the maturity of shrews begins in April (Heikura 1984), which increases their mobility and makes them more accessible for aerial predators. Shrews are also much more vociferous than voles, particularly in spring (Heikura 1984). Therefore, boreal owls locating prey by auditory clues can hunt shrews even in May and June, when the dense vegetation cover protects voles against aerial predators.

Because field and sibling voles only rarely reproduce in winter, and winter reproduction is confined to the increase phase of the vole population cycle (Hansson 1984, Norrdahl and Korpimäki 2002b), the abundance of these voles (the main prey of boreal owls) in the field probably becomes reduced from their seasonal peak in autumn until their next reproductive season, which in our study area usually starts in late May (Norrdahl and Korpimäki 2002b). This decrease is reflected in a decreasing number of *Microtus* voles in the diet of boreal owls. Among the alternative prey, bank voles are more important than shrews and birds. When the availability of *Microtus* voles decreased, boreal owls first shifted to bank voles, and when their availability became reduced at the end of May, the owls finally shifted to capturing shrews and small birds. The availability of birds increased especially in June, when there were plenty of nestlings and juveniles which were quite easily captured by owls. Therefore, the order of seasonal dietary shifts was the same as for the between-year variation of the most important prey groups (see Chapter 5.6).

Our results from boreal owls were consistent with many earlier investigations of birds of prey, in which a decrease in the proportion of vole prey and an increase in bird prey in the course of the breeding season have been documented. This trend has been reported, for example, in the sparrowhawk (Sulkava P. 1964, 1972), the pygmy owl (Kellomäki 1977) and the Eurasian kestrel (Korpimäki 1985b, 1985d) in western Finland, and in the long-eared owl in southern Sweden (Nilsson I. 1981) and the Netherlands (Wijnandts 1984). The proportion of field voles, which are the main food of barn owls in Scotland, was lowest in the three spring months (March–May), increased during the summer, and peaked in late autumn and early winter. This seasonal change largely followed the annual changes in vole abundance in the field (Taylor I. 1994). A similar pattern for barn owls has also been recorded many times in temperate areas of Europe (review in Taylor I. 1994). The proportion of shrews (the main alternative prey to voles in much of Europe) in the diet of barn owls correspondingly increased in winter and spring in comparison to summer (Taylor I. 1994). Also in studies on the diet of Eurasian kestrels in Scotland (Village 1982) and of long-eared owls in southern Sweden (Nilsson I. 1981) and the Netherlands (Wijnandts 1984), the importance of *Microtus* voles decreased continuously during the first half of the year and the occurrence of birds in the diet was largely confined to the June–July period. The most striking difference was that *Microtus* voles were caught by boreal owls less in winter than in the breeding season. This dissimilarity was probably due to the much deeper snow cover in western Finland in comparison with the more southern areas of temperate Europe. This theory is also supported by the fact that in areas with even deeper snow cover, such as southern Norway (Jacobsen and Sonerud 1993) and northern Sweden (Hörnfeldt et al. 1990), the importance of bank voles moving more on the snow surface was much greater in the winter diet of boreal owls than in our study area. In general, bank voles and other *Myodes* spp. might be the main foods of boreal owls in snowy areas of Eurasia, but more comprehensive studies on the diet composition of boreal owls in winter in other areas are badly needed. In addition, more information is still needed on the food resources used by female boreal owls to lay eggs at the time when vole populations are in their seasonal low phase and the voles are also protected by the deep snow layer in late winter and early spring.

5.8.   Are prey selected by species, sex and body condition?

Prey animals face a conflict between the demands of being active and of avoiding predators (see reviews in Lima and Dill 1990, Caro 2005). Prey selection is the primary subject of predator–prey interactions, because predators usually kill some prey species more often than others, and this may noticeably change the composition of prey communities (Sih et al. 1985, Kuno 1987). A mechanistic approach to prey selection proposes that the decision of prey choice is influenced by the behaviour of both predator and prey – not just by the predator, as optimality models assume (Sih 1993). Therefore, the refuges, crypticity, moving activity and anti-predator behaviour of the prey may also influence the encounter rate between prey and predator. The selection of some prey items is simply a result of varying encounter rates and capture success, rather than a result of an active decision by the predator to attack or not to attack (Sih 1993). Prey preference is a vague ecological term because, in a strict sense, it means a deliberate choice when two prey types are available simultaneously (Stephens and Krebs 1986). However, mammal- and bird-eating predators usually encounter their prey sequentially. Many researchers, including us, have used prey preference to mean that a prey type is taken more frequently than expected from its proportion in the field (e.g. Erlinge 1981, Nilsson I. 1981, Korpimäki 1992a).

Traditionally it has been assumed that predators mainly kill substandard individuals in poor condition, such as sick, injured, inexperienced, young or aged individuals, because the dominant individuals force subordinates into poorer habitats where the predation risk is high (Errington 1946, 1956; review in Temple 1987). Therefore, the detrimental effects of predators on prey densities may remain insignificant. This 'Erringtonian' view of prey selection is partially supported by the finding that wood mice with asymmetrical hind-leg bones fall victims to tawny owls more often than wood mice with symmetrical hind-leg bones. Thus, asymmetry in motion traits that are apparently important for evading avian enemies increased the predation risk of mice (Galeotti et al. 2005). Alternatively, predators may selectively capture one or other sex and/or various age classes of prey, and therefore markedly influence prey population dynamics. The sex ratio of the prey population may also change if predators preferentially take members of one sex. If females are the predominant sex in the diet of a predator, the intrinsic growth rate of the prey population may decline (Longland and Jenkins 1987).

Predators may selectively kill either large prey individuals, usually males or old specimens (e.g. Morse 1980), or small prey individuals, often females or young specimens (e.g. Marti and Hogue 1979, Rohner and Krebs 1996). Most studies on avian predators mainly feeding on small mammals have shown that males are the most vulnerable prey category (e.g. Southern and Lowe 1968, Lagerström and Häkkinen 1978, Korpimäki 1981, 1985d, Halle 1988, reviews in Christe et al. 2006, Taylor I. 2009), but a preference for young females has sometimes also been found (Longland and Jenkins 1987, Dickman et al. 1991). In a minority of studies conducted on birds of prey, no clear sex or size preferences were recorded (e.g. Boonstra and Krebs 1977, Castro S. and Jaksic 1995). Selective predation is usually due to the predator's habitat selection or to behavioural differences between species or social groups within the prey population (Dickman 1992).

To find out whether there is selectivity in the prey capture of boreal owls by prey species, either sex, or substandard individuals disproportionately to their occurrence in the field, we snap-trapped small mammals in the main habitat types of owl home ranges and simultaneously identified prey items cached by the same owls in their nest-boxes. The study was carried out in the neighbourhood of 21 nest-boxes occupied by owls in 1992, when vole densities declined during the breeding season (Fig. 1.1). Sibling voles were the preferred prey, followed by field voles, bank voles and common shrews. A similar order of preference was obtained on the basis of prey remains accumulated at the bottom of nest-boxes in the late nestling stage (Koivunen et al. 1996b). This was consistent with the contention that predation risk seems to increase with the size of the prey (Kotler et al. 1988). There were no obvious differences in the sex ratio between sibling, field and bank voles captured by the owls and those trapped by us in the field at the same time. However, owls captured significantly more male common shrews than were available in their territories (Koivunen et al. 1996b). Voles and shrews captured by owls tended to be lighter than those caught in our snap-traps, although the difference was significant only for female field voles. Furthermore, the female field voles and male common shrews found in the stores of owls were shorter than those trapped by us in the same territories (Koivunen et al. 1996b). There was a tendency for stored prey items to have more internal fat than prey items available in the field (Fig. 5.17). This tendency was significant for male field voles and female common shrews and for pooled data from both sexes of field voles (Koivunen et al. 1996b). In addition, great horned owls *Bubo virginianus* tended to capture snow-shoe hares in above-average condition (Rohner and Krebs 1996).

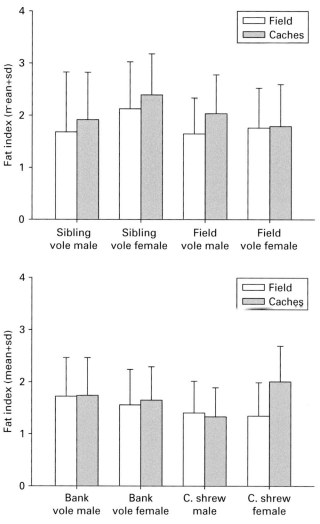

Fig. 5.17.   Mean (s.d.) fat indices for male and female sibling and field voles (upper panel), and for bank voles and common shrews (lower panel) in the field (white bars) and in the larders of boreal owls (grey bars) in 1992 (data from Koivunen et al. 1996b).

We also identified, sexed and weighed prey items stored by boreal owls in their nest-boxes, and compared the characteristics of these prey with those of small mammals trapped in our study area during 1985–92. For sibling, field and bank voles, owls captured more males than females, but we did not find that owls preferred one sex of common shrews (Koivunen et al. 1996a). Our long-term data also indicated that the male bias towards sibling and field voles in the diet of owls was highest in the low phase of the 3-year vole population cycle, and decreased through the increase and decline phases (Fig. 5.18), which may explain why we did not find obvious male bias in the study of Koivunen et al. (1996b), which was carried out only in the decline phase of the vole cycle.

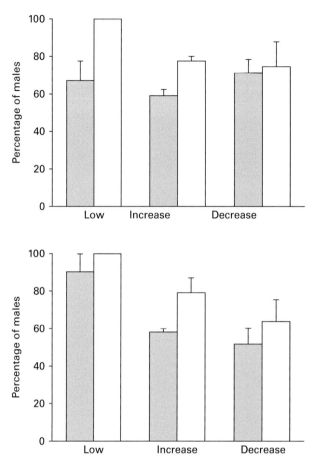

Fig. 5.18.   Mean (95% confidence limit) sex ratio of field voles (upper panel) and sibling voles (lower panel) captured by boreal owls (white bars) and snap-trapped in the field (grey bars). Phases of the vole cycle are separated into low (pooled data from 1987 and 1990), increase (1985, 1988) and decrease (1986, 1989 and 1992) phases (data from Koivunen et al. 1996a).

The yearly mean body mass of three vole species and common shrews stored by the owls was also compared with the yearly mean body mass of small mammals snap-trapped in the field during 1985–92. These comparisons revealed that, for all four prey species, stored prey individuals of both sexes were significantly lighter than those trapped in the field. This difference was consistent in every year, with the exceptions of sibling vole males in 1989, and common shrew males in 1987 and females in 1987 and 1990 (Koivunen et al. 1996a). It is also noteworthy that, although all the trap types used by trappers might be selective in regard to the species, sex, age and body size of captured small mammals, dissimilarities in trappability when using Finnish metal snap-traps appeared not to bias our results (Koivunen et al. 1996a, 1996b).

On the basis of small-mammal trapping in the vicinity of nest-boxes and prey animals simultaneously identified in the stores of owl nests, sibling voles were the preferred prey because they were taken more frequently than expected from their proportion in the field.

A similar preference order was obtained on the basis of prey remains. This preference may be due to sibling voles' patch-forming way of life and preference for a short-grass habitat (Norrdahl and Korpimäki 1993, Koivisto et al. 2007), which may make them more vulnerable to boreal owls that are able to search for high-density prey patches and repeatedly visit these patches until they are depleted. In our study area, field voles mainly occupy high-grass open habitat (Norrdahl and Korpimäki 1993, Koivisto et al. 2007), and the tall ground vegetation cover probably shelters them to some extent from avian predators, although boreal owls locate their prey by auditory cues. In contrast to the sibling and field voles occupying open habitats, bank voles and common shrews also inhabit sheltered coniferous forests, and are therefore less vulnerable to owl predation. In addition, the larger herbivorous sibling and field voles are clumsier when escaping predators than the omnivorous slender bank voles. Hawk owls also preferred to capture *Microtus* voles and avoided *Myodes* voles in the boreal forests of Yukon, Canada (Rohner et al. 1995). Barn owls, in turn, took more field voles and less wood mice than were available in the field, and this was suggested to be a result of the greater ability of wood mice to detect the owls and their greater agility in escaping catches (Taylor I. 2009).

The prey preferences of boreal owls were similar to those previously reported for vole-eating Eurasian kestrels and long-eared owls in our study area (Korpimäki 1985d, 1992a). However, these earlier results were based on pooled data, where the diets of several pairs breeding in our large study area and determined from analyses of pellets and prey remains collected at nests were compared with the available numbers of small mammals calculated from snap-trappings conducted in four main habitat types of the study area and then extrapolated to a much larger area. The same method has also been used in most previous studies on the diet selection of avian predators subsisting on small mammals (e.g. Nilsson I. 1981, Halle 1988, Mappes et al. 1993; but see Taylor I. 2009). This method was coarse and did not consider the prey choice of birds of prey at the individual or pair level. Also, the effects of habitat distribution within individual home ranges were not considered. In our earlier studies and in the study of Koivunen et al. (1996a) on boreal owls, the only replicates were between years, whereas in the study of Koivunen et al. (1996b) the captured and available prey items were simultaneously measured in 21 different home ranges, including both good and poor home ranges.

Small passerine birds are an important alternative prey for boreal owls when the densities of the main prey, voles, decline (Table 5.5; see also Korpimäki 1988b). Censuses of breeding birds and records of prey items used by the owls on the same home ranges in five years (1974–8) indicated that tree-nesting and hole-nesting forest birds were killed by the owls more than their availability in the field indicated, whereas secretive ground-nesting forest and open-country birds were killed less than available in the field. Among the most common bird species breeding in our study area, the fieldfare, chaffinch, bullfinch, song thrush, redwing, pied flycatcher and crested tit were captured substantially more often, and the willow warbler *Phylloscopus trochilus*, yellowhammer, robin, tree pipit, meadow pipit *Anthus pratensis* and garden warbler *Sylvia borin* were captured less often than would be expected from their availability in the field (Korpimäki 1981).

Boreal owls caught more male sibling, field and bank voles than were expected on the basis of their availability in the field according to snap-trap captures. This suggests that male voles are more vulnerable to owl predation, probably because they are more mobile and noisy in spring when searching for mates and defending territories. Accordingly, male sibling and field voles radio-tracked in the increase and decrease phases of the vole cycle moved substantially more than female conspecifics in late April to June (Norrdahl and Korpimäki 1998). Male-biased mobility was also stated to be the main reason for the predominance of field vole males in the diet of Scottish barn owls in the breeding season, because in winter, when the voles were not breeding, the sex ratio of field voles in the pellets was close to unity (Taylor I. 2009). The selection for male voles by hunting owls in spring was thus the result of augmented detection and encounter rates of owls that use auditory cues when hunting rather than 'optimal' foraging based on the larger body mass and energy content of male voles. This interpretation was also supported by the fact that there was a male predominance in chaffinches and yellowhammers found in the food stores of breeding boreal owls in our study area (Korpimäki 1981) and a male bias in blackbirds found in the food stores of tawny owls in Switzerland (Christe et al. 2006). Male songbirds are also more vociferous and active in spring when attracting mates and defending territories. Alternatively, but less probably, the higher vulnerability of male voles may be related to differences in microhabitat use between the sexes. For example, young female house mice that used open vegetation more than adults were more vulnerable to barn owl predation (Dickman et al. 1991), but house mice are an introduced species in Australia (Singleton et al. 2005) and in most studies a majority of male small rodents has been recorded in the diet of barn owls in the breeding season (Christe et al. 2006, Taylor I. 2009).

Boreal owls tended to capture lighter and shorter voles and common shrews than those present in the field according to snap-trappings (Koivunen et al. 1996b). Animals both in larders of owls and trapped in the field were over-wintered adult individuals, and thus variation in body length was not substantial. Although the differences in mean body size were relatively small, even these could, for example, affect the spacing behaviour and dominance status of field voles (e.g. Nelson 1994, 1995). Because all the small mammals included in our study (Koivunen et al. 1996b) were over-wintered, this difference was not due to the owls capturing juveniles that had recently left the nest but had not yet entered our snap-traps. Prey items stored by owls in their nest-boxes may not accurately reflect the prey actually caught by the owls if they consume the small prey items at the capture site and transport the larger prey items to their nests (i.e. the load-size effect; see Sonerud 1992a). If the load-size effect was really biasing our data on the prey captured by the owls, the real size difference between the killed and available prey would have been even larger. However, field data have shown that the model described above did not work in predicting the load-size effect of birds of prey subsisting on small mammals in the field (Korpimäki et al. 1994, Taylor I. 2009). Alternatively, owls may also consume larger prey items at the capture site and transport the smaller ones to the nest. However, male owl parents are presumed to maximise their food delivery rate on the basis of energetic efficiency (Kacelnik 1984). The mean mass of the prey has an important effect on the energy costs of breeding via the rate of delivery of food. Prey items that are too small or large are likely to increase the energy costs of food delivery for the males. For example, the range of prey sizes for male owls was 0.5–115 g (Table 5.5; see also Korpimäki

1988b). In our study, the heaviest vole trapped in the field weighed 40 g, which was far below the delivery capacity of boreal owls. Therefore it is unlikely that boreal owls preferentially ate larger voles at the capture site and transported smaller ones to the nest.

It is well known that a number of activities related to reproduction, such as mate-searching activities and pregnancy, can increase the predation risk for individuals (Magnhagen 1991 and references therein). In small mammals, the level of predation risk may depend on their visibility, audibility and smell, and all of these are increased in the reproductive season. Ural owls preyed proportionally more on reproductively active field voles than expected on the basis of snap-trappings in the field (Karell et al. 2010). Both in the aviary and in the field, we studied whether the maturity of female field voles affected their behaviour, and consequently their risk of predation by boreal owls. In an aviary, we recorded the behaviour of mature and immature voles when owls were present or absent, but did not find any obvious differences in behaviour or vulnerability between mature and immature female voles. In the field, we compared the maturity status of female field voles snap-trapped in boreal owl home ranges with those caught by breeding owls in 1992 and 1994. In accordance with the results from the aviary experiment, there were no obvious differences in vulnerability between mature and immature individuals. This suggests that mature and immature female field voles are equally exposed to boreal owl predation (Koivunen et al. 1998a).

There are at least two alternative, but not mutually exclusive, explanations for the tendency of boreal owls to capture small individuals, because larger individuals are normally stronger in competitive interactions (Grant P. 1972). One possible explanation is that small, subordinate individuals may behave differently from large, dominant ones, and may therefore be more exposed to owl predation. This seems unlikely, however, because dominant individuals probably move more when defending their territories than subordinate ones do (see Nelson 1994, 1995).

If dominant individuals force subordinate ones to disperse from safe to unsafe habitats, the subordinates may be more vulnerable to owl predation. Therefore, we suggested that large, dominant vole individuals may occupy safe habitats with dense vegetation cover, where owl predation risk is minimal, and force subdominant individuals into less-safe habitats. But we found that these individuals were not necessarily in poor physiological condition, because the prey mammals caught by owls tended to have more internal fat than conspecifics caught by our snap-traps (Koivunen et al. 1996b). Therefore, large, dominant voles probably consumed more energy when defending their territories and searching for mates (males) or producing young (females). Small voles could instead take more time to eat and accumulate internal fat reserves.

Little attention has been focused on identifying the exact mechanisms causing the vulnerability of small prey individuals to owl predation. This may be because proper experiments concerning behavioural differences in small mammal populations are laborious, especially in the field, where it is difficult to control for all the behavioural and other ecological factors that may act on the prey population. However, a great deal of useful information can be obtained by using aviary experiments, which provide opportunities to examine the behaviour of small mammals with and without the risk of avian predation. Therefore, these findings were further tested in an aviary and in the field.

In an aviary experiment, we studied whether body size or habitat familiarity of field voles affected boreal owl predation risk (Koivunen et al. 1998b). Large field voles occupied the good habitat more than small ones did, both in the presence and absence of boreal owls. Furthermore, habitat-familiar individuals inhabited the good habitat more than those that were unfamiliar with the habitat when an owl was present in the aviary. Large field voles also moved less than small ones when an owl was not present, whereas there was no such a difference in moving when an owl was present. Habitat-familiar individuals moved less than unfamiliar ones when owls were present. Boreal owls captured 2 of 17 large voles, 8 of 17 small voles, 1 of 17 familiar voles, and 6 of 17 unfamiliar voles. Small voles were found to be more vulnerable to owl predation than large voles, and habitat-unfamiliar voles were exposed to owl predation almost significantly more than habitat-familiar voles (Koivunen et al. 1998b). Our field data were consistent with our aviary data: larger field voles were more frequently found in good habitats than in poor habitats. This suggests that large field voles may have priority in inhabiting sheltered habitats.

We are well aware that the abnormal spatial scale of our enclosure experiment may restrict the application of the results to field conditions. Nonetheless, our results from aviary and field circumstances are parallel, suggesting that body size may reflect social status within field vole populations. Large field voles may be stronger and will therefore choose to occupy safe habitats with dense vegetation cover, where owl predation is minimal. Furthermore, habitat familiarity may contribute to decreased owl predation risk because individuals that know their home range need to move less in searching for feeding and refuge sites than habitat-unfamiliar ones. We suggested that the mechanistic approach to diet selection (Sih 1993) may help to explain the vulnerability of small voles to avian predators, because we found distinct differences in habitat use and movement activity between small and large field voles. Furthermore, habitat familiarity may play a central role in avoiding risky habitats (Koivunen et al. 1998b) and thus boreal owl predation.

To summarise, sibling voles were the preferred prey of owls, followed by field voles, bank voles and common shrews. For each of the three vole species, owls captured more males than females, and smaller and lighter individuals. However, within each vole species and sex, individuals captured by the owls tended to have more internal fat than those available in the field (Koivunen et al. 1996a, 1996b). We suggested that large, dominant vole individuals may occupy safe habitats with dense vegetation cover, where owl predation risk is minimal, and force subdominant individuals into less-safe habitats. However, these individuals are not necessarily in poor physiological condition. It has previously been reported that boreal owls took field voles more, and grey-sided voles *Myodes rufocanus* less, than expected from their abundance in the field according to snap-trap captures, particularly late in the breeding season (Hörnfeldt et al. 1990)

5.9.   Impact of owls on prey populations

The possible limiting, or even regulatory, impacts of predators on prey population densities are a central question of population dynamics, but have largely remained unanswered until recent experimental studies, where the densities of vertebrate predators have been manipulated (see reviews in Salo et al. 2007, 2010). The impact of predators on prey populations largely depends on whether and how predators respond to changes in prey densities. Therefore, to answer the question on predator limitation of prey populations, the first step is to document the numerical and functional (dietary) responses of predators to prey densities (Solomon 1949). The numerical response is due to the mobility of predators (immigration/emigration due to dispersal), reproductive potential (fecundity), generation time, and possible time lags between density fluctuations of predator and prey populations. The functional response is due to the degree of food specialisation, availability of alternative prey, and competition and predation between various predator species (Andersson and Erlinge 1977, Erlinge et al. 1983, Korpimäki and Norrdahl 1991a, 1991b; review in Valkama et al. 2005).

5.9.1.   Numerical and functional responses

A study on the numerical and functional responses of boreal owls and their impacts on small mammal populations was carried out during 1977–87 in an area covering 100 km^2 in the core of our long-term study area and including 71–90 nest-boxes suitable for boreal owls (Korpimäki and Norrdahl 1989). With respect to the numerical response of boreal owls, the yearly breeding density varied from 1 to 26 nests (mean 9.1) and the yearly number of non-breeding males from 0 to 10 (mean 4.3) per 100 km^2 during 1977–87 (Table 5.9). The yearly breeding density increased in line with increasing abundance indices of *Microtus* voles in the current spring (Pearson correlation, $r = 0.62$, $n = 11$, $p = 0.04$). Supposing that the number of breeding owls tracked the abundance indices of *Microtus* voles with a time lag of 6–7 months, only an almost significant correlation was obtained ($r = 0.53$, $n = 11$, $p = 0.09$).

Table 5.9. The number of nests and non-breeding males of boreal owls, and the mean number of fledglings produced per nest and the total number of fledglings produced in the 100 km^2 study area during 1977–87 (data from Korpimäki and Norrdahl 1989).

Year	No. of nests	No. of non-breeding males	No. of fledglings Mean	Total
1977	23	8	3.5	81
1978	2	–	3.4	7
1979	7	3	2.4	17
1980	6	3	1.0	6
1981	3	3	3.2	10
1982	14	5	4.0	56
1983	6	10	0.8	5
1984	2	–	1.5	3
1985	11	5	3.5	39
1986	26	8	3.4	87
1987	1	2	3.0	3

Comparing nest-holes that were occupied by male owls before the start of breeding (determined by the point stop method; see Chapter 3.8) with the number of nests revealed that, on average, 47% of males during 1977–87 were non-breeding, although they possessed nest-holes (Table 5.9). Assuming that there is no year-to-year variation in the efficiency of the point stop method, the number of non-breeding males in the study area varied in close accordance with the abundance of available bank voles ($r = 0.77$, $n = 11$, $p = 0.004$). Presuming that there was a time lag of 6–7 months between the fluctuations in the number of non-breeding males and abundance of bank voles, the correlation was not significant ($r = 0.44$, $n = 11$, $p = 0.18$). The yearly mean number of fledglings per pair increased significantly with the abundance indices of *Microtus* voles in the current spring ($r = 0.72$, $n = 11$, $p = 0.01$). The abundance of bank voles and common shrews was not obviously correlated to the mean production of fledglings per pair ($r = 0.30$, $n = 11$, $p = 0.37$ and $r = 0.50$, $n = 11$, $p = 0.12$, respectively). There was a close positive correlation between the total number of fledglings produced in the 100 km^2 study area and abundance indices of *Microtus* and bank voles during 1977–87 ($r = 0.72$, $n = 11$, $p = 0.012$ and $r = 0.64$, $n = 11$, $p = 0.04$, respectively).

Boreal owls rapidly responded numerically to density changes of their main prey and were thus able to track without obvious time lags the high-amplitude cyclic population fluctuations of their vole prey. The rapid numerical response was due to the high degree of mobility, vole-supply-dependent adult survival and recruitment of young to the future breeding population, large reproductive potential and early maturity (Korpimäki and Norrdahl 1989, Korpimäki 1994b). In our study area, adult male boreal owls are mostly site-tenacious after their first breeding attempt (Korpimäki 1987e, 1993a), whereas most of the adult females and juveniles show long-distance dispersal and are thus nomadic (Korpimäki and Hongell 1986, Korpimäki et al. 1987). The survival of males on their home ranges and the number of immigrating females are positively related to the vole supply in the previous autumn and winter (Korpimäki 1987e, Hakkarainen et al. 2002). The owls have a large reproductive potential, as a clutch often contains six or even eight eggs (Korpimäki 1987a, Korpimäki and Hakkarainen 1991) and, because of simultaneous polygyny and successive polyandry, a considerable proportion of the population produces two clutches per season in good vole years (Korpimäki 1989b, 1991b, Korpimäki et al. 2011). The mean number of fledglings produced per pair also varied widely between years and was positively related to the vole supply (Table 5.9; see also Korpimäki 1987a). The survival of fledglings depends on the phase of the vole cycle, so that the fledglings produced in the increase phase survive and are later recruited to the breeding population twice as well as those produced in the other phases (Korpimäki and Lagerström 1988). Boreal owls mature as yearlings (see Korpimäki and Hongell 1986) and, especially in good vole years, a large proportion of both females and males of the population are first-year breeders (Korpimäki 1988c, Laaksonen et al. 2002). These factors led to a 21-fold year-to-year variation in the number of owls at the end of the breeding season (Table 5.9) and a very rapid numerical response to an increase in the vole population, thus permitting nearly synchronous population fluctuations of voles and boreal owls.

A rapid numerical response to vole cycles without obvious time lags has also been recorded for long-eared, short-eared and hawk owls, as well as for Eurasian kestrels, in our study area (Korpimäki 1985e, 1994b, Korpimäki and Norrdahl 1991b). All these avian predator species have at least three features in common with boreal owls that facilitate a rapid numerical response.

- First, the high degree of mobility, which results in a rapid influx of juvenile and even adult newcomers into the area with increasing densities of voles and a rapid outflow from the area with decreasing densities of voles.
- Second, a large reproductive potential, indicated by the large clutches and broods at high densities of prey.
- Third, a weak territoriality (Korpimäki 1992c), which means that breeding densities can be high if nest-sites are not a limiting factor.

Strong territoriality was one of the main factors that limited breeding densities and thus time-lagged the numerical responses of great horned owls to 10-year population fluctuations of snow-shoe hares (Rohner 1995, 1996). A considerable time lag of 1–3 years has also been documented for other avian predators that remain resident in their territories when densities of main prey populations decline. These include common buzzards (Erlinge et al. 1983) and tawny owls (Jedrzejewski et al. 1996) subsisting mainly on small mammals, and gyrfalcons *Falco rusticolus* (Nielsen 1999) and goshawks (Tornberg et al. 2005) subsisting mainly on grouse species. In mammalian predators with more limited mobility, a considerable time lag in the numerical responses to changing main prey densities has most often been found in our study area (Korpimäki et al. 1991) and also elsewhere (Jedrzejewski et al. 1995, O'Donoghue et al. 1997, 1998a).

The density of *Microtus* voles in spring was the most important factor determining the diet composition of boreal owls. At peak densities of these voles, their proportion in the diet was 60–80% of prey weight, and at low vole densities their proportion remained at 5–10%. The functional response curve of owls to changing numbers of these voles was close to linear (Fig. 5.19). Holling (1959) distinguished three different functional response curves. At low prey densities, type 1 has a linear, type 2 a convex and type 3 a sigmoid (concave) curve. All curves level off at high prey densities, as predators become satiated. Theoretically, only the sigmoid curve has stabilising potential, whereas in the two other curves the percentage taken by predators either remains constant (type 1) or declines (type 2) with increasing prey numbers (Murdoch and Oaten 1975, Taylor R. 1984). The functional response curve of boreal owls to varying numbers of *Microtus* voles was close to type 1, but the curve did not level off at high vole densities. This indicated that a boreal owl pair takes a constant proportion from the *Microtus* voles available and that the owls did not become satiated, although vole populations showed 45-fold fluctuations during the study period. This interpretation is further supported by the finding that the largest yearly proportions of *Microtus* voles of prey weight were far from 100% (max. 80%). A close to linear (type 1) functional response to fluctuating densities of voles has also been recorded for long-eared and short-eared owls, as well as Eurasian kestrels, in our study area (Korpimäki 1985e, Korpimäki and Norrdahl 1991b). The probable reason is that when *Microtus* vole densities are high, all the small – and even larger – avian and mammalian predators, including Ural owls, eagle owls and red foxes, shift to using these voles as their main foods (e.g. Korpimäki et al. 1990, Dell'Arte

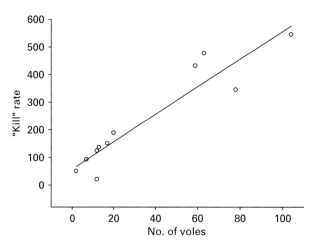

Fig. 5.19.   The number of *Microtus* voles taken by a boreal owl pair in the breeding season (the 'kill' rate) plotted against the total number of these voles in the 100 km² study area during 1977–87. (Note that 10 = 10 000, 20 = 20 000, 30 = 30 000, etc. on the x-axis) (data from Korpimäki and Norrdahl 1989). Linear regression, $r = 0.94$, $p < 0.0001$, regression equation $y = 4.99x + 56.66$, $r^2 = 0.88$.

et al. 2007). These dietary shifts probably induced high inter- and intraspecific interference and food competition between predators (Korpimäki 1987c, Hakkarainen and Korpimäki 1996) which, in turn, decreased chances that predators cannot subsist only on voles and did not become satiated at high densities of voles.

The functional response curves of avian predators are usually convex (type 2, Murdoch and Oaten 1975) (for examples, see Keith et al. 1977, Rohner et al. 2001 for great horned owls to snow-shoe hares; Jedrzejewski et al. 1996 for tawny owls to voles and mice; Rohner et al. 1995 for hawk owls to voles; Salamolard et al. 2000 for Montagu's harriers to voles; Lindén and Wikman 1983 for goshawks to forest grouse), but are sometimes also sigmoid (for example, hen harriers to grouse; Redpath and Thirgood 1999). The linear functional response curve of boreal owls seemed to differ from this general pattern. The reason was that the owls shifted rapidly (i.e. without time lags) to alternative prey (bank voles, common shrews and birds) with decreasing numbers of *Microtus* voles (the preferred prey, see Chapter 5.8), so that very few of these voles were taken in the low phase. Functional responses of resident predators normally lag well behind the changes in prey numbers (Ryszkowski et al. 1973, Keith et al. 1977, Erlinge et al. 1983, O'Donoghue et al. 1998b, Rohner et al. 2001).

5.9.2.   Prey consumption and predation rates

The total prey biomass consumed by boreal owls in one breeding season averaged 161.7 kg during 1977–87 (Table 5.10). Most of the prey was consumed by breeding adults, followed by their young and non-breeding males. The proportion of prey eaten by breeding adults was highest in 1984 and lowest in 1987. The percentage for young owls of the prey consumption was largest in 1978 and smallest in 1983. Non-breeding males consumed most in 1983 and 1986 compared with their consumption in other years.

Table 5.10. The total prey weight (kg) consumed by boreal owls in the breeding season, the percentage of prey weight consumed by breeding adults (A), young (Y) and non-breeding males (NM) from this consumption, and the number of most important prey species or groups consumed in the 100 km^2 study area during 1977–87 (data from Korpimäki and Norrdahl 1989).

Year	Prey weight (kg)	Percentage consumed by			Number consumed[a]			
		A	Y	NM	*Microtus* spp.	Bank voles	Common shrews	Birds
1977	410.9	52	39	9	10 950	4610	2430	1290
1978	32.4	57	43	–	300	600	480	270
1979	112.5	58	30	12	950	2960	2390	540
1980	81.4	68	15	17	550	1000	1230	1220
1981	63.8	44	31	25	370	850	2190	840
1982	264.1	49	42	9	4820	4470	5020	670
1983	111.9	50	9	41	1130	1960	1810	970
1984	24.5	76	24	–	100	300	1350	230
1985	202.5	51	38	11	4740	3670	2190	180
1986	450.6	54	38	8	14 160	4050	2450	200
1987	24.5	38	24	38	20	300	970	420
Mean	161.7	54	30	16	3460	2250	2050	620
s.d.	152.7	11	11	14	4870	1740	1190	410

[a] The number of prey animals (NPA) consumed by boreal owls in the breeding season was estimated as follows (see Korpimäki and Norrdahl 1989 for further details):

$$\text{NPA} = \frac{((\text{CA1}+\text{CY1})\times\text{PPA1})+((\text{CA2}+\text{CY2})\times\text{PPA2})}{\text{MWPA}}, \text{ where}$$

CA1 = consumption (g) of adults in the early part of the breeding season = (2 × number of breeding pairs + number of non-breeding males) × daily food requirement (38 g) × length of the season (71 days),
CA2 = consumption (g) of adults in the late part of the breeding season = (2 × number of breeding pairs + number of non-breeding males) × daily food requirement (38 g) × length of the season (51 days),
CY1 = consumption (g) of young in the early part of the breeding season = mean number of fledglings produced per pair × number of breeding pairs × mean daily food requirement of young from hatching to post-fledging period (30.5 g) × length of the season (14 days),
CY2 = consumption (g) of young in the late part of the breeding season = mean number of fledglings produced per pair × number of breeding pairs × mean daily food requirement (30.5 g) × length of the season (51 days),
PPA1 and PPA2 = percentage of prey animals' biomass in the diet in the early and late part of the breeding season, and
MWPA = mean weight (g) of prey animals.

During 1977–87, *Microtus* voles were the most frequently consumed prey item of boreal owls, followed by bank voles, common shrews and small birds (see Appendices 1–2 in Korpimäki and Norrdahl 1989). In good vole years, *Microtus* voles were the most frequent prey group eaten, while in poor vole years the consumption of bank voles, common shrews, and even birds, largely exceeded the consumption of *Microtus* voles. The proportions of main prey, *Microtus* voles, by prey weight varied from 0% to 82% in the early part and from 3% to 74% in the late part of the breeding season (see Appendices 1–2 in Korpimäki and Norrdahl 1989). The proportion of these voles in the diet in these two parts of the breeding season markedly increased with increasing numbers of these voles available in the study area during 1977–87 (Korpimäki and Norrdahl 1989). A similar relationship was also evident for the bank vole in the late part of the season. In contrast, the abundance of common shrews in the field was not obviously correlated with the proportion of these prey items in the diet. There was also a significant tendency for boreal owls to take more common shrews and birds when there were less *Microtus* voles in the field (Korpimäki and Norrdahl 1989).

The number of *Microtus* voles in the study area was the most important factor determining the diet composition of breeding boreal owls (see also Chapter 5.6. and Korpimäki 1988b). When the number of *Microtus* voles taken by a boreal owl pair in the breeding season (i.e. the 'kill' rate) was plotted against the total number of these voles in the 100 km^2 study area during 1977–87, the functional response curve was very close to linear (Fig. 5.19). The concave curve with the quadratic equation explained the functional response only slightly better ($r^2 = 0.91$). The yearly 'kill' rate of bank voles was also closely related to the number of *Microtus* voles in the study area during 1977–87 ($r = 0.80$, $n = 11$, $p = 0.003$) but not so obviously to their own numbers ($r = 0.14$, $n = 11$, $p = 0.68$). The 'kill' rate of birds increased with decreasing numbers of *Microtus* voles ($r = -0.65$, $n = 11$, $p = 0.03$), and the 'kill' rate of common shrews tended to do so ($r = -0.56$, $n = 11$, $p = 0.08$). This was obviously due to the fact that owls shifted to shrews and birds at low densities of main prey.

The mean predation impact (or predation rate = the percentage of prey items taken by boreal owls out of the total number available in the study area) on *Microtus* voles, bank voles and common shrews was quite similar during 1977–87 (mean ± s.d. $7 \pm 5\%$, $6 \pm 3\%$ and $5 \pm 3\%$, respectively). However, there was large year-to-year variation in the predation rates; the predation rate of *Microtus* voles tended to vary more than that of bank voles and common shrews. The owl predation impact on *Microtus* and bank voles was also significantly larger in good vole years than in poor ones, but a similar difference was not found for common shrews (Korpimäki and Norrdahl 1989). When the predation rates of *Microtus* voles, bank voles and common shrews were plotted against the total number of *Microtus* voles available during 1977–87, the correlations for voles were significant for *Microtus* voles (Fig. 5.20) and almost significant for bank voles ($r = 0.60$, $n = 11$, $p = 0.05$) but not for shrews ($r = 0.19$, $n = 11$, $p = 0.57$). The curvilinear regression with quadratic equation appeared not to explain the relationship for the predation rates of *Microtus* voles and their numbers in the field better ($r^2 = 0.43$) than the linear regression.

Folivorous *Microtus* voles are slow and clumsy and are thus more vulnerable to predators than are granivorous bank voles. *Microtus* spp. occupying open habitats with low spring vegetation cover are also more exposed to avian predators than are bank voles in woodland with higher spring vegetation cover (Chapter 5.8). Thus, *Microtus* voles seem to suffer from heavier predation than bank voles, and the predation impact on bank voles in temperate Europe is usually inversely density-dependent on the predation impact on *Microtus* spp. (e.g. Goszczynski 1983, King 1985, review in Oksanen T. et al. 2000). However, the mean predation impact of boreal owls on bank voles was similar to that on *Microtus* voles and the yearly predation rate tended to be positively dependent on the density of *Microtus* voles. This was due to the 'flexible' switching of boreal owls to bank voles when the number of *Microtus* voles decreased. This switching possibly promotes

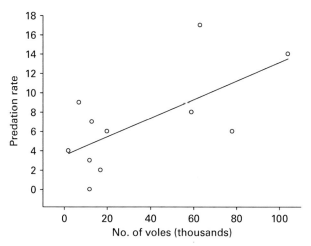

Fig. 5.20. The predation rate of *Microtus* voles plotted against the total number of these voles available in the $100\,km^2$ study area during 1977–87. (Note that $10 = 10\,000$, $20 = 20\,000$, $30 = 30\,000$, etc. on the *x*-axis) (data from Korpimäki and Norrdahl 1989). Linear regression, $r = 0.66$, $p = 0.028$, regression equation $y = 0.10x + 3.53$, $r^2 = 0.43$.

synchrony between population fluctuations of *Microtus* and bank voles, the populations of which fluctuate in close synchrony in our study area (Fig. 1.1).

When the numerical and functional responses of boreal owls and other birds of prey, including long-eared owls, short-eared owls and Eurasian kestrels, were pooled, the predation impacts of these four species on voles was positively and directly density-dependent. A larger proportion of voles was consumed with increasing vole densities (Fig. 5.20; see also Korpimäki and Norrdahl 1991a). This indicated that the boreal, long-eared and short-eared owls, as well as Eurasian kestrels, dampen vole cycles by truncating population peaks (Korpimäki and Norrdahl 1989, 1991a). The following factors seemed to promote the dampening impact of boreal owls and other birds of prey on vole populations: rapid and wide numerical and functional responses of owls and kestrels to changes in vole densities, the spatial heterogeneity of our study area, and breeding densities of owl and kestrel populations only slightly limited by territoriality (Korpimäki and Norrdahl 1989). The most important factor was the rapid numerical response of boreal owls to cyclically fluctuating main prey densities due to their high degree of mobility and large reproductive potential, as time lags tend to have substantial destabilising effects in predator–prey interactions (Murdoch and Oaten 1975). For boreal owls, it is of course much easier to track population densities of main prey that fluctuate in only 3-year cycles than, for example, for great horned owls attempting to track snowshoe hare densities with 10-year density fluctuations. Boreal owls consumed plenty of *Microtus* spp. only when they were abundant, which indicates a low catching efficiency of these voles by boreal owls in the low phase of the cycle. The owls switched to alternative prey mainly occupying forests (bank voles, shrews and small birds) when the abundance of the main (preferred) prey in farmland decreased. Farmlands interspersed through forest probably improved this 'flexible' switching. Moreover, after snowmelt, *Microtus* voles are more vulnerable to boreal owls and other avian predators, as the old vegetation cover is flattened and no new vegetation has yet grown (Chapter 5.7). Also, in general, spatial heterogeneity of the study system tends to stabilise predator–prey interactions (Kuno 1987). The weak territoriality of boreal owls and other vole-eating nomadic avian predators, in comparison with many resident birds of prey including, for example, Ural, tawny and eagle owls (Korpimäki 1992c), means that territorial behaviour had only a slight limiting influence on boreal owl densities.

The predation impact of boreal owls substantially differed from that of many resident predators, which usually showed negative density dependence: a low predation rate in the increase and peak phases of the cycle and a high rate during the decrease and low phases (see e.g. Jedrzejewski et al. 1996, Rohner et al. 2001 for avian predators; Pearson 1966, Ryszkowski et al. 1973, Goszczynski 1977, Korpimäki et al. 1991, Jedrzejewski et al. 1995 for mammalian predators). However, our correlative results from boreal owl predation impacts on vole densities were not necessarily proof of the 'control' of vole populations by boreal owls, as there were no data on vole densities in the absence, or at experimentally reduced densities, of owls. This was later studied by reducing the breeding densities of birds of prey (mainly boreal owls and Eurasian kestrels) in our study area (Norrdahl and Korpimäki 1995a, 1996).

We studied the effects of a reduction of breeding densities of boreal owls and Eurasian kestrels on voles and shrews during 1989–92 to find out whether they have a regulating or limiting impact on their prey populations. We removed potential breeding sites from five manipulation areas (about 3 km^2 each), while control areas had nest-boxes in addition to natural cavities and stick-nests. The densities of small mammals were monitored by snap-trapping in April, June and August, and densities of mammalian predators (the least weasel, the stoat and the red fox) by snow-tracking in early spring and late autumn. The yearly mean number of owl and kestrel nests was 0.2–1.0 in reduction areas and 3.0–8.2 in control areas. Breeding raptors alone did not limit prey populations in the long-term, but probably caused short-term changes in the population dynamics of both the main prey, *Microtus* voles, and an alternative prey, the common shrew. The densities of an alternative prey, the bank vole, decreased in raptor reduction areas, most probably due to increased least weasel predation pressure resulting from the reduced breeding densities of owls and kestrels (Norrdahl and Korpimäki 1995a).

Three-to-five-year population oscillations of northern small rodents are usually synchronous over hundreds of square kilometres. This regional synchrony could be due to similarity in climatic factors affecting reproduction and survival of voles (Moran 1953), or due to nomadic predators, such as boreal owls, reducing the patches of high prey density close to the average density of a larger area (Ydenberg 1987, Korpimäki and Norrdahl 1989). In the reduction experiment of breeding densities of owls and kestrels, we also studied whether these avian predators concentrate in high prey density areas, and whether this decreases spatial variation in prey density. Hunting birds of prey concentrated in high prey density areas after their breeding season in August, but not necessarily during their breeding season (April–June) when they were constrained to hunt in the vicinity of the nest. The experimental reduction in breeding owls and kestrels increased the variation in prey density between reduction and control areas but not within areas.

The difference in variation between predator reduction and control areas was largest in the late breeding season of birds of prey, and decreased rapidly after the breeding season (Norrdahl and Korpimäki 1996). These results appear to support the hypothesis that the geographical synchrony of population cycles in voles may be driven by nomadic predators concentrating in high prey density areas. Predation by owls and kestrels and climatic factors are apparently complementary rather than exclusive factors in contributing to the geographical synchrony of population cycles of voles.

We have also conducted large-scale replicated field experiments where the densities of avian predators (boreal and long-eared owls and Eurasian kestrels) and mammalian predators (least weasels and stoats) were reduced in four large (each 2.5–3.0 km^2) unfenced areas through 3-year vole cycles (Korpimäki and Norrdahl 1998, Korpimäki et al. 2002). In addition, we excluded avian and mammalian predators by fencing four 1-ha farmland areas, and also provided supplemental food for voles during two winters (Klemola et al. 2000, Huitu et al. 2003a). While only least weasel densities were substantially reduced in the decline phase of the vole cycle, this treatment was not able to prevent the decline of vole densities in the course of the summer (Korpimäki and Norrdahl 1998). In this sense the results were similar to our avian predator reduction experiments (Norrdahl and Korpimäki 1995a), which did not have obvious long-term positive effects on vole densities. The probable reason is that the reduction of one predator species (the least weasel) or group (owls and kestrels) relaxes inter- and intraspecific food and interference competition between predators, and the remaining predators are thus able to compensate for the experimentally reduced mortality imposed by another predator group.

The large-scale experimental reduction of all main avian and mammalian predators increased fourfold the autumn density of voles in the low phase, accelerated the increase twofold, increased the autumn density of voles twofold in the peak phase, retarded the initiation of decline of the vole cycle, and prevented the summer decline of vole populations (Korpimäki and Norrdahl 1998, Korpimäki et al. 2002). Experiments conducted with voles (cycle period 4–5 years) gave similar results in northern Norway (Ekerholm et al. 2004). Based on our experiments, demographic models predicted changes from regular multi-annual cycles to annual fluctuations with declining densities of specialist predators, whereas the reduction of generalist predators tended to increase the period and the amplitude of these fluctuations (Korpimäki et al. 2002). Food supplementation in winter increased vole population growth and subsequently prevented the crash of vole populations, but only in the absence of predators (Huitu et al. 2003a). These novel results largely solve the puzzle of the 3-year high-amplitude population cycles of northern voles. The most likely hypothesis is that these population cycles are driven by delayed density-dependent losses to predators but that vole populations which have succeeded in escaping regulation by predators are limited in growth by a shortage of winter food. The delayed density-dependent mortality that is necessary to drive 3–4-year vole cycles is mainly imposed on vole populations by specialist mammalian predators such as least weasels and stoats, which are considerably less mobile than avian predators. On the other hand, avian predators, including boreal owls, mainly decrease the amplitude of vole cycles, because their main predation impact on voles is directly dependent on vole densities (Korpimäki and Norrdahl 1989, 1991a, Korpimäki et al. 2004).

6 Life-history of the boreal owl

6.1. Morphological characteristics under fluctuating food conditions

6.1.1. Factors affecting body size of recruits

Body mass of female boreal owls differs greatly depending on the phase of the vole cycle: in poor vole years, females are much lighter than in good vole years, whereas the body mass of males remains at a relatively constant level over years (Fig. 2.9; see also Korpimäki 1990a). Body mass of females is largely influenced by the present vole abundance in the field, which largely contributes to the male's ability to provide food to the incubating and brooding female. During autumn migration, females are only 4% heavier than males, but are approximately 30% lighter than during the breeding season (Hipkiss 2002) when they carry large fat deposits and can be twice as heavy as males (Korpimäki 1990a). Owing to extremely large quantitative and qualitative variations in food resources (Chapter 5.6), we also expected that the development of the offsprings' phenotypic traits during their first months of life might be influenced by the direction of future changes in vole abundance. It has been shown that environmental variation, along with maternal effects, may hasten the development of avian offspring and may also modify their future morphological and breeding characteristics (Boag 1987, Boag and van Noordwijk 1987, Richner 1989).

To study whether the body measurements of recruits were affected by parental age or vole abundance in the year of their birth, we pooled data from three adjacent large study areas: our study area in the Kauhava region, the Seinäjoki region and the Kokkola region (Table 6.1). The relatively short distances between the three study areas (approx. 40–170 km) and the large number of nest-boxes (300–500) within the large study areas (1000–1300 km^2) ensured that at least some offspring recruited to one of our local breeding populations or were met as migrants in coastal bird observatories during their autumn migration (Fig. 3.1). Data up until 1993 included body size measurements from 183 recruits that were born and re-trapped in the three study sites (see Hakkarainen et al. 1996c).

Surprisingly, the large annual variation in the abundance of the main prey, as measured by the phase of the vole cycle (low, increase and decrease), did not have long-lasting effects either on the body mass or on the wing length of recruits. Furthermore, parental age did not have any obvious effects on the body dimensions of recruits (Hakkarainen et al. 1996c). In contrast, the laying date appeared to be significantly associated with the wing length of recruits: eggs laid early in the season produced long-winged offspring, whereas

Table 6.1. Classification of data on the basis of the birth and the recruitment site of owls and the season (breeding time or autumn migration) of the recruitment (see map in Fig. 3.1 for the location of the different study areas). Data from population studies conducted in the Kauhava and Seinäjoki study areas during 1979–93 and in the Kokkola region during 1986–93 (Hakkarainen et al. 1996c).

Birth and recruitment site	No. of recruits
1. Born in Kauhava, recruited as breeder in Kauhava	45
2. Born in Kauhava, recruited as breeder outside Kauhava	10
3. Born in Kauhava, recaptured in autumn in Kokkola	25
4. Born in Seinäjoki, recruited as breeder in Seinäjoki	10
5. Born in Seinäjoki, recruited as breeder in Kauhava	12
6. Born in Seinäjoki, recaptured in autumn in Kokkola	10
7. Born in Kokkola, recruited as breeder in Kokkola	8
8. Born in Kokkola, recruited as breeder in Seinäjoki	3
9. Born in Kokkola, recaptured in autumn in Kokkola	45

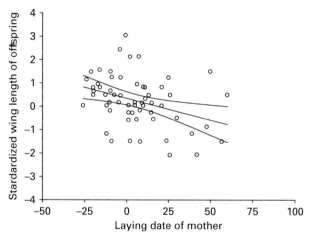

Fig. 6.1.   Standardised wing length of male and female recruits against the laying date of their mother (regression line with 95% confidence limits, $F = 8.37$, df = 60, $p = 0.005$) (Hakkarainen et al. 1996c).

late-laid eggs produced short-winged offspring (Fig. 6.1). Therefore, laying date, which is a reliable estimate of the hatching time of owlets (Korpimäki 1981), significantly explained the wing length of offspring in the next breeding season. This may be due to the fact that owlets born early in the season have more time for a post-juvenile moult, or because early-born offspring may receive more parental care because of good parental or home range quality. In particular, the time for a post-juvenile moult may be important for developing owlets, because they have to moult all their feathers before gaining their adult plumage in the late summer (März 1968). In male recruits, the body mass was independent of the date of birth within the season. In contrast, the body mass of breeding female recruits was significantly positively related to the laying date of their mother. This positive trend was largely due to the fact that the late clutches in the increase phase of the vole cycle produced well-fed and fat recruits, because in that phase food abundance substantially improved throughout the first year of life, whereas offspring born in the decrease phase experienced a decline in food abundance until the next year. Accordingly, our results suggest that the conditions experienced during an individual's early life may result, at least to some extent, in lasting morphological differences.

6.1.2.   Heritability of morphological characteristics

Heritability plays an essential role in the process of natural selection, as characters beneficial for fitness are transferred to the next generation (see Falconer 1981, Endler 1986). High heritability is beneficial if offspring will be living in environments similar to those that their parents have encountered. The situation may differ, however, if there is large temporal or spatial variation in an environment. Under such conditions, the importance of phenotypic plasticity and low heritability has been emphasised. In species such as the boreal owl, which subsists on annually varying food resources, the offspring will often face a different environment from their parents. Food conditions may even change so drastically that parent-like traits may be harmful for the offspring, while different characteristics from those of their parents may be beneficial. If the fitness of an individual is associated with spatial variation in an environment, then offspring may be able to shift to the most appropriate habitats, where they will perform best. However, if there are detrimental temporal long-term changes related to a certain character, then in extremely bad times offspring would reproduce inadequately or would need to wait for a better time to reproduce. Therefore, we predicted that boreal owls living under annually changing food conditions would show low heritability for morphological and breeding traits.

The data on recruits presented in Table 6.1 enabled us to measure the heritability of morphological characters (wing length and body mass), which was measured by father–son and mother–daughter regressions. Doubling the regression coefficients along with standard errors gave estimates of heritability between parents and offspring (see, e.g., Falconer 1981). A high parent–offspring similarity (i.e. high regression coefficient between those traits) describes high heritability. However, environmental factors may contribute considerably to trait variation, along with genetic factors, and it is often hard to separate environmental and heritable influences from each other (see Alatalo et al. 1989, Gebhardt-Henrich and van Noordwijk 1991). Therefore, it is important to take into account environmental variation in studies on heritability, as individuals encountering similar environmental conditions may resemble each other independently of their relatedness. The boreal owl is a good study subject in this context, as the young disperse widely between their birthplace and first breeding site (Korpimäki et al. 1987, Chapter 9.2). Adult females also disperse widely between successive breeding seasons, whereas adult males are resident after their first breeding attempt (Löfgren et al. 1986, Korpimäki 1987e, Chapter 9.3). These age- and sex-related differences in dispersal behaviour are likely to lessen the confounding effect of a common environment on the heritability estimates. Furthermore, body size measurements (wing length and body mass) of recruits were independent of the phase of the vole cycle, as shown in Chapter 6.1.1. Hence, the large variation in food abundance did not have an obvious influence on our heritability estimates. Sibling competition may also bias the heritability estimates, as members of the same brood may obtain different amounts of food due to intense between-sibling competition (Falconer 1981). In the boreal owl, however, the wing length and body mass of recruits were not obviously associated with clutch size, indicating a low degree of sibling competition at the time of hatching. Furthermore, size-assortative mating may confound the interpretation of parent–offspring regressions (Falconer 1981, Boag and van Noordwijk 1987), but in boreal owls, mating seemed to be random in relation to the body size of mates within pairs (Korpimäki 1989b). Instead, the start of egg-laying within the season was known to be related to the wing length and the body mass of recruits, having the potential to affect heritability estimates in this species. Therefore, the laying date effect was included in the heritability analyses, if appropriate (for further details, see Hakkarainen et al. 1996c).

We found a significant mother–daughter regression for wing length (Table 6.2, Fig. 6.2). This relationship was, however, lowered after the removal of the laying date effect in 27 cases, where the start of egg-laying was known ($r = 0.30$, df $= 25$, $p = 0.40$). Some of the decline in significance level, however, may be attributable to a smaller sample size, as the regression coefficient still remained quite high, corresponding to a heritability of 0.60. The mother's wing length did not significantly correlate with the wing length of her son, as it did between the wing length of the father and his offspring (Table 6.2, Fig. 6.2). Furthermore, neither paternal nor maternal heritability in body mass was significant (Table 6.2). In other words, the only morphological character that showed relatively high heritability was the wing length of the mother and daughter.

Our heritability estimates for morphological characters were relatively low compared with many other bird species (Boag and van Noordwijk 1987). This finding is consistent with the idea that low heritability may promote adaptability to largely varying environmental conditions. If the heritability for body size was high, owls could only reproduce 'copies' of their own phenotype, which would perform properly only 3–4 years later,

Table 6.2. Heritability estimates (± s.e.) for wing length and body mass of boreal owls from parent–offspring regressions (Hakkarainen and Korpimäki 1995).

	Mother				Father			
	Wing length	df	Body weight	df	Wing length	df	Body weight	df
Daughter	0.76[***] (0.20)	79	0.00 (0.28)[a] −0.09 (0.14)[b]	76[a] 34[b]	−0.04 (0.46)	36	−0.08 (0.67)[a] 0.26 (0.78)[b]	22[a] 11[b]
Son	0.24 (0.16)	84	−0.12 (0.11)	83	0.10 (0.30)	49	−0.13 (0.34)	48

Note. Regression coefficients of single parents and their standard errors are doubled. Because females are heavier in the breeding season than during the autumn migration, their body weights were analysed separately. Significance level: [***] two-tailed $p < 0.001$.
[a] Breeding females.
[b] Migrating females.

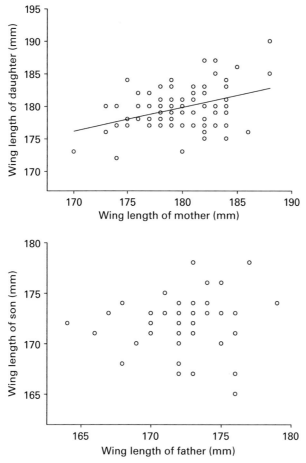

Fig. 6.2.   Regression of the wing length of the daughter on the wing length of her mother (upper panel) and the wing length of the son against the wing length of his father (lower panel) (Hakkarainen et al. 1996c).

when the same phase of the vole cycle was reached again. Low heritability and high non-additive genetic variance in morphological characters could instead result in the production of various phenotypes. This might increase the probability that under rapidly changing environmental conditions there would be some individuals which would survive because of their 'fitting' phenotype (see Endler 1986). This prediction parallels the idea proposed by Laaksonen (2004): parents may ensure a safe bet by producing diverse young, some of which will succeed. Therefore, the maintenance of phenotypic variation is important under varying environments and can be considered as a trait subject to natural selection and evolutionary change (Levins 1968, Caswell 1983).

Higher heritability values seemed to be obtained for maternal wing length than for paternal wing length. The father–son regression for wing length was significantly lower than the mother–daughter regression for the same trait (Hakkarainen et al. 1996c). The coefficient of determination was 16% higher in mother–daughter regressions than in father–son regressions of wing length (interaction term $F = 2.83$, $p = 0.04$; Fig. 6.2). Extra-pair paternity may lessen the heritability estimates of father–son characteristics (e.g. Alatalo et al. 1989, Møller 1989b), as the frequency of extra-pair copulations leading to fertilisation has been found to be high in many avian species (Birkhead and Møller 1992). Hence, male boreal owls trapped at their nests may not have fertilised all of their offspring, leading to insignificant paternal heritability in wing length. This may not be a valid explanation, however, in our boreal owl population: DNA fingerprinting of blood samples of boreal owl families revealed only a few cases of extra-pair paternity and no cases of extra-pair maternity (Chapter 7.3.1). This suggests that there may be real sex-related differences in the heritability of wing length, although further studies on this subject are needed.

6.2.   A puzzle: why are female owls larger than males?

In many species, the sexes differ in size. For example, in reptiles, mammals and birds, males are usually larger than conspecific females. This sexual size dimorphism has been the topic of several studies (e.g. Ralls 1976, Shine 1988), and plausible mechanisms causing sexual size dimorphism have been proposed ever since Darwin (1871). Size differences between the sexes have been thought to be the product of natural or sexual selection, or both (e.g. Hedrick and Temeles 1989). Of these, sexual selection is most emphasised. Variation in the body size of males, for example, may result in fitness differences between males if mating success in inter-male competition is enhanced by large body size (see Clutton-Brock and Iason 1986). Owing to intense male–male competition, males may evolve to be larger than females (i.e. intrasexual selection). Alternatively, large male size may evolve through female choosiness if females prefer large males as mating partners (i.e. intersexual selection). It has also been suggested that sexual size dimorphism may have evolved to reduce food niche overlap and intersexual food competition within a pair (Peters and Grubb 1983, Temeles 1985).

   Although male vertebrates are usually larger than conspecific females, in most owls and diurnal raptors females are substantially larger than males. The reason(s) for this reversed sexual size dimorphism (RSD) have remained a long-lasting puzzle, although a plethora (>20) of hypotheses have been put forward (e.g. Andersson and Norberg 1981). The hypotheses for RSD can be divided into more general groups based on their different roles in natural and sexual selection: those that favour large female or small male size. Here we concentrate on the five main hypotheses that have most often been proposed for the evolution of RSD in birds of prey. Most of these hypotheses stress the importance of female largeness instead of male smallness as a cause for the evolution of RSD – a general research and publication bias found for other organisms as well (Blanckenhorn 2000).

6.2.1.   Hypotheses for reversed sexual size dimorphism (RSD)

The large size of females in owls, hawks and falcons has been explained by the following hypotheses: (1) starvation, (2) reproductive effort, (3) female dominance, and (4) sexual selection.

1. According to the *starvation hypothesis* (Korpimäki 1986d, Lundberg 1986, Mendelsohn 1986), the probability of starvation increases under unpredictable food conditions, when selection could favour large female size with extra body reserves to increase survival and breeding prospects when food is scarce. This hypothesis could be valid, especially for boreal owls, because they breed very early in the season when wintery conditions prevail.

2. The *reproductive effort hypothesis* proposes that large females may produce more offspring and/or provide better parental care than small females can. Reproductive role division may evolve size differences between the sexes, and is mainly thought to result from selection pressures acting on female fecundity (e.g. Ralls 1976). Accordingly, egg production and incubation efficiency are suggested to be enhanced with increasing body mass of females (e.g. Reynolds R. 1972).

3. The *female dominance hypothesis* suggests that females choose small males because they are easier to dominate compared to large males, especially at the time of mating and pair formation. Hence, female largeness compared to male size could play an important role in the maintenance of the pair bond, resulting in increased food provisioning by the male (e.g. Mueller 1986).

4. The *sexual selection hypothesis* states that RSD may be due to sexual selection operating in reverse to the 'normal situation' (i.e. males larger than females): females may compete for and/or prefer small males at pairing, which may result in RSD. This would be a valid explanation in birds of prey because males invest in their breeding attempts to a great degree, which may result in inter-female competition for males (Newton 1986). For example, in the Eurasian kestrel it has been shown that the energy expenditure of breeding females is only 75% of that of males (Masman et al. 1989), which suggests that male birds of prey invest even more in offspring than females do. This may be because most male birds of prey are primarily responsible for feeding the young, which is energetically very costly (Masman et al. 1989). Hence, large females could be better at competing for good-quality males, especially because the parental effort of male birds of prey seems to be considerable. Alternatively, selection may favour small male size if smallness is significantly associated with mating success or reproductive output. Hence, both female choosiness and the importance of male foraging ability may contribute to RSD.

However, some hypotheses for RSD are instead based on factors that promote small size in males.

5. The *small male hypothesis* states that males have become smaller for more efficient foraging or territorial defence (Hakkarainen and Korpimäki 1991, Massemin et al. 2000). Most of these hypotheses are based on predatory habits, such as specialisation towards relatively agile and/or rare prey. Small male size might have evolved, in particular, because males have a major role in the provisioning of food for the incubating female and nestlings. Hence, if small male size enhances foraging ability, reproductive success and fitness within a pair, the proportion of large males will decrease in the future breeding population.

Although some evidence has been found for most RSD hypotheses, no clear consensus has yet emerged concerning the relative importance of these factors in explaining the evolution of RSD among birds of prey. So far only one comprehensive synthesis on the mechanisms of RSD has been conducted (Krüger 2005). This is probably because it is very difficult to test different hypotheses on RSD in the field. Generally a comparative approach at species level has been used, and there have been few studies on RSD studying selection for body size within a species in their real environments. During the last two decades, however, some field studies have focused on RSD and body size in a single species (e.g. Hakkarainen and Korpimäki 1991, 1993, 1995, Hakkarainen et al. 1996a, Catry et al. 1999, Tornberg et al. 1999, Massemin et al. 2000, Phillips et al. 2002).

In studies on the evolution of RSD, it is also important to take into account temporal variation because, in fluctuating environments, there may also be variation in the extent to which a phenotype is suited to the environment, so that the phenotypic correlation for some traits may have opposite trends in different environments. While one form of a trait might be beneficial in some conditions, a different form might be optimal in other conditions (see, e.g., Stearns 1989). An individual coping with a fluctuating environment also regularly faces different environmental conditions, such as variations in food quality and abundance, when the optimal phenotype obviously changes. In this context, a classic example comes from the medium ground finch *Geospiza fortis* in the Galapagos Islands. In this species, the environmental conditions are mostly stable, but in some years long drought and rainy seasons induce changes in the seed composition of some plant species. This raises a problem of directional selection for bill structure of the medium ground finch: during drought periods, when only large seeds are available, a large beak is favoured, whereas during the rainy periods, with an abundance of small seeds, a small beak is beneficial (Grant B. and Grant P. 1989). Accordingly, quantitatively and qualitatively varying resources may cause different phenotypes of a forager to evolve (Stephens and Krebs 1986). In the case of boreal owls, both the abundance (vole cycles) and the quality (main vs. alternative prey) of food resources vary considerably over years, as in many other owl species living in the Northern Hemisphere (Mikkola 1983). Therefore, comprehensive and accurate measurements of selection pressures on the body size of owls should be made, as this serves as a good example of changing selection pressures under fluctuating food conditions.

6.2.2. The starvation, reproductive effort, and female dominance hypotheses

Female boreal owls are approx. 30–40% heavier and have on average 5% longer wings than males in the mid-nestling period (Table 2.1, Fig. 6.3; see also Korpimäki 1990a), although in the autumn the degree of RSD in body weight is much less (Hipkiss 2002). With regard to body mass, the degree of RSD within breeding boreal owl pairs is one of the highest among the European and North American owl species studied so far (Korpimäki 1986d, Lundberg 1986). In the boreal owl, as in other birds of prey, males do most of the hunting for the whole family from before egg-laying until the offspring are half-grown. In fact, boreal owl males usually provide for their families until the chicks are independent, while females only begin to help in chick feeding when the young are at least 3 weeks old

Fig. 6.3.   Boreal owl females (on the right) are heavier and larger than males (on the left). Photo: Erkki Korpimäki.

(Korpimäki 1981). Hence, the different roles of the sexes during breeding may cause contrasting selection pressures on morphological characteristics depending on sex.

We trapped 473 females and 423 males (a total of 483 breeding pairs) at their nests during 1981–90. This period covered three different vole cycles, including regular low, increase and decline vole years. Based on these long-term field measurements (see Hakkarainen and Korpimäki 1991) and direct observations of food provision of male owls during one vole cycle (Hakkarainen and Korpimäki 1995), we examined the factors promoting RSD in boreal owls.

According to the starvation hypothesis, females laying early in the season under adverse weather and unpredictable food conditions should be larger than late-laying females, when environmental conditions are more constant (Korpimäki 1986d). Therefore, pairs breeding early in the season should be more dimorphic. At that time, deep snow cover may prevail for the whole incubation period and temperature may drop to as low as −25°C, when prey delivery rate by males may decrease considerably, especially during stormy nights. Despite that, no clear evidence for the starvation hypothesis was found. In low and declining vole years, no correlations between body mass of females and their laying date were found, or between the body mass differences within a pair and laying date. Furthermore, long-winged females did not breed earlier than short-winged females when the timing of breeding was analysed separately for young and old females. Accordingly, body mass differences within a pair or female largeness were not associated with the initiation of breeding. In contrast, during the increase phase of the vole cycle the heaviest females and more dimorphic pairs in body mass were the earliest breeders.

These results are generally in disagreement with the predictions of the starvation hypothesis because the benefit from female largeness did not appear in the low vole years, when the starvation risk was highest. Furthermore, the significant results for body mass in the increase phase of the vole cycle probably reflect the hunting ability of the males, rather than female largeness, because in most owl species variation in female body mass is extremely large, owing to changes in the provisioning ability of males. Therefore, structural body size of females should be measured by wing length or other stable structural measures, such as tarsus length, instead of body mass. Accordingly, we interpreted our results on body size of females mostly on the basis of wing length, which is stable throughout the breeding season. Using wing length as an estimate of body size, there seemed to be no clear advantage to large female size in initiation of breeding, and hence no clear support for the starvation hypothesis was detected (Hakkarainen and Korpimäki 1991).

Body mass of females was significantly positively associated with clutch size in low, increase and decrease vole years. In 3+-year-old females, however, no significant benefit from heavy body mass was found except in low vole years. Because the body mass of females changes greatly during the breeding season, it is better to use wing length instead of body mass when interpreting the results for body size (Chapter 6.2.3). If this was done, in both young and old female boreal owls wing length was not associated with the number of eggs or fledglings produced. Furthermore, the incubation efficiency of females, measured by mass and wing length in relation to hatching success (proportion of hatched eggs in relation to those laid) was not significant: the largest females were not the most successful incubators. This suggests that large female size does not increase incubation efficiency. However, our analyses revealed that long-winged and heavy females produced larger eggs than did short-winged and light females, but only in the decline phase of the vole cycle (Hakkarainen and Korpimäki 1993). With respect to egg size, this was the only obvious evidence for the reproductive effort hypotheses. In conclusion, although some support for the benefits from large female size was found, our long-term correlative data do not give any strong or explicit support for the hypothesis that RSD would have evolved through selection favouring high breeding success of large females.

According to the various role-differentiation hypotheses (e.g. the niche differentiation hypothesis, the female dominance hypothesis), a large size difference between the sexes (i.e. the degree of RSD) should improve the breeding success of owl pairs. For example, if parents of different sizes prefer different types of prey, then intraspecific competition within a pair could be diminished. Some evidence for the dietary separation between the boreal owl sexes was obtained in Idaho, USA. In winter, northern flying squirrels (body mass 140 g) were frequently captured by female boreal owls, representing 45% of their prey weight, but only one specimen was captured by a male (Hayward et al. 1993). This is a fascinating observation, and intersexual dietary separation definitely needs further study. However, the degree of RSD within boreal owl pairs in our study area was not of great importance to breeding success as measured by the number of eggs or fledglings produced. Therefore, the relative size difference between parents appeared not to substantially contribute to RSD in boreal owls. Although there is some evidence for the niche differentiation hypothesis, it and the female dominance hypothesis were not supported by our long-term data.

6.2.3.   Body size, fledgling production and provisioning rate of males

It appears that the selective pressure on male size varied depending on the phase of the vole cycle. Based on our 10-year correlative data covering three vole cycles, it was discovered that light and long-tailed males were more successful than heavy, short-tailed males in their breeding attempts in poor vole years, whereas heavy and short-tailed males were more successful than light, long-tailed males in good vole years, as measured by the number of fledglings produced. These results for males also held when young (1–2-year-old) and older (2+-year-old) males were analysed separately. These results held true especially when fledgling production was examined with respect to the loading index of flying area, i.e. $100 \times$ (body mass/wing $\times$ tail length). This index combines the effects of wing and tail length in relation to body mass and describes pooled wing- and tail-loading (g/cm^2). We found that in low vole years, males with a low loading index of flying area were able to raise more fledglings than those with a high loading of flying area ($r_s = -0.56, n = 38, p < 0.001$; see Fig. 6.4). The opposite was true in decrease vole years: heavy males with respect to flying area performed better than those with low wing and tail loading ($r_s = 0.23, n = 210, p < 0.01$). In the increase vole years, the number of fledglings produced was not associated with the loading index of flying area.

To test more male size effects under cyclic food conditions, we also made direct observations on male provisioning rate in the field during one vole cycle consisting of a

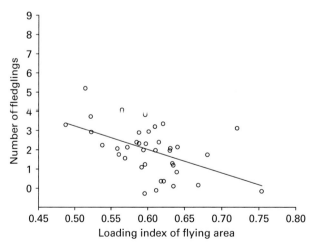

Fig. 6.4.   The correlation between the loading index of flying area [$100 \times$ (body mass/wing $\times$ tail length)] of males and the number of fledglings produced in the low phase of the vole cycle (Hakkarainen and Korpimäki 1994a).

Fig. 6.5.   Male boreal owl delivers a bank vole to the nest-box where the female is incubating or brooding. Photo: Benjam Pöntinen. (For colour version, see colour plate.)

low, increase and decrease vole year (Fig. 6.5). Nightly 4-hour prey delivery sessions of a total of 57 males were counted (14 in the low vole year 1990, 22 in the increase vole year 1991, and 21 in the decline vole year 1992). Each male was observed during a 4-hour period at night (22:00– 02:00 hours). Observations were made from a hide near the nest-hole when nestlings were 2 weeks old. At that time, females brood their offspring in the nest-box and males alone provision the whole owl family. In accordance with our correlative field data, our field observations revealed that in the poor vole year, males with a low loading index of flying area fed their chicks significantly more than those with a high loading of flying area (Fig. 6.6). The reverse appeared to be true in the decline vole year, when heavy males with respect to flying area provisioned their offspring more than those with low wing and tail loading. In the increase vole year, such a relationship was not found. (Fig. 6.6; see also Hakkarainen and Korpimäki 1995).

We suggest the following ecological explanations for the contrasting phenotypic correlations for food provisioning and fledgling production of male boreal owls. These are based on the hunting efficiency of males depending on changes in prey size and abundance over the vole cycle. In poor vole years, when breeding is costly due to food scarcity, an increase in hunting areas and small prey size, light males are obviously more economical fliers and efficient hunters than heavy males. In poor vole years, owing to the scarcity of the main prey, males have to make longer hunting trips from their nests, and these high energy costs of flight are partly compensated for by lightness of the male, which probably improves flight and hunting performance. Hence, it is probably energetically advantageous for males to have low wing and tail loading, because the lift of flying increases with decreasing body size. Therefore, in low vole years much energy is required for flying and capture attempts. This is further supported by our unpublished radio-tracking data which showed that in low vole years males hunted as much as 3–4 km away from their nests, whereas in the increase and decline vole years males usually hunt within 1 km of the nest (see Chapter 4.1.3).

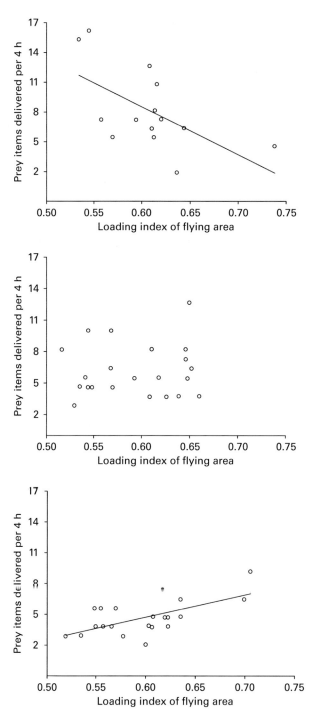

Fig. 6.6.   The number of prey items delivered during 4-hour observation periods in relation to the loading index (g/cm^2) of flying area [100 × (body mass/wing × tail length)] of males in the low (upper panel), increase (middle panel) and decline (lower panel) vole years (Hakkarainen and Korpimäki 1995).

In addition to large hunting areas in poor vole years, in those years prey size is also unprofitable. Males therefore have to capture all the prey items they encounter, as there are limited opportunities to select optimal prey size, owing to the scarcity of food. Accordingly, in low vole years, small shrews (approx. 7 g) represent about two-thirds of prey items, while in good vole years *Microtus* voles (approx. 25 g) are the main prey (Chapter 5.6). Hence, in poor vole years about four shrews are needed to satisfy the daily food requirements of one nestling, whereas in good vole years just one *Microtus* vole satisfies these requirements. This, in turn, increases the number of hunting trips and the energetic costs of hunting during poor vole years compared with those of good ones, and obviously favours small and efficient male size. Furthermore, the vulnerability of different prey types varies greatly in the course of the 3-year vole cycle. In low vole years, birds represent about one-third of prey items, whereas in good vole years mostly voles are used as prey. This probably increases the costs of hunting in low vole years, because it is generally known that birds are more difficult to catch than mammalian prey (hunting success 13% vs. 23%; Temeles 1985).

In poor vole years, the prey is of relatively small size and scarce, and is difficult to catch. This all apparently increases selection for small and efficient male size, especially as males are mainly responsible for feeding the young and the incubating female. In good vole years, however, small male size may be disadvantageous, because selection may favour large male size in catching the large voles available at that time. Accordingly, in good vole years, when large reproducing *Microtus* voles are common in the field, large males captured larger voles than those that were concurrently snap-trapped in their territories. Lighter male owls captured relatively small voles (Fig. 6.7). In good years, heavy male boreal owls may have more strike power in hunting large voles, which may weigh about 50% of their body mass (for further details, see Hakkarainen and Korpimäki 1995).

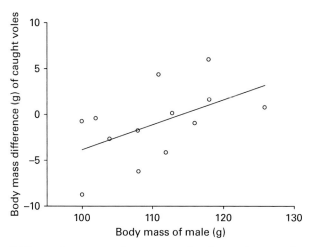

Fig 6.7.   Difference in prey mass between stored and trapped prey against the body mass of males in the decline vole year 1992. The difference in prey mass was calculated as mean body mass of *Microtus* and bank voles in the nest minus mean body mass of those voles concurrently snap-trapped in the main habitat types in the vicinity of the same nest (Hakkarainen and Korpimäki 1995).

To summarise, our explanation for RSD in the boreal owl and other birds of prey highlights the availability of prey, especially from the energetical point of view of the male. This explanation, based on energetic costs between the sexes during the breeding season, seems to be comprehensive and valid for other species demonstrating RSD, such as skuas. This hypothesis needs to be tested more intensively at an individual level in the field, and also include other animal groups displaying RSD. Our 'small efficient male' hypothesis (Hakkarainen and Korpimäki 1991) provides an explanation for both the direction and degree of RSD. The direction is based on the division of duties, in which the male is the main food provider, while the high energetic costs of males in their duties compared with those of females in the breeding season determine the degree of RSD. Although we underline the importance of male smallness to the evolution of RSD, at the same time we cannot say that factors favouring large female size do not exist. Instead, we conclude that small male size is probably of greater importance, because we did not find strong evidence that large female size was beneficial in the course of the vole cycle: it was only in the decline phase of the vole cycle that long-winged and heavy females seemed to produce larger eggs than short-winged and light females. Furthermore, the number of breeding attempts and longevity contributes more to life-time breeding success than does the number of offspring produced in any one breeding attempt (see Newton 1989, Korpimäki 1992b, Laaksonen et al. 2004). Hence, male owls that are also able to breed in poor vole years may increase their lifetime repro-ductive success and fitness in comparison with males that can only breed in good food conditions. This may be a great selection advantage, because in poor vole years only about one-tenth of males manage to breed.

6.3.   Selection for small males and large females: a synthesis

Under fluctuating food conditions, a fixed body size may be an inappropriate phenotypic expression to cope with the varying environmental conditions. Therefore, at certain stages of a temporally heterogeneous environment, the optimum body size may never be attained, while an attained optimum may prove costly under other environmental conditions. Probably owing to such large variation in ecological environment, the heritability of morphological characters in boreal owls was relatively low, as shown in Chapter 6.1.2. Low heritability may be beneficial if offspring will face different environ-mental conditions than their parents, when better adaptability may be achieved by producing a wide spectrum of offspring: then the probability that at least one of the offspring is well adapted to future changes in ecological environment will be assured. In contrast, the production of parent-like phenotypes would be costly, because in boreal owls even differing phenotypes will be favoured during the same year, for example, at the time of the crash of vole populations.

Small males were the best food providers in low vole years (Fig. 6.6). It is particularly important for northern vole-eating predators to survive through the regular 'lean' periods (i.e. poor vole years). This may explain why in boreal owls, like most other birds of prey, males have evolved to be smaller than females. During 'lean periods', large males probably die more often than small males, which induces selection for smallness of males. There is also increasing support for the hypothesis that small males are more successful when food is difficult to catch (e.g. Temeles 1985) or scarce (Hakkarainen and Korpimäki 1991, 1995). However, we found that large males were favoured in good vole years, as phenotypic correlations with fledgling production and feeding efficiency of different-sized males changed over between the years of food scarcity and abundance (Fig. 6.6). This indicates that a fixed phenotype may not be appropriate in greatly varying environmental conditions. These results suggest that if studies are made in different years, and there is seasonal or other periodic selection, then generalisation to other seasons would be dangerous (Endler 1986).

A plausible explanation for RSD, however, is not so simple, because the two sexes show different optima with respect to body size: large females and small males are favoured at certain times, depending on the stage of fluctuations in food abundance. Hence, RSD is probably not simply the outcome of a single selective force acting on either of the sexes. Instead, directional selection probably increases female largeness to some extent, but especially decreases male body size in food-limited conditions. The intensity of selection pressure on male body size probably varies over time, depending on the phase of the vole cycle, because even contrasting selection pressures on male size were found. We suggest that this large temporal variation has the potential to create counterbalancing selection on body size and probably maintains body size and RSD in boreal owls at a relatively constant level. In accordance with our results, the small male hypothesis has also been supported by two reviews on the origin and maintenance of RSD (for birds of prey, see Krüger 2005; for other animals, see Blanckenhorn 2000). At present, this seems to be the most plausible explanation for the evolution of RSD in birds of prey: small males are efficient in the provision of food. Support for the hypothesis that large females compete for smaller males, or produce early and large clutches/eggs was not found (Hakkarainen and Korpimäki 1991, 1995, Krüger 2005), which further emphasises the importance of small size in the evolution of RSD. Furthermore, we have also found that in Eurasian kestrels, small males were preferred by females (Hakkarainen et al. 1996a) and performed best in poor vole years (Massemin et al. 2000), which gives further support for the connection between small male size and RSD in that ecologically quite similar species.

In the future, there is a need to study the association between body size measurements, RSD and lifetime reproductive success, although such data are difficult to obtain. A significant connection between lifetime reproductive success and body size measurements, however, would reveal the real fitness effects of body size, and hence a plausible mechanism for RSD. Temporal and spatial variations in environmental quality should also be taken into account, because selection pressure on body size may vary greatly over time, as was shown by our study on contrasting phenotypic correlations between male size and feeding frequency. Our studies on male size effects under cyclic food conditions are in accordance with the comprehensive literature published on the evolution of body size. Many studies on phenotype–environment interactions concern spatially varying environments, such as comparisons between poor and good habitats. These interactions, however, can also be studied in a single environment, if an individual encounters regular temporal changes in its ecological environment, for example in the abundance of the primary food (see Stearns 1976).

7 Mating and parental care

7.1. Pair bonds and divorce rate

Some of the most popular study questions in behavioural ecology have long been the evolution of mating systems and the factors affecting mate choice. 'Mating system' means the way in which individuals acquire mates including the forming and duration of pair bonds, how many mates they have, and patterns of parental care for each gender (Davies 1997). Pairs may be formed only during the breeding season, or the male and female may remain together all year. Successive breeding with the same mate (mate fidelity) and successive use of the same territory (site fidelity) are widespread, particularly among long-lived bird species (e.g. Black 2001, Naves et al. 2007, Jeschke and Kokko 2008) including Ural and tawny owls (Saurola 1987, 1989, Korpimäki 1992c). Successful reproduction is usually associated with the higher probability of pair reformation, while failure is followed by an increased probability of divorce.

We trapped (ringed or re-trapped) breeding boreal owl females at 1224 nests during 1976–2009 and males at 954 nests during 1979–2009 (see Table 1). The annual proportion of nests where parent female owls were trapped varied from 80% to 100% during 1980–2001 and 2006–9, and the corresponding proportion was 60–100% for parent male owls during 1981–96, 1999–2000 and 2007–9. We recorded a total of 923 pair bonds during 1979–2009. The vast majority (>99%) of pair bonds of boreal owls last only for one breeding season or part of it (see Korpimäki 1989b and Chapter 7.3.7); in only three cases did the pair bond last more than one breeding season. These three pairs bred in the same box in successive years, but in one case (out of these three) there was a one-year 'divorce' and subsequent 'remarriage' in the pair bond. The male (ring no. 264345) and female (ring no. 1553) first bred together in box no. 366 in 1987 and again in 1989. The male of this pair bred in two neighbourhood nest-boxes with two separate females in between these two pair-bondings, in 1988 (i.e. he was polygynous; see Korpimäki 1989b and Chapter 7.3.3). He was polygynous again in 1989, when the same female as in 1987 was the first egg-laying mate of this polygynous male. These scanty data suggest that it is probably the nest-site rather than the mate that makes the pair bond of boreal owls last for more than one breeding season.

If the members of the pair survive over the winter and change mates in the next breeding season, they are said to have 'divorced'. There are no fewer than 11 hypotheses that have been proposed to explain why socially monogamous bird species divorce, but one important reason for the splitting of a pair bond has been suggested to be poor reproductive success (Dhondt 2002). The divorce rate is usually defined as the verified number of pairs separated in cases where both members of the pair were recorded alive in the subsequent breeding season. The divorce rate of boreal owls in our study area was very high (87%, 13 out of 15 cases), particularly in comparison with the divorce rate of tawny owls (12%) and that of Ural owls (3%) in southern Finland (Saurola 1987). The main reason is probably that boreal owl females are highly mobile and may disperse up to 500–600 km even after successful breeding attempts, whereas males are site-tenacious after successful breeding attempts in our study area (Korpimäki et al. 1987, 1993a). This leads to the situation where the members of the pair only very seldom meet again on the home range of the male, even when they are alive in the next breeding season.

7.2.   Assortative mating

Assortative mating is the tendency to mate with partners of the same age or breeding experience. Possible mechanisms for assortative mating by age in boreal owls include choice by differences in plumage colour or timing of settling on territories between age classes, and choice by differences in nest-hole quality or in food quantity provided by the male during courtship (Korpimäki 1989b).

The age composition of the parent boreal owls in our breeding population varies markedly during the course of the vole cycle. A marked variation in the age composition of parent owls in relation to abundance of the main food has also been documented in a German population (Schwerdtfeger 1991). Very few yearling (1-year-old) males are able to breed in the low phase of the cycle; in the increase phase most breeding males are 2+ years old; and in the decline phase most breeders are either yearling or 2+-year-old males (Table 7.1; see also Korpimäki 1988c, Laaksonen et al. 2002). Therefore, it is important to analyse the pair bonds by parental age in the three different phases of the vole cycle. In the increase phase of the vole cycle, 2-year-old and 2+-year-old males

Table 7.1. Number of pair bonds among three age classes (1 year old, 2 years old, and 2+ years old) of boreal owls in the low, increase and decrease phases of the vole cycle (pooled data from 1981–97).

Cycle phase	Male age	Female age 1	2	2+	Total	Percentage of age class of males
Low	1	3	3	3	9	13.4
Low	2	8	11	10	29	43.3
Low	2+	6	14	9	29	43.3
Increase	1	10	11	19	40	14.7
Increase	2	15	20	44	79	29.0
Increase	2+	27	47	79	153	56.3
Decrease	1	92	27	35	154	44.4
Decrease	2	28	12	22	62	17.9
Decrease	2+	65	25	41	131	37.8
Total		254	170	262	686	

mated more often with the oldest (2+-year-old) females than would be expected by chance (Table 7.1). In the decline phase, both yearling (1-year-old) and 2+-year-old males mated more often with yearling females than expected by chance; while there was no obvious indication of assortative mating by age in the quite scanty data from the owls breeding in the low phase of the cycle.

In our study population, boreal owls show two tactics of breeding dispersal: they shift nest-hole within the study area (residents), or they move into the area from elsewhere (immigrants) (Korpimäki 1987e, 1988d). Adult males usually shift nest-hole only within their own home range, whereas adult females frequently change home range and even disperse away from the study area altogether (Korpimäki 1987e). Therefore, differences in mating by age class in the increase and decrease phases of the vole cycle might be caused by different arrival times of adults and yearlings. It is difficult to examine the arrival times of night-active boreal owls but, in all probability, immigrants arrive later than residents that probably stay over winter on their home ranges. We found that 77% of 22 resident males mated with 1+-year-old females, while the corresponding percentage was only 51% for 47 immigrant males. A similar tendency was apparent in 15 resident females, none of which paired with first-year males, while the corresponding proportion was 14% for 81 immigrant females (pooled data from 1985–6; Korpimäki 1989b).

There are no clear differences in plumage colour between first-year and older boreal owls, although 1-year-old owls are usually slightly darker than older ones (Fig. 2.1). Out of the four nest-box types available in our study area, boreal owls clearly avoided only the largest board boxes, which are accessible to Ural owls (Chapter 4.3.3), although their number in the field has been low (Korpimäki 1987e). Therefore, dissimilarities in plumage colour and nest-hole quality as mechanisms for assortative mating can probably be discounted in our study area. However, we do not have any information on whether hooting characteristics vary between different-aged male boreal owls and whether these hoots are signals of male quality that could possibly lead to assortative mating. There is some evidence that the hoots of tawny and little owls provide information on caller quality (Galeotti 1998, Hardouin et al. 2007, 2008).

Adult (>1-year-old) resident females had more time for mate choice than the adult immigrants had, and they evidently chose at least 2-year-old males in the increase phase of the vole cycle, when the size of the breeding owl population was rapidly increasing. These older males probably occupied high-quality territory, because the longevity of males is greater in good territories than in poor ones (Korpimäki 1988d, 1992b). Older, experienced males may also be skilful hunters and they know the patches of main and alternative prey on their home ranges. Therefore, older males would probably provide plenty of food for their partners, and thus the choice by food quantity during the courtship period seems most plausible. This view is further supported by the fact that resident males mated with older females more than immigrant males of the same age class did (Korpimäki 1989b).

In the decline phase of the vole cycle, main food abundance is still good-to-moderate during the mating and courtship periods in late winter and early spring. Under these conditions, first-year males also succeed in pairing and they mainly mate with first-year females. On the other hand, males are competing for the smaller number of unpaired females arriving in the area at any one time, because there are many bachelor males in boreal owl populations (see Chapter 7.3.2). Therefore, males may accept the first female they encounter, irrespective of age, which may explain the high proportion of first-year females as partners of 2+-year-old males in the decline phase of the vole cycle. However, if these older males have a choice, it would be adaptive to reject first-year females due to their poor breeding performance (Korpimäki 1988c, Laaksonen et al. 2002).

To conclude, different timing of settling on home ranges between age classes, and choice by differences in food quantity provided by the male during the courtship period are the most likely explanations for the assortative mating by age class found for boreal owls in the increase and decrease phases of the vole cycle. However, our data from night-active boreal owls did not allow us to assess the relative importance of these two factors. In partially migrating Eurasian kestrels in Scotland, assortative mating seemed largely due to the different arrival times of yearlings and adults (Village 1985). There is a need to study the settling times of different-aged male boreal owls on their home ranges in different food conditions and then to find out the arrival times of different-aged females in the area. One important question is how many males do these females visit and court before making their final choice?

7.3.   Mating systems and parental care

Mating systems are the result of individual behaviours: individuals compete and attempt to maximise their reproductive output (Emlen and Oring 1977). Parental care increases the survival and fitness of offspring (Clutton-Brock 1991) but may be costly to parents if it decreases the resources they allocate to self-maintenance and reduces their survival and future reproductive success (Roff 1992, Stearns 1992). Therefore, mating systems should not be considered as cooperative adventures in which females and males rear offspring in agreement, but rather as each individual attempting to maximise its own reproductive success, even at the expense of its mate (Trivers 1972; see reviews in Parker et al. 2002, Houston et al. 2005). Therefore, there is an apparent intersexual conflict over parental care in animals (e.g. Hinde and Kilner 2007, Olson et al. 2008), in which both parents have important duties during the reproductive season. Thus, the costs and benefits of caring versus desertion need to be considered for each gender.

In most birds, one male and one female form a pair and raise a brood together, and they thus typically have bi-parental care and monogamy (Lack 1968). The main advantage proposed for monogamy is that males and females produce the most offspring if both parents help to raise a brood (Lack 1968). However, many individual-level population studies have revealed that regular social polygyny occurs in at least 10% of bird species from at least ten orders (review in Bennett and Owens 2002). In these cases it is the male that partially or entirely deserts his offspring and re-mates in the early, or sometimes the later, phases of the breeding cycle. In addition, social polyandry has been documented in at least 11 different families, including fewer than 5% of all bird species (Bennett and Owens 2002). Therefore, in birds, offspring desertion by females is far less common than desertion by males (e.g. Clutton-Brock 1991). Uni-parental male care and 'classical polyandry', in which one female is paired with two or more males with separate nests in one season, is usually possible only in precocial birds, for example in waders, where the young are relatively mature and mobile from the moment of hatching (see reviews in Oring 1983, Andersson 2005). Multi-nest sequential polyandry in birds with altricial chicks (which are dependent on their parents for food), where males are either unwilling or physiologically unable to perform the majority of incubation, is rare and has been reported only in a handful of species (reviews in Wiktander et al. 2000, Wiebe 2005).

In species with bi-parental care, both sexes benefit from the reproductive effort of their partner and are suggested to somewhat decrease their own reproductive costs (Clutton-Brock 1991, Parker et al. 2002; but see Jones et al. 2002). When the benefits of desertion are assumed to exceed the benefits of care, parents are expected to abandon their offspring (Houston et al. 2005). A main benefit of offspring desertion is that it allows faster re-mating for a new breeding attempt (Korpimäki 1989b, Szekely et al. 1996, Roulin 2002). By shortening the time period between two annual breeding events in seasonal environments, desertion could be beneficial in terms of fitness. Desertion may also allow individuals to reallocate some resources from reproduction into body maintenance, moult or migration (e.g. Szekely et al. 1996, Currie et al. 2001). However, offspring desertion can be costly, due to the loss of a high-quality mate or territory and a reduction in reproductive success at the first nest (Eldegard and Sonerud 2009). The former may be mitigated by re-mating opportunities (Emlen and Oring 1977) and the latter by the ability and choice of the partner to rear offspring on its own.

In birds of prey, the division of breeding duties between the sexes is more marked than in most other birds, and they thus have obligatory bi-parental care. The duties of female boreal owls include the production and incubation of eggs and brooding the young until they are about 3 weeks old (Fig. 7.1). Male owls, in turn, provide nearly all the food for the whole family from before egg-laying until the offspring become independent after the post-fledging period (Korpimäki 1981, Eldegard and Sonerud 2009, Zárybnická 2009b) (Fig. 7.2). Female boreal owls stay in the vicinity of the nest-box or in it for some 2–3 weeks before the start of egg-laying, put on weight, and are solely provided for by their mates during this courtship feeding period (Korpimäki 1989b, 1991b). Female owls thus concentrate on incubation and brooding, whereas males specialise in hunting. Thus, both owl parents have important roles in parental care, which has been suggested to promote monogamy (Wittenberger and Tilson 1980) through increased offspring production by both females and males.

7.3.1.   High paternal investment prevents cuckoldry?

We trapped (ringed or re-trapped) breeding boreal owl females at 1227 nests during 1976–2009 and males at 954 nests during 1979–2009 (Table 7.1). Single-locus DNA fingerprinting confirmed that the male parents trapped at the nests were genetic sires of their offspring. Some evidence of mismatching paternity was revealed in only four nests (6%) and no evidence of mismatching maternity was detected in a total of 65 broods sampled during 1991–4 and analysed later (E. Korpimäki and H. Hakkarainen, unpublished data). Similarly, no evidence of extra-pair fertilisations among boreal owls in

Fig. 7.1.   Female boreal owls lay eggs every second day. The incubation of eggs and brooding of young until they are approx. 2–3 weeks old is the duty of females. Photo: Benjam Pöntinen. (For colour version, see colour plate.)

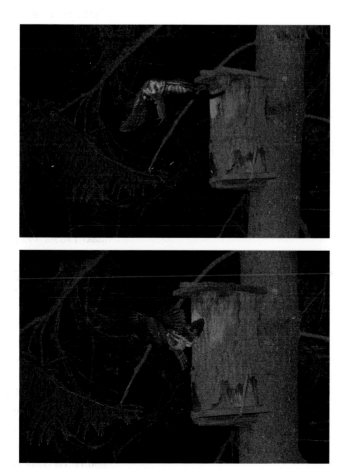

Fig. 7.2.   Male parents provide for their families from before the egg-laying until the young fledge at the age of 30–33 days. In this case the young were about to fledge. Photos: Pertti Malinen. (For colour version, see colour plate.)

either the boreal forest of Alaska or the Rocky Mountains of Idaho, Montana, Wyoming and Colorado were found in blood samples collected from 32 broods including 109 nestlings (Koopman et al. 2007b). Although the number of samples was quite low, the geographical coverage of sampling was quite extensive. Moreover, no evidence for extra-pair fertilisations has been detected in many other owl populations studied so far, including little owls *Athene noctua* (Müller et al. 2001), long-eared owls (Marks et al. 1999), eastern screech owls *Megascops asio* (Lawless et al. 1997), flammulated owls *Otus flammeolus* (Arsenault et al. 2002), and burrowing owls *Athene cunicularia* (Korfanta 2001). Extra-pair fertilisations have been documented at low rates (2–5% of broods) in two owl species: barn owls and tawny owls (Roulin et al. 2004, Saladin et al. 2007), as well as in some diurnal raptors including Eurasian kestrels and lesser kestrels *Falco naumanni* (reviews in Korpimäki et al. 1996, Negro et al. 1996).

These results show that the vast majority of female owls show genetic monogamy, which is unusual among other avian families (reviews in Birkhead and Møller 1992, Bennett and Owens 2002). Male birds can adopt two alternative, but not mutually exclusive, means of paternity assurance. In mate guarding, males closely follow their mates during the fertile period prior to and during egg-laying, and attempt to prevent copulation with another male. Alternatively, males may display a very high frequency of within-pair copulation, by which males can at least augment their chances of paternity (Birkhead 1987, Birkhead and Møller 1992). One can speculate that the possibility for extra-pair copulations and fertilisations could be high among night-active birds, including boreal owls. During the female's fertile period prior to and during egg-laying, males should try to ensure paternity by closely guarding their partners in the darkness of night, but at the same time they need to provision their mate to allow her to gain weight and maximise egg production. Thus, males have to compromise between leaving the nest vicinity and the female to collect food for their mate, or staying close to her to assure paternity.

We suggest that the main reason for successful paternity assurance and high genetic monogamy of boreal owl males is that their parental investment in the offspring is considerably higher than that of female parents. Male owls provide food for their mates during at least the 2-week courtship feeding period before egg-laying, while females stay immobile in the vicinity of the future nest-hole and put on weight to produce eggs (Korpimäki 1981, 1989b). Bi-parental care – egg-laying, incubation of eggs and brooding of the young by females, as well as hunting and food provision by males – is obligatory until the offspring can keep themselves warm at the age of approx 3 weeks in the adverse conditions of northern boreal forests. Thereafter, females can start hunting and may help the males in providing prey for the offspring, but their contribution to chick feeding is normally low (Korpimäki 1981, Zárybnická 2009b). A recent radio-tracking study revealed that about 70% of parent female boreal owls deserted their brood in the late phase of the 4-week nestling period, while their mate continued to care for their joint offspring in the nest. He alone fed the offspring for an additional 6–7 weeks during the post-fledging dependency period outside the nest-hole (Eldegard and Sonerud 2009). Eggs are laid every second day, incubation of the first egg lasts on average 29.2 (± 1.7, s.d.) days, that of the last egg on average 26.6 (± 1.3) days, first-hatched chicks stay in the nest-hole for an average of 32.6 (± 2.3) days, and last-hatched chicks for 30.3 (± 1.4) days (Korpimäki 1981).

This all means that the period of high parental investment of male boreal owls in successful nests lasts for 4.5–5 months. It is probable that male owls provide considerably less care if females take part in extra-pair copulations, which probably leads to lowered reproductive success or even to total nest failure. Because the costs of extra-pair copulations of females through reduced fitness are likely to be high, female owls may not jeopardise future parental care from their mates by soliciting extra-pair copulations from intruders. The frequency of intruders is not known for night-active owls, but apparently these intrusions by extra-pair males happen because of the existence of bachelor males in the boreal owl population (see Chapter 7.3.2). In diurnal raptors, the within-pair copulation frequencies are high (for example, 207–230 per clutch in Eurasian kestrels, 326 in lesser kestrels, and 690 in American kestrels *Falco sparverius*; review in Korpimäki et al. 1996), but the copulation frequencies of night-active owls remain to be studied in the wild. Despite this, we suggest that boreal owls successfully use both frequent within-pair copulation and mate guarding as complementary paternity protection, which in all probability assures their paternity. It may also be true that female owls do not normally solicit extra-pair copulations from intruding males in the same way that Eurasian and lesser kestrels do (Korpimäki et al. 1996, Negro et al. 1996). Intruding male owls might also be juvenile, inexperienced individuals, and female boreal owls have been shown to choose older, experienced males when these are available (Korpimäki 1989b). Therefore, high paternal investment is probably the most important factor preventing extra-pair fertilisations in boreal owls.

7.3.2. Existence of bachelors: lack of mating partners?

In animal populations, the breeding system depends on the ability of one sex to acquire mates by associating with them directly or by defending territories and other resources for breeding (e.g. Reynolds J. 1996, Davies 1997). Most population studies on birds have focused on the traits of breeding individuals, while the existence of an 'underworld' portion of the population, the non-breeding individuals, has received considerably less attention. Evidence for the existence of non-territorial floaters in the population is the rapid replacement of territory owners after removal experiments or after territory owners have died naturally (reviews in Newton 1992, 1998), or observations of radio-tracked individuals which did not breed (Rohner 1996, Kenward et al. 2000). The probable reason for neglecting non-breeding individuals is that such studies are laborious to perform, although it may be imperative to study non-breeders in order to better understand the life-history and population characteristics of animals. Most long-term studies have shown that a minority of breeding individuals may produce the majority of offspring in the future breeding population (reviews in Newton 1989 for birds; Clutton-Brock 1988 for many animal taxa including birds). Therefore, even intensive studies may be biased towards 'fit' individuals in the population, while less-'fit' individuals are only occasionally met as breeders. To better understand population ecology, the mechanisms of non-breeding should be examined more closely.

In our boreal owl population, on average 24% of males were bachelors, 66% of males were monogamous and 10% polygynous in good vole years (Table 7.2). Bachelor males

Table 7.2. The numbers of available nest-sites, boreal owl nests, bachelor males and nests in which parent owls were trapped, and the mating status of males trapped in the core of the study area (870 km^2) in the Kauhava region. Data from Korpimäki (1991b).

Year	No. of nest-sites	No. of nests	No. of bachelor males	Nests in which parent owls were trapped		Mating status of males		
				Females	Males	Monogamy	Bigyny	Trigyny
1985	400	42	11	32	33	30	3	–
1986	400	81	15	70	66	58	8	–
1988	440	92	22	82	82	69	12	1
1989	440	133	40	107	105	92	11	2
Total	1680	348	88	291	286	249	34	3

hoot in the vicinity of and deliver courtship prey items to their nest-holes but do not breed. Thus, bachelor males possess home ranges and nest-boxes (i.e. they essentially differ from non-territorial floaters), but still do not breed. The existence of non-breeding bachelors and breeding with two mates simultaneously in boreal owl populations has also been found in Germany (Schwerdtfeger 1984, 1993) and Sweden (Carlsson B.-G. et al. 1987, Carlsson B.-G. 1991). Because the bachelor males, at least in our study area, possessed home ranges with suitable nest-holes for breeding, they appeared not to be non-territorial floaters, the existence of which has quite often been documented in other bird species, including birds of prey (e.g. Rohner 1996, Newton 1998, Kenward et al. 2000). This raises the intriguing question of why a high proportion of males remain as bachelors while, at the same time, quite a large number of males are able to attract two – and some even three – mates.

Non-breeding may occur because non-breeders are immature or somehow physiologically incapable of breeding, or because of a lack of the food resources and mating partners needed to breed. There is, however, a lack of experimental evidence on whether bachelor males possessing territories and nest-sites are able to breed when supplemented with extra food or provided with mating partners. Therefore, we sought to examine the reason for non-breeding in male boreal owls. We asked whether this was due to (1) low food resources in the territory, leading to reduced prey delivery rates by the male to the female, so that she was unable to accumulate sufficient resources for egg-laying, (2) a lack of females, or (3) the males being physiologically incapable of breeding. To answer these questions we first conducted a supplementary feeding experiment in the nest-boxes of bachelor males during the courtship period. We predicted that if limited food resources were the cause of non-breeding, the experimental bachelor males provided with adequate supplementary food should attract more females and attempt to breed more often than unfed control males. Second, we performed a mate addition experiment by transporting females to the territories of bachelor males. We predicted that if a lack of mating partners is the cause of non-breeding, the experimental addition of females should increase the proportion of bachelor males that attract a female and start to breed in comparison with control bachelor males without female addition. If so, physiological constraints of breeding would also be excluded. We also estimated habitat and nest-box quality and determined whether these characteristics differed between breeding and non-breeding males.

In early March 1988, 1989 and 1991, non-paired males were located by counts of hooting owls with playbacks in the vicinity of each nest-box ($n = 470$) for 10 minutes per a recording point at nights (19:30–06:00 hours) in good weather conditions. In late March and early April, we checked all the nest-boxes. If the nest-box was not occupied where a hooting male had been observed, the box was visited regularly to ensure that the male was not paired and did not breed. These nest-boxes were visited from the mean laying date of the population (in 1988 this was 10 April ± (s.d.) 20 days, $n = 93$; in 1989 it was 14 March ± 10 days, $n = 131$; and in 1991 it was 29 March ± 20 days, $n = 97$) until the end of the laying period (approx. 1 June; Korpimäki and Hakkarainen 1991). Males were classified as non-paired, if some marks of occupancy (such as fresh courtship small mammals in the nest-box) were observed, but no eggs had been laid.

A previous supplementary feeding experiment in boreal owls, conducted prior to and during egg-laying, showed that pairs receiving supplemental food advanced their laying by 6 days and laid 0.9 more eggs than unfed control pairs (Korpimäki 1989a). We used the same method by placing dead rooster chicks (mean weight of each ± s.d. = 40.1 ± 3.0 g) in the nest-box of non-paired males in 1988 ($n = 12$) and 1989 ($n = 13$). In 1988 (mean laying date 10 April), the supplementary feeding was performed every 2–3 days in each nest from 15 April to 11 May. In 1989 (mean laying date 14 March), supplementary food was provided every 2–3 days from 31 March to 27 April. During each visit, three rooster chicks were placed at the bottom of the nest-box. The amount of food provided to the experimental males was much greater than the estimated daily food requirement, 38 g, of an adult boreal owl (Korpimäki 1981). During the supplementary feeding period of 1988, the mean biomass of the chicks consumed per territory by non-paired males was 213.9 (s.d. 198.9) g. Only 2 of the 12 non-paired males receiving food supplements did not consume any of the chicks provided. In contrast, during the supplementary feeding period of 1989, the mean consumption per non-paired male was only 61.7 (s.d. 96.3) g, and only 5 of the 13 males consumed the extra food. In early spring of the decline vole year of 1989, food might have been so abundant that there was little need for owls to consume supplementary food. The rest of the non-paired males in the population (18 in 1988 and 29 in 1989) were used as non-food-supplemented controls.

In the increase vole year 1991, 33 males were detected to be non-paired. During the late laying stage of the population (mean = 4 May, s.d. 6 days), we transported a female to the territory of randomly chosen non-paired males ($n = 16$) and placed her inside the nest-box. The females were captured from early nests in the mid-nestling phase, when most females would typically abandon the brood to re-mate and breed again (Korpimäki 1989b, Eldegard and Sonerud 2009). The nest-boxes of the other 17 non-paired males were visited at the same time but no females were provided, and these were used as controls. In early June, when all breeding attempts have usually been initiated (Korpimäki 1987f, Korpimäki and Hakkarainen 1991), the occupancy of both female-addition and control nest-boxes was determined.

Food supplementation did not increase the breeding prospects of unpaired males. In 1988, only one of 12 supplemented unpaired males started to breed, while none of the control group (18 males) did so. Similarly, in 1989, none of the 13 supplemented males or 29 control males started to breed. Territory grades did not differ between supplemented and control males in either 1988 ($2.3 \pm$ s.d. 1.4 vs. $2.6 \pm$ s.d. 1.4, respectively) or 1989 ($2.3 \pm$ s.d. 1.4 vs. $1.9 \pm$ s.d. 1.6, respectively). Territory quality, as estimated by breeding frequency of owls during an earlier 10-year period, was significantly (1988, 1991) or almost significantly (1989) lower among non-paired males than among breeding males in the three experimental years (Hakkarainen and Korpimäki 1998).

Nest-box quality (poor-quality boxes, diameter of the entrance hole 13–18 cm vs. good-quality boxes, 8–9 cm diameter) did not differ between non-paired and breeding males in 1988 (proportions of poor-quality nest-boxes were 0% and 9% for non-paired and breeding males, respectively) or in 1989 (11% and 6%, respectively). In 1991, however, a higher proportion of non-paired males were in lower-quality nest-boxes (32% vs. 7%, respectively).

Of the 16 females transported to the territories of non-paired males, three (19%) initiated breeding, while no breeding attempts were observed in control territories ($n = 17$; Fisher exact test, two-tailed $p = 0.10$). All three of the males who bred with a transplanted mate fledged chicks (clutch sizes were 4, 5 and 6, and the numbers of fledglings 4, 3 and 5). These values represent relatively good breeding success for boreal owls (Korpimäki and Hakkarainen 1991) despite the exceptionally late laying dates in those nests: 13 May, 4 May and 8 May. The territory quality in those three nests (grades 1, 2, 3) represented average values for the territory quality of boreal owls.

Even in good vole years, approximately one-quarter of males remained unpaired. Our supplementary feeding experiment suggests that lack of food resources did not prevent those males from breeding. In contrast, males appeared to be able to breed successfully when females were transplanted to their nest-boxes. This also shows that non-breeding males were not physiologically incapable of reproducing. The addition of mates, however, increased the breeding probability of bachelor males only slightly. The number of re-mating females among the transplanted females may have been low because some females probably did not stay in their new nest-boxes; they may disperse up to 200 km between two breeding attempts within a year (see Chapter 7.3.7). In addition, because of the late timing of the female addition experiment, some transplanted females may have been unwilling to re-mate.

An explanation for the availability of females limiting breeding by males could be a male-biased sex ratio (e.g. Emlen and Oring 1977). This may be because boreal owl females have a high probability of mortality during autumn migration, exacerbated following the crash phase of the vole cycle (Korpimäki and Hongell 1986, Korpimäki et al. 1987). Therefore, there is probably a lag period before the female breeding population is recovered, because of their long dispersal distances (even over 500 km; Korpimäki and Hongell 1986, Löfgren et al. 1986, Sonerud et al. 1988). Most males, in contrast, are site-tenacious throughout their life after the first breeding attempt (Korpimäki and Hongell 1986, Korpimäki 1993a), and therefore are continuously present in the mating population. The operational sex ratio, i.e. the ratio of males to females among individuals ready to mate

(Emlen and Oring 1977), may also be male-biased because in good vole years about 10–20% of males are able to acquire two, or even three, females.

Habitat quality, as indicated by the frequency of breeding occupancy, may also play some part in decreasing the breeding prospects of bachelor males. In all three years, breeding males tended to occupy higher-quality territories than unpaired males. High-occupancy, high-quality territories sustain higher vole populations than low-quality territories (Hakkarainen et al. 1997a). This suggests that non-paired males are forced to occupy lower-quality territories where prey density is relatively low. Therefore, lack of food resources in the territories of bachelor males would account for their lower breeding prospects. This interpretation, however, was undermined by the fact that, despite our supplementary feeding, only one bachelor male was able to attract a mate and breed. In 1991, in addition to poor habitat quality, non-paired males occupied territories with poor-quality nest-boxes, which are more vulnerable to larger predators and adverse weather conditions. Our female addition experiment in the same year, however, showed that in spite of poor habitat quality or nest-box quality, bachelor males seemed to increase their breeding probability when given additional females.

We conclude that the relatively high number of bachelor males in boreal owl populations is probably due to the lack of females in the mating population. In addition, some traits associated with habitat quality (e.g. food resources, nest-box quality and predation risk) may also play a minor role in non-breeding of owl males possessing nest-holes. In the future, there is a need for attempts to study the body condition and age of these bachelor males. One of the most interesting questions is whether the boreal owl males starting their reproductive lifespan at the age of 1+ year are already present in the population as yearling bachelors. If so, they may gain experience in good hunting patches and refuges that may improve their future survival and reproductive success.

7.3.3.   Polygyny in good food conditions

Of a total of 721 males trapped at their nests during 1981–97 (Table 1), sixty-nine (9.6%) were trapped at two (66 males) or three (3 males) nests and were defined as polygynous (Fig. 7.3). DNA fingerprinting also confirmed that polygynous males were the genetic fathers of the offspring that they were provisioning with food. The annual proportion of polygynous males of all males breeding in our study area varied from 0% to 18% during 1981–97 and was closely positively correlated with the abundance index of voles in the current spring (Fig. 7.4). Moreover, polygyny cases were not recorded in any of the poor vole years: 1981, 1983, 1984, 1987, 1990, 1993 and 1997. On the Swedish side of the Bothnian Bay, Baltic Sea, the corresponding proportions were 9% in 1982, 0% in 1983, and 14% in 1984 (Carlsson B.-G. et al. 1987). These annual proportions of polygynous males are probably underestimates, because males are more reluctant to enter the trap on their second or third nest, but our polygyny frequencies are comparable between years because our trapping effort of parent males was consistently high during 1981–97.

In the core area (870 km^2) of our boreal owl study, 87% of breeding males were monogamous, 12% bigynous, and 1% trigynous in the pooled data from four years. These included two increase years of vole abundance (1985 and 1988) and two decrease

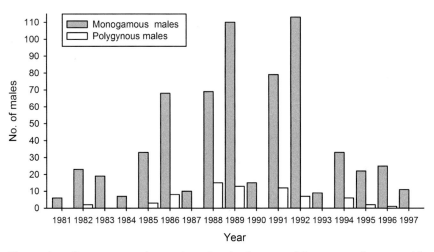

Fig. 7.3.   The number of monogamous (grey columns) and polygynous (bigynous or trigynous, white columns) males trapped at the nests during 1981–97 in the Kauhava region (data from Korpimäki 1991b and unpublished).

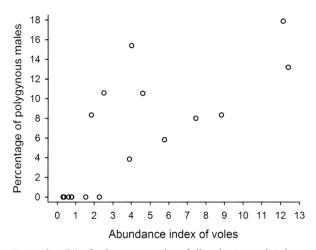

Fig. 7.4.   Proportion (%) of polygynous males of all males trapped at the nests in the Kauhava region plotted against the spring abundance index (individuals per 100 trap-nights) of voles (*Microtus* and bank voles pooled) during 1981–97 (Spearman rank correlation, $r_s = 0.80$, $p < 0.001$) (data from Korpimäki 1991b and unpublished).

years of vole abundance (1986 and 1989), when the censuses of hooting males, and thus mating options, were carried out (Table 7.2). The proportions of polygynous males were 9% in 1985, 12% in 1986, 16% in 1988, and 12% in 1989.

Primary nests of polygynous males were 1351 ± 864 m (mean $\pm$ s.d.) from secondary nests, but only 812 ± 359 m (850 m) from the closest vacant nest-hole (upper panel in Fig. 7.5). There were 1–4 vacant nest-holes between the primary and secondary nests in 70% (26 out of 37) of cases of polygyny. In the Umeå study area, two nests of polygynous

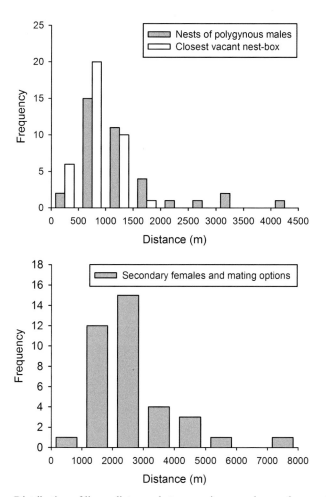

Fig. 7.5. Distribution of linear distances between primary and secondary nests of polygynous males (mean ± s.d. 1351 ± 864 m, median 1050 m, range 520–4360 m) (grey bars) and between the primary nest and closest vacant nest-box (812 ± 359 m, 850 m) (white bars). The difference was significant using the Wilcoxon matched-pairs signed-ranks test, $z = 4.37$, two-tailed $p < 0.001$), pooled data from 1985–6 and 1988–9. The number of polygynous males was 37 (upper panel). The distribution of linear distances between secondary females and closest mating option was either the closest bachelor male or nearest unmated male that later attracted a mate and bred successfully (grey columns in lower panel) (data from Korpimäki 1991b).

males were 650–2300 m (median 950 m) apart (Carlsson B.-G. et al. 1987). The home range size of monogamous boreal owl males is 150–230 ha in northern Europe (see Chapter 4.1.3), indicating that polygynous males tend to space out their different nests. In an extreme case, a male had his two nests 4360 m apart, which probably meant much flying between these two nests during the short nocturnal hunting and chick feeding periods. In addition, there was one nest of a monogamous male in between these two nests of the polygynous male, and the occupancy of a monogamous breeding male in between the nests of bigynous males has also been observed elsewhere (Carlsson B.-G. 1991).

Table 7.3. Characteristics of territories and breeding performance of primary and secondary female boreal owls compared with monogamous females laying at the same time (Korpimäki 1991b).

| | Polygynous | | | | | | Monogamous[a] | | | | | |
| | Primary females | | | Secondary females | | | Simultaneous to primary | | | Simultaneous to secondary | | |
	Mean	s.d.	*n*	Mean	s.d.	*n*	Mean	s.d.	*n*	Mean	s.d.	*n*
Grade of territory	3.2	1.5	37	3.1	1.5	37	3.1	1.4	37	2.9	1.4	37
No. of prey items												
– egg-laying and incubation	1.9	1.5	37	1.9	1.5	37*	3.7	5.8	37	2.0	1.7	37
– hatching and nestling	2.1	1.9	37*	1.3	1.4	37***	4.4	4.0	37**	2.9	3.2	37
Laying date[b]	22.3	14.0	37***	38.2	20.9	37	22.4	14.8	37	38.4	21.5	37
Clutch size	6.2	1.4	37	5.7	1.0	36	6.3	1.0	37	5.9	0.9	37
No. of fledglings	3.9	1.9	37***	2.1	1.6	37	4.4	1.4	37***	3.8	1.6	37
Fledgling mass (g)	129.6	15.6	25**	118.8	16.1	25	132.8	12.9	23*	129.0	13.6	23

Note. Pooled data from 1985–6 and 1988–9; *n* = number of nests. The Wilcoxon matched-pairs signed-ranks test was used for statistical comparisons between primary and secondary females, between primary and simultaneously laying monogamous females, and between secondary and simultaneously laying monogamous females.
Only statistically significant differences between columns area shown (*two-tailed $p < 0.05$, **$p < 0.01$, ***$p < 0.001$).
[a] Simultaneous monogamous females were chosen irrespective of their distance from polygynous nests as follows: the number of polygynous and monogamous females from different years was the same (3 from 1985, 8 from 1986, 13 from 1988, 13 from 1989), and the laying dates of monogamous females were close (± 2 days) to those of primary and secondary females.
[b] 1 March = 1, 2 March = 2, 3 March = 3, etc.

The time between the start of egg-laying in primary and secondary nests of polygynous males ranged from 1 to 43 days (mean 16 days; Table 7.3). The corresponding time interval ranged from 2 to 31 days (median 18.5 days) in 1982 and from 4 to 42 days (13.0) in 1984 in the Umeå study area (Carlsson B.-G. et al. 1987).

Our data set also included four cases of trigyny, in which the same male was trapped in three nests (Table 7.4). The same male (ring no. 284008, nicknamed the 'Sultan of Sippolankorpi') was trigynous in both 1988 and 1989 (Fig. 7.6). He initiated his breeding lifespan as a yearling in 1986, when he was monogamous and raised five fledglings of the six-egg clutch. He was again monogamous in 1987 and reared three fledglings of the five-egg clutch. His first two mates started to lay eggs within 3 days in 1988 and produced seven- and eight-egg clutches, respectively, of which seven + seven young hatched. The third female (1 year old) initiated egg-laying 11 days later than the primary female (2 years old) and 8 days later than the secondary female (1 year old), laid six eggs that all hatched and produced four fledglings. Altogether, this male raised 14 fledglings in 1988. In 1989 his first (1 year old) and second (1 year old) females initiated egg-laying on the same day. Both females laid four-egg clutches, of which two and four young hatched, while his third female (1 year old) started to lay eggs 10 days later and laid six eggs that

Table 7.4. The details of four trigyny cases of males recorded in our study area in 1988, 1989 and 2005.

	Primary nest (P)	Distance P–S	Secondary nest (S)	Distance S–T	Tertiary nest (T)	Distance P–T
Case 1[a]						
Ring no.	284008	900 m	284008	2200 m	284008	1200 m
Laying date	20 March 1988		23 March 1988		1 April 1988	
Clutch size	7		8		6	
No. of fledglings	5		5		4	
Case 2[a]						
Ring no.	284008	1500 m	284008	2100 m	284008	900 m
Laying date	9 March 1989		9 March 1989		19 March 1989	
Clutch size	4		4		6	
No. of fledglings	2		0		2	
Case 3						
Ring no.	15321	960 m	15321	1600 m	15321	760 m
Laying date	7 March 1989		10 March 1989		21 March 1989	
Clutch size	7		8		4	
No. of fledglings	4		3		3	
Case 4						
Ring no.	80249	3700 m	80249	3900 m	80249	1800 m
Laying date	22 March 2005		31 March 2005		1 April 2005	
Clutch size	7		6		8	
No. of fledglings	3		5		6	

[a] The 'Sultan of Sippolankorpi'.

Fig. 7.6.   The home range of the male that was trigynous in two successive years ('Sultan of Sippolankorpi') included the old-growth forest of Paukunmäki on the southern side of Sippolankorpi, Kauhava. Photo: Pertti Malinen.

all hatched. The male raised four fledglings in 1989. His last breeding attempt was recorded in 1990 with a 2-year-old female, which laid five eggs, of which four hatched but none fledged. All in all, this 'Sultan of Sippolankorpi' male appeared to be extremely productive, because he paired with at least nine different females in 5 years. His mates laid a total of 51 eggs, mainly on the basis of food provided by this male during the courtship feeding period, of which 46 hatched, 26 fledged, and 3 later recruited to the breeding population. This is the only male owl known to us which has been recorded as being trigynous in two breeding seasons.

The second trigynous male (ring no. 15321) was recorded as breeding only in 1989, when he was 3 years old and all three mates were yearlings (Table 7.4). His first two mates started to lay eggs within 3 days in 1989, produced seven- and eight-egg clutches, of which all the eggs hatched. His third female initiated egg-laying 14 days later than the primary female and 11 days later than the secondary female, laid and hatched four eggs and reared three fledglings. Altogether, this male raised 10 fledglings in one breeding season.

The third trigynous male (ring no. 80249) was found in 2005 (Table 7.4). He was 3 years old, his second mate was 3 years old and his third mate 1 year old (the age of the primary female remains unknown). The second female started to lay eggs within 9 days of the first, and they produced seven- and six-egg clutches, all of which hatched. The third female, who started egg-laying 10 days later than the primary female and 1 day later than the secondary female, laid and hatched eight eggs. This male reared 14 fledglings in one breeding season. He also bred in 2006 (six eggs but no fledglings) and again in 2008 (six eggs and six fledglings). Altogether, this male reared 20 fledglings during his lifetime, although his three nests were on average 3.1 km apart, which is at least twice as far the corresponding distance for two other trigynous males (1.5 km and 1.1 km, respectively).

Supplementary feeding during the pre-laying, laying and incubation periods probably induced a trigyny case in boreal owls (Carlsson B.-G. and Hörnfeldt 1989), in which the linear distance between the boxes of the primary and secondary, the secondary and tertiary, and the primary and tertiary was 2400 m, 2500 m and 850 m, respectively. The time interval between the initiation of the primary and secondary clutches was 5 days, and 25 days between the secondary and tertiary ones. The male was 3+ years old, and the primary, secondary and tertiary females were 3, 3+ and 3+ years old, respectively. They laid eight, six and seven eggs, and reared three, five and two fledglings, respectively. The pooled number of fledglings produced by this male was ten (Carlsson B.-G. and Hörnfeldt 1989). There has also been a case of trigyny recorded in saw-whet owls. Three nests of the male were very close to each other: two were in paired boxes 15 m apart and the third was 130 m away (Marks et al. 1989). The abundance of small rodents was probably unusually high in the Snake River Birds of Prey Area, Idaho in that year (Marks et al. 1989). Three simultaneous nests of three different boreal owl females have also been discovered in the same tree, but in this case the identity of the male(s) remained unknown (Chabloz et al. 2001).

In addition to the two owl species of the genus *Aegolius*, there are reports of occasional polygyny in seven owl species in Europe, including the snowy owl, tawny owl, scops owl *Otus scops*, barn owl, great grey owl, hawk owl and the long-eared owl (reviews in

Korpimäki 1988e, Marks et al. 1989). There are also a few recent records of polygyny in pygmy owls (E. Korpimäki, unpublished), short-eared owls (E. Korpimäki unpublished) and Ural owls (P. Saurola, unpublished) in Finland. Among diurnal raptors, hen harriers *Circus cyaneus*, marsh harriers *C. aeruginosus*, and Montagu's harriers *C. pygargus*, common buzzards and Eurasian kestrels have frequently been recorded to be polygynous (review in Korpimäki 1988e), but the boreal owl is the only owl species known to be regularly polygynous. A common feature of all of these species is that they use small mammals as their main foods and polygyny usually occurs in conditions of food abundance.

The annual proportion of polygynous males of all boreal owl males breeding in our study area was higher in years of vole abundance than in those of vole scarcity (Fig. 7.4), which evidently suggests a close link between food abundance and polygyny. Food abundance is the main factor enabling the male to feed two or even three females, so that they can put on weight and produce eggs. Polygyny cases were not recorded in any of the poor vole years, which also supports this interpretation. Supplementary feeding prior to and during the egg-laying period probably also induced a case of trigyny in boreal owls (Carlsson B.-G. and Hörnfeldt 1989). This is consistent with the prediction that supplementary feeding prior to and during the egg-laying periods can increase the number of males in harems of birds of prey (Korpimäki 1988e).

Food abundance and number of nest-holes in the home range are probably the crucial factors determining the marital status of boreal owl males, but it is also probable that male quality is at least a contributing factor. All three trigynous boreal owl males were at least 3 years old, and their primary and secondary females started to lay eggs nearly simultaneously (within <5 days) >900 m apart. Because males are in conflict whether to guard their mate so as not to lose paternity, or to leave their mate in order to hunt for themselves and feed their mate properly, bigynous males need to be very good hunters. Trigynous males would be even more effective hunters and food providers for their mates, because their tertiary females started to lay eggs 8–11 days later than the secondary females. In the future, there is a need for radio-tracking studies on the allocation of time given by males to mate guarding, hunting and prey delivery as well as hooting to attract additional mates, because these interesting behaviours happen at night and are thus very challenging to study. Individual identification of hooting males indicated that males continued to sing at a new nest-box after having mated with a primary and a secondary female (Carlsson B.-G. 1991).

7.3.4.   Polygyny by female choice or constraint

There are several hypotheses or models that have been developed to explain the evolution of territorial polygyny by female choice. These models can be tentatively divided into two categories. '*Compensation models*' are derived from the polygyny threshold model (e.g. Verner 1964, Orians 1969), which suggests that a female breeds more successfully as a secondary female (i.e. pairing with an already-mated male) in a good territory than by joining an unmated male in a poor territory (Verner 1964, Orians 1969). The prediction usually derived from this model is that polygynous females should breed as well as simultaneously laying monogamous females. However, several studies on, for example, passerines and birds of prey have shown that secondary females (i.e. the second mate of the male) raise fewer fledglings than primary females (i.e. the first mate of the male; reviews in Korpimäki 1988e, Alatalo and Lundberg 1990). The compensation may also take the form of increased survival of secondary females or their young (Wittenberger 1976), or of attractiveness of offspring fathered by high-quality males (Weatherhead and Robertson 1979). In addition, nest predation may lower the polygyny threshold so that the influence of predation should be considered when studying the validity of compensation models (Bensch and Hasselquist 1991, Sonerud 1992b).

'*Non-compensation models*' suggest that female choice is constrained in some way and that females choosing already-mated males have poorer fitness than monogamously mated females. The 'no better option' hypothesis was first put forward to explain the common occurrence of harems of up to six females mating with one male in hen harriers (Simmons et al. 1986a, 1986b, Redpath et al. 2006) and marsh harriers (Altenburg et al. 1982). This hypothesis says that a lack of alternative mating options could mean that the only alternative to polygyny is not breeding (Newton 1979). The 'deception' hypothesis (Alatalo et al. 1981) suggests that males of polyterritorial species 'deceive' females into polygyny by hiding the fact that they are already mated. Thus, females choose males on the basis of territory quality, but they are not able to compensate for the costs of sharing the male. In addition, searching for unmated males may be costly (Stenmark et al. 1988), which may reduce the mating options available for females.

Primary data showed that secondary female boreal owls raise only 36% of the fledglings reared by all monogamous females, and 41% of those produced by monogamous females laying concurrently in our study area (Korpimäki 1989b). These very low numbers of fledglings indicate that secondary female owls make a bad mate choice, particularly because many bachelor males capable of breeding are available (Table 1; also see Chapter 7.3.2). The low productivity of secondary females was also obvious in the Umeå study area, where bigynous males were able to fledge only 49% of the eggs laid by their second females, while the corresponding proportion was >75% for the eggs of females of monogamous males laying simultaneously. In addition, the first females of these bigynous males raised 83% of eggs laid to fledglings (Carlsson B.-G. et al. 1987). Probably one of the best ways to study why females choose already-paired males in polyterritorial species is to compare the options available to individual females. Therefore, we analysed the apparent options available to females that paired with already-mated males. We also asked whether these females would have bred more successfully if they had paired with bachelor males available at the same time.

There were no obvious differences in territory quality and clutch size between the primary and secondary females, but primary females produced significantly more and heavier fledglings than secondary females (Table 7.3). Primary and secondary females stored equal numbers of prey items in their nest-boxes per visit during the egg-laying and incubation periods, but the prey stores of primary females were significantly larger during the hatching and nestling periods (Table 7.3). Territory quality of primary and secondary females was very similar to that of monogamous females laying simultaneously. There were no obvious differences in the clutch size and number and body mass of fledglings produced between the primary females and monogamous females starting concurrently. This was also the case for the clutch size of secondary females, but they produced significantly fewer and lighter fledglings than monogamous females laying at the same time. At all stages of the breeding season, the mates of monogamous males had significantly larger food stores than primary females starting simultaneously (Table 7.3). This held also for the number of prey items stored by mates of monogamous males and secondary females during the hatching and nestling periods, but not during the egg laying and incubation periods.

We compared the territory quality and breeding success of secondary females with those of the unmated males available in two ways. First, the grade of the territory and breeding performance of secondary females were compared with those they would have if they mated with bachelor males (see Table 7.2 for numbers). These males were first recorded in the vicinity of nest-holes during listening surveys (i.e. during the settling period of females) and also possessed their nest-holes in the early breeding season, as fresh prey items (i.e. 'courtship gifts') were present in the hole. We used the following procedure when choosing mating options. The first-laying secondary female was 'allowed' to mate with the closest bachelor male and thereafter the second one 'got' the closest vacant male, and so on. The clutch size and number of fledglings of females that would have 'paired' with these bachelors were estimated based on the grade of territories, assuming that these males were monogamous. We used the mean values of the data collected during 1977–86 extracted from table 2 in Korpimäki 1988d). This data set included four good vole years and six poor ones. Second, we compared the territory quality, laying date, clutch size and number and body mass of fledglings of secondary females and later-started females of monogamous males. According to listening surveys, these males were already present on their territories during the settling period of the secondary females, and were also available to them in the early breeding season (i.e. 'courtship gifts' were present in the nest-hole). The procedure when choosing these options was the same as described above.

Secondary nests and the closest bachelor males were on average 2827 m (± 1390 m) apart (Fig. 7.5, lower panel). The territorial call of male boreal owls is audible to humans up to 3–4 km away (Holmberg 1979, authors' own observations), but the bilateral ear asymmetry of boreal owls substantially improves their auditory localisation (Chapter 6.1). Thus, female boreal owls are probably able to locate the territorial calls of males from 3 km away. Using this criterion, 62% of bachelor males were so close to the secondary nests that they should have been detected by the females that mated with already-paired males. Secondary nests and the closest unmated male that later attracted a mate and bred successfully were on average 5688 m (± 4382 m) apart (Fig. 7.5, lower panel). Of these males, 35% were <3 km from the secondary nests. Pooling the two data sets (Fig. 7.5, lower panel) reveals that the mean distance between the secondary females and the closest mating option was 2464 m (± 1288 m). Of these, 76% were closer than 3 km from the secondary females.

One could argue that an excess of high-quality nest-boxes available in our study area led to artificial conditions, so that the distances recorded between secondary nests and mating options would differ from the 'natural' situation. The density of cavities made by black woodpeckers in south Finnish old forests is 0.5–1.5/km^2 (Pouttu 1985, Virkkala et al. 1994), which probably equals the 'natural' situation in our study area before modern forestry destroyed most of the cavities. The density of nest-boxes (approx. 0.5/km^2) used

Table 7.5. Characteristics of territories and breeding performance of secondary females compared with their closest option to mate with a bachelor male and a monogamous male that were available for the secondary female when she paired (Korpimäki 1991b).

	Secondary females			Mating options: Bachelor males[a]			Monogamous males		
	Mean	s.d.	*n*	Mean	s.d.	*n*	Mean	s.d.	*n*
Grade of territory	3.1	1.5	37	2.6	1.3	37	2.7	1.3	37
No. of prey items									
– egg-laying and incubation	1.9	1.5	37				2.4	2.1	37
– hatching and nestling	1.3	1.4	37				1.9	1.8	37
Laying date	38.2	20.9	37				47.9	23.9	37
Clutch size	5.7	1.0	36	5.8	0.3	36***	5.8	0.9	36
No. of fledglings	2.1	1.6	37*	2.9	0.4	36***	3.5	1.4	37
Fledgling mass (g)	118.8	16.1	25				123.1	12.6	25

Notes. Pooled data from 1985–6 and 1988–9; *n*, laying date and statistical comparisons between secondary females and their mating options are as in Table 7.3.
[a] Clutch size and number of fledglings produced by possible 'mates' of bachelor males were estimated based on the grade of their territories and assuming that they were monogamous.

in our study area is substantially lower. There is thus no solid evidence that 'unnatural' numbers of nest-boxes were provided.

Both bachelor males and unmated males that paired later occupied slightly, but not significantly, poorer territories than mates of secondary females (Table 7.5). However, only one bachelor male possessed a territory that was never used for breeding in a 10-year period (1977–86). Other territories provided for a breeding pair at least in good vole years (Korpimäki 1988d) such as 1985–6 and 1988–9. The clutch size of secondary females and the estimated clutch size of females pairing with bachelor males were similar, but secondary females had significantly fewer fledglings (Table 7.5). Considering only those bachelor males that were <3 km from the secondary nests does not change the results (Table 7.5). It should also be noted that the estimates of breeding success are conservative, as the data used for estimation also included six low vole years, when reproductive success is generally poor (Korpimäki 1981, 1987a, Korpimäki and Hakkarainen 1991).

The mates of monogamous males that were unpaired when secondary females joined already-mated males laid on average 10 days later (Table 7.5). Thus, secondary females should have had more than a week to choose a mate in order to avoid polygyny. By doing so they would have raised significantly more fledglings, as did the mates of these males. This difference also held when only those monogamous males that were closer than 3 km from the secondary nests were compared (Table 7.5). The better breeding success is probably due to the higher feeding rates of monogamous males during the hatching and nestling periods, as at these stages the prey caches of the females of these males were almost significantly larger than the prey stores of the secondary females. On the other hand, there was no difference in clutch size and number of prey items in food stores during the egg-laying and incubation periods. Fledging mass seems to be a good estimator of future survival of offspring (Korpimäki and Lagerström 1988). Fledglings of secondary females tended to be slightly, but not significantly, lighter than those of the females of the monogamous males (Table 7.5), suggesting that their future survival is lower.

Our comparisons above would be biased if the females mated to the monogamous males were of higher quality than secondary females (Davies 1989). Primary females (31% of them were yearlings) tended to be older than secondary females (46%), while there was no obvious difference in comparison with females of the monogamous males that later bred successfully (51%). In addition, there were no obvious differences in the wing length and body mass of primary females, secondary females and females of the monogamous males (mean wing length ± s.d. 179.6 ± 3.7 vs. 179.2 ± 3.1 vs. 179.4 ± 3.2 mm; mean body mass 168.6 ± 18.4 vs. 165.8 ± 13.2 vs. 166.5 ± 16.5 g). This suggests that the poor breeding success of secondary females is not due to their poor quality.

7.3.5.   Nest predation lowers the polygyny threshold?

In Eurasian coniferous forests, boreal owls co-exist with pine martens, which also use tree-holes for roosting and nesting. In southern Norway, pine martens took 47% of boreal owl nest contents (Sonerud 1985a). This high predation rate led Sonerud (1992b) to question our conclusion that polygynously mated female boreal owls make a poor mate choice (Carlsson B.-G. et al. 1987, Korpimäki 1989b, 1991b). He developed a model, the first assumption of which was that polygynous males allocate all their feeding effort to secondary nests if predators destroy primary nests (Sonerud 1992b). To rigorously test this assumption, one either needs to remove the primary nest of polygynous male owls to find out whether this induces males to switch feeding to their secondary families, although this is questionable for ethical reasons, or to observe the behaviour of males after the natural failure of the primary nest. However, to our knowledge, this has not yet been done. The second assumption of the model was that pine martens almost always prey on boreal owl nests before hatching, and thereafter secondary families may get all the food collected by polygynous males (Sonerud 1992b). However, 10 out of 30 nest predation cases by pine martens happened after hatching (pooled data from 1974–92; Korpimäki 1994a). The time interval between the laying dates of primary and secondary clutches of polygynous males is sometimes only a few days and averages 16 days (Table 7.3). Therefore, in 10–20% of predation cases, the secondary females could not gain, because the polygynous males would switch to feeding them only after their young were 1–2 weeks old, and most of their young would die within a few days after hatching (Carlsson B.-G. et al. 1987, Korpimäki 1989b).

Sonerud's (1992b) model (equations 1–2) showed that predation rates of boreal owl nests should be 60–90% before secondary females could achieve the number of fledglings of simultaneously mating monogamous females, and 74% if they were to achieve the reproductive success of later-mating monogamous females. In his equations, the assumption was that, after predation of primary nests, fledging success (number of fledglings per number of eggs produced) of secondary females is higher than that of the females pairing monogamously at the same time. This was suggested to be because bigynous males are better hunters and have better territories than males mating monogamously at the same time. The latter appears to be an incorrect assumption, as there was

Table 7.6. Number of fledglings relative to monogamous females for primary and secondary female boreal owls and the overall frequency of nest predation during 1985–92[a] in the core of the study area (870 km^2), the Kauhava region. Number of females in parentheses. Data from Korpimäki (1994a).

Year	Primary females[b]	Secondary females[b]	Pooled no. of females	Frequency of nest predation
1985	0.64 (3)	0.42 (3)	33	0.05
1986	1.22 (8)	0.55 (8)	66	0.05
1988	1.25 (13)	0.75 (13)	82	0.03
1989	0.95 (13)	0.38 (13)	105	0.03
1991	0.67 (12)	0.46 (12)	78	0.03
1992	0.86 (7)	0.86 (7)	100	0.02
Mean	0.93	0.57		0.03
S.d.	0.26	0.19		0.01

[a] No polygyny was recorded in 1984, 1987, 1990 and 1993, when breeding density was low, associated with scarcity of the main food (voles).
[b] Including nests that produced at least one fledgling or where nestlings starved to death and excluding total nest losses by predators.

no difference in territory quality between polygynous and monogamous males pairing simultaneously (Korpimäki 1991b). To test this assumption, data from provisioning rates of polygynous and synchronously mating monogamous males feeding experimentally randomised broods are needed.

To test how much random nest predation might lower the polygyny threshold, data on fledglings reared by three female categories (primary, secondary and monogamous) and frequency of nest predation are needed (Bensch and Hasselquist 1991). In addition, one has to exclude the cases where reproductive success might have been affected by nest predation. Therefore, when calculating breeding success, only nests that produced at least one fledgling or where nestlings starved to death were included. The relative reproductive success of primary females, averaged over 6 years, was close to that of monogamous females, but secondary females produced only 57% of the fledglings raised by monogamous females (Table 7.6). There was a wide between-year variation in the relative breeding success of both primary (range 0.64–1.25) and secondary females (0.38–0.86), but the yearly relative success of primary females was not closely related to that of secondary females (Spearman rank correlation, $r_s = 0.37$, $n = 6$). The mean predation rate over 6 years was low (3.5%, Table 1), and its incorporation in equation (2) of the model of Bensch and Hasselquist (1991) gives the breeding success of secondary females relative to that of monogamous females as 0.59. The inclusion of the highest predation rate recorded in boreal owls (47%; Sonerud 1985a), increases the reproductive success of secondary females to 74% of that of monogamous females, but females still suffer an obvious fitness cost when they choose already-mated males. Therefore, random nest losses may not be able to lower the polygyny threshold sufficiently, and 'non-compensation' models are probably needed to explain the poor reproductive success of secondary boreal owl females.

In the models of Sonerud (1992b) and Bensch and Hasselquist (1991), the breeding success of secondary females was estimated by the number of fledged young. However, this may not be the best measure. The first-year survival of owlets varies depending on fledging condition (Korpimäki and Lagerström 1988), and this trait is probably crucial for female fitness. The fledging mass of offspring of secondary females is lower than that of offspring of simultaneously or later-pairing monogamous females (Table 7.3). The poor phenotypic quality of the few fledglings produced by secondary females indicates that future offspring survival is probably poor. If the fitness of boreal owl females of different mating status could be estimated, not by the number of fledglings, but by the number of offspring subsequently recruited to the breeding population, the fitness of secondary females would be even worse than estimated by the number of fledglings alone.

7.3.6.   Cheating in mate alluring, and sexual conflict in parental care?

Our main findings on polygyny of boreal owls can be summarised as follows.

- Polygyny is mostly polyterritorial, although the closest females of the harems may also nest at opposite corners of a single home range.
- Polygyny is sometimes nearly simultaneous, and males usually have to feed their two, or even three, females and their chicks at the same time.
- The breeding success of secondary females is poor (only 55% of the number of fledglings raised by monogamous females laying simultaneously), though there are no marked differences in clutch size. Thus, secondary females appear to make a bad choice.
- The majority of secondary females would raise more and heavier fledglings if they mated with a bachelor male nearby. This indicates that they do not make the 'best of a bad job' and is in disagreement with the 'no better option' hypothesis.
- Differences in the quality of territories and females cannot explain the very poor reproductive success of secondary females.
- High predation rates of owl nests by martens during the evolution of mate choice are not likely to explain the poor reproductive success of secondary female owls. The models based on predation risk could only work in environments where marten predation rates are 'unnaturally' high (>90%).
- The quality (in terms of fledging mass) of the few young produced by secondary females is poor. So the future success of their offspring may be poorer, rather than better, than that of monogamous females laying concurrently, arguing against the compensation hypotheses on higher phenotypic quality of offspring of polygynous males (e.g. Weatherhead and Robertson 1979, Wittenberger 1976).
- Primary and secondary females had similarly sized prey stores during the egg-laying and incubation stages, but the food stores in the primary nests were substantially larger later on. The food stores of primary females were consistently smaller than those of monogamous females laying at the same time, but a similar difference was recorded for secondary females only during the hatching and nestling periods. Based on these differences and on the fact that the clutch sizes are similar, polygynous males probably feed their secondary females during courtship, egg-laying and incubation stages at the expense of their incubating primary females. After the young of their first nest hatch, they are not able to provide for their two families and they preferentially feed their

primary family. According to our radio-tracking of hunting trips and prey deliveries of two bigynous males (in 1985 and 1986), the ratio of prey deliveries between primary and secondary nests was approx. 5:1 at the time when young in both nests had hatched. Polygynous males started to hunt in the vicinity of the primary nest after sunset, and delivered 4–5 prey items to the young of the primary nest for 1.5–2 hours. Thereafter, around midnight, they flew to the vicinity of the secondary nest, hunted there for 0.5–1 hour and delivered 1–2 prey items to their secondary families. Thereafter, they returned to hunt in the vicinity of the primary nest and delivered most prey items to this nest during the early morning hours. Therefore, low provisioning rates of polygynous males induced the high nestling mortality, and thus poor reproductive success, of the secondary females. In another study, a radio-tracked bigynous hawk owl male allocated most prey deliveries to the young of the primary nest. A lack of food resulted in the failure of the secondary nest, although the secondary female was still begging for food from the bigynous male after deserting the eggs of the secondary nest (Sonerud et al. 1987).

Boreal owls start to breed so early (Chapter 8.1) that incubating and brooding secondary females cannot leave their nests to hunt, as their eggs and young would probably become chilled during cold nights. Therefore, secondary female owls are not able to compensate for the low provisioning rates of their mates. Partial compensation has been found for secondary females in harems of hen harriers in Scotland, where larger prey animals were available along with voles, and gamekeepers also removed most nest predators (Redpath et al. 2006) so that secondary females were able to leave their nests to hunt. In contrast, secondary female hen harriers suffered from a substantially reduced breeding success in New Brunswick, Canada. No larger prey items than voles were frequently available and the abundance of nest predators was high (Simmons et al. 1986a, 1986b). On the other hand, the performance of male boreal owls during short nights is crucial for their breeding success and female owls are probably unable to raise a brood without the male's feeding support. Thus, unlike passerine birds and hen harriers in favourable conditions, polygynous male boreal owls cannot transfer their share of the parental care to their mates, as their nesting attempts would most probably fail.

The polygyny threshold model predicts that if a female has a choice, she should settle polygynously only if she performs better than the female that pairs later with the monogamous male she rejected (Davies 1989). Our comparisons of mating options available to secondary females argue against this prediction. Moreover, it should be noted that the number of bachelor males revealed by listening surveys is probably an underestimate, rather than an overestimate, as the males do not call all the time and their presence was checked only once by the listening method. Bachelor males may be of such poor quality that they were not able to feed a female during courtship, and the females therefore avoided them. However, 97% of these males occupied territories that provided for a breeding pair at least in good vole years (Korpimäki 1988d). In addition, supplementary feeding experiments showed that food-supplemented bachelor males did not start to breed more frequently than the non-supplemented ones. In contrast, transplantation experiments of females into the nest-boxes of bachelor males showed that the lack of a mate, rather than a poor ability to provide food, was the reason for not breeding (Chapter 7.3.2).

The strongest evidence that better mating options were available to secondary females is that the females of monogamous males settling later raised more and heavier fledglings. However, there are at least three factors that may confound these comparisons. First, the males may not have been present when secondary females settled in the area, but the results of listening surveys indicated that this was not so.

Second, better males may have taken over the territories of these males in the course of the settling period, and thereafter were able to attract a female. As it was not possible to capture bachelor males, this possibility cannot be ruled out. It seems unlikely, however, because males do not shift territories between successive breeding attempts (Korpimäki 1987e, 1993a).

Third, the costs of searching for a mate due to a lack of time and energy may have constrained female choice, and females may thus have accepted the first mate encountered. Although there are no data from radio-marked female boreal owls visiting and choosing mates, indirect evidence from singing activity of males indicates that females visit more than one male during the mating period (Carlsson B.-G. 1991). For example, female pied flycatchers *Ficedula hypoleuca* seem to check on average four (range 1–9) males before they make a choice (Slagsvold et al. 1988). As these flycatchers are long-distance migrants and the prospect of successful breeding declines sharply throughout the breeding season, females are selected to choose quickly (Korpimäki 1993a). In contrast, the settling and breeding periods of boreal owls are long, and breeding success does not decrease abruptly with later laying date (Korpimäki 1987f, Korpimäki and Hakkarainen 1991). Especially in the increase phase of the vole cycle, as in 1985 and 1988, when vole numbers increase rapidly in the course of the breeding season, the survival of late fledglings is also high (Korpimäki and Lagerström 1988). It seems likely that female boreal owls are not much constrained by time when making their choice, in particular in the increase phase of the vole cycle. Polyandrous females move on average 22.9 ± 48.2 km between successive breeding attempts (Chapter 7.3.7). Thus, secondary female boreal owls probably checked the mating options available nearby, but because these males occupied slightly poorer territories, they seem to have rejected them. This difference in territory quality between polygynous and bachelor males indicates that females may search for a mate and may not accept the first mate encountered.

Females of vole-eating birds of prey seem to use provisioning rates during courtship feeding as a cue for mate choice (e.g. Altenburg et al. 1982, Simmons 1988, Palokangas et al. 1992). The number and quality of nest-holes defended by males may also be of particular importance to female choice (Carlsson B.-G. 1991). Therefore, it is likely that territory quality (probably intercorrelated with male quality; Korpimäki 1988e) affects the female's choice via provisioning rates. At the early stages of the breeding season, the males on good territories are probably able to feed even their secondary females at least as well as bachelor males on poor territories. This interpretation is supported by the fact that there were no differences in the clutch size and size of prey caches during egg-laying and incubation between secondary females and females of the monogamous males. However, the cue for mate choice used by females is not a good estimate of future breeding success, as secondary females presumably cannot predict that males will provide little food during the nestling period. The finding that most secondary females had an unmated male nearby when entering the harems and that they would have done better by choosing these males gives further support to the deception model. Therefore, our interpretation is that polygyny in boreal owls appears to be best explained by the deception hypothesis.

Why, then, can boreal owl males acquire females by 'deceptive' behaviour? The most important factor is probably the cue that females use when choosing. Already-paired males on high-quality territories are able to 'imitate' the provisioning rates of unmated males in good vole years. These rates are mostly good estimates of future prey delivery rates of males and reproductive success, and thus females may 'accept' polygyny. Primary females cannot prevent males from trying to attract more mates, as they are either inactive and close to the nest putting on weight during pre-laying (Korpimäki 1986d), or tied to the nest because of a risk of their eggs becoming chilled. Thus, the aggression of primary females towards other females (e.g. Breiehagen and Slagsvold 1988) probably cannot explain why polygynous males space out their nests. Secondary females usually do not follow their mate when he leaves the vicinity of the nest. He hunts far away and may be absent for a long time if hunting success is poor. During hunting trips, he can also provide for the primary female. In addition to the spacing out, the three-dimensional forest habitat, dark nights and silent flight may help to hide his marital status. Moreover, it may not even be necessary for already-paired male owls to conceal their marital status perfectly. The defence of two or more nest-holes and similar or better provisioning rates than those of unmated males during the pre-laying and laying periods may be the only 'tricks' that are needed to 'deceive' females into polygyny. Monoterritorial male hen harriers may acquire members for their harems by 'deceitful' food provisioning during courtship (Simmons 1988). In this open-country raptor, harem females are presumably aware of the marital status of their partners, but because these males provide food at a higher rate than unmated males available at the same time, females seem to 'accept' polygyny.

Alternatively, polygynous male boreal owls perhaps only work at the 'optimal level' in the early breeding season, but because this is not enough to feed two or more families in the late season, they partly desert the secondary nests. This interpretation is unlikely, however, as polygynous males feed their secondary mates at the expense of incubating primary females. Although males fed their secondary females during the courtship feeding, egg-laying and incubation periods at the expense of incubating primary females, primary families were highly favoured later on, which resulted in the poor reproductive success of secondary females. This, polyterritoriality and a lack of polygyny in poor vole years suggest that already-paired males on high-quality territories may 'imitate' the food provisioning rates of unmated males in good vole years, thereby 'deceiving' females into polygyny.

As high-quality mated boreal owl males are likely to be able to match the provisioning rates of bachelor males available at the same time (Korpimäki 1991b), polygynous males gain substantial fitness benefits, because they produce more fledglings in one breeding season than monogamous males starting their breeding simultaneously. Average fledgling production of trigynous males was 9.3, that of bigynous males 6.0 and that of monogamous males 4.4. Males that are polygynous at least once during their lifetime also live longer and in their lifetime they produce twice the number of fledglings as monogamous males, although they do not have markedly better territories (Korpimäki 1992b). This suggests that male boreal owls seem to have been quite successful in the battle of the sexes, as polygynous males gain, but polygynously mated females lose. On the other hand, female boreal owls can also desert their first brood at the late nestling stage and remate polyandrously, thereby entirely transferring the burden of raising the young to their mate (Korpimäki 1989b, Eldegard and Sonerud 2009, Zárybnická 2009b).

7.3.7.   Polyandry by brood desertion increases fitness of females

Female boreal owls concentrate on incubation and brooding, whereas males specialise in hunting and provisioning their families. Prey deliveries by male birds of prey are obligatory prior to egg-laying until the independence of the young, while maternal care is probably less crucial after the offspring are half-grown. Therefore, female birds of prey have been shown to desert their offspring more often than males do (Beissinger and Snyder 1987, Korpimäki 1989b, Kelly and Kennedy 1993, Eldegard and Sonerud 2009, Zárybnická 2009b). However, we are aware of only a few studies, with a minimum number of records showing that deserting females do re-mate (Marks et al. 2002, Roulin 2002) and on the benefits of offspring desertion for female fitness.

A total of 14 polyandrous female boreal owls were documented in Europe up to 1986. These females deserted their first brood in the late nestling period, moved an average of 5.0 ± 3.7 km, re-mated and bred again (review in Korpimäki 1989b). Recently it has been found that some 70% of radio-tagged female boreal owls deserted their first brood and that male owls continued to care for the offspring alone (Eldegard and Sonerud 2009). Therefore, we analysed long-term data from our boreal owl population supplemented with corresponding nationwide Finnish ringing data to find out the timing of brood desertion, and the occurrence and fitness consequences of sequential polyandry. We first aimed to document whether females that deserted their brood re-mated and became polyandrous. If so, we determined the inter-nest distances of polyandrous females and examined how the frequency of polyandry varied relative to fluctuating densities of main prey. Finally, our most important question was whether polyandrous females produced more offspring than did monogamous females that nested at the same time.

We trapped (ringed or re-trapped) female boreal owls at 1135 nests during 1981–2009 early in the nestling period (Table 1). Supplementary nationwide ringing data from Finland during 1981–2009 included a total of 11 610 females ringed or re-trapped at their nests (Table 7.7). Double-brooded females were defined as females that were trapped or re-trapped at two nests with a different male parent so that the first nest was successful until the young were at least 3–4 weeks old. Dispersal distances of these females were estimated as the linear distance between the nest-boxes that they used in successive breeding attempts. Females trapped for the first time at the nests that failed in the early stages of the breeding season were omitted from the data, because these re-traps included re-nesting attempts, not real brood desertion and sequential polyandry cases.

In the nationwide data, annual numbers of boreal owl nests in which the female parent was trapped or re-trapped varied from 92 to 1203 (mean $\pm$ s.d. 400 ± 259) during

Table 7.7. Annual number of monogamous (Mono) and polyandrous (Poly) females ringed or re-trapped, and number of boreal owl broods ringed in Finland outside our study area during 1981–2009 (Korpimäki et al. 2011).

Year	Mono	Poly	No. of broods
1981	92	0	185
1982	158	0	364
1983	222	0	460
1984	146	0	330
1985	393	3	1043
1986	696	1	1755
1987	217	0	422
1988	644	2	1967
1989	1199	4	3412
1990	183	0	363
1991	907	1	2013
1992	643	2	1439
1993	141	0	292
1994	605	1	1636
1995	320	0	819
1996	471	1	1446
1997	383	1	1153
1998	331	0	992
1999	376	1	1184
2000	316	0	516
2001	133	0	350
2002	368	1	995
2003	765	0	2214
2004	124	0	283
2005	354	0	1325
2006	477	0	1431
2007	153	0	374
2008	303	0	1067
2009	470	2	1794
Total	11 590	20	31 624

1981–2009. The average (± s.d.) proportion of females ringed or re-trapped in nests in which chicks were ringed was 39.6% (± 8.2%) with an annual range of 26–61% during 1981–2009 (Table 7.7). Therefore, the ringing effort of female parents was relatively high nationwide, which increased the re-trapping probability of double-brooded females even at long distances.

Of 1135 females trapped at their nests in our study area over 29 years, 12 (1.1%) were paired with two males in successive breeding attempts (Fig. 7.7) and were thus sequentially polyandrous. Ten of the 12 polyandrous females made both of their breeding attempts within our study area (maximum linear distance moved 35 km; Fig. 7.8). One female first bred south of our study area and made her second breeding attempt in our study area (linear distance moved 95 km in 1991; see Korpimäki 1993b), while the other

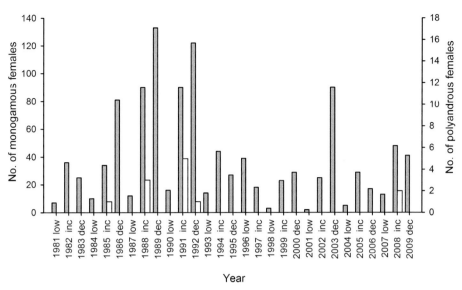

Fig. 7.7.   Number of monogamous (grey bars) and polyandrous (white bars) boreal owl females in the three phases (low, increase and decrease) of the 3-year vole cycle in the Kauhava region, western Finland during 1981–2009. Note the different scale for monogamous and polyandrous females (left and right y-axis, respectively) (Korpimäki et al. 2011).

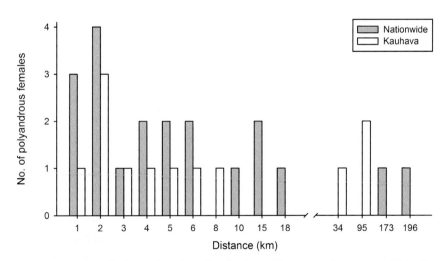

Fig. 7.8.   Inter-nest distances (km) of polyandrous females in the Kauhava region, western Finland (white bars) and in the Finnish nationwide ringing data (grey bars) (Korpimäki et al. 2011).

female first bred successfully in our study area and moved 95 km southwards to breed again in 2008. In addition, one probable case of simultaneous polyandry in which two males (one yearling and one 1+-year-old male) were trapped simultaneously at the same nest in 1986 when feeding young were recorded in our study area. Of 11 610 females trapped at their nests elsewhere in Finland, 20 (0.2%) females were trapped at two nests and were thus defined as sequentially polyandrous (Table 7.7). Overall, polyandrous females searched for mates over relatively long distances (mean ± s.d. 22.9 ± 48.2 km, median 4.5 km, range 1–196 km; Fig. 7.8).

The 12 polyandrous females in our study area initiated egg-laying of their first annual breeding attempt on average 14 days earlier and produced on average 0.4 egg more than monogamous females of the study population (mean ± s.d.: 15 March ± 6 days, $n = 12$ vs. 29 March ± 18 days, $n = 373$ for laying date; 6.3 ± 0.6, $n = 12$ vs. 5.9 ± 1.2, $n = 369$ for clutch size). These 12 polyandrous females started egg-laying in their two nests at intervals of 54–68 days (mean ± s.d. 59.7 ± 4.0 days). The incubation period of the first egg averages 29 days and the nestling period of the first-hatched young averages 33 days (Korpimäki 1981). If we assume conservatively that the time needed to search for and pair with a new mate, and putting on mass to lay a new clutch is at least 10 days, polyandrous females should have deserted their first brood when the oldest young were 21 days old on average. In one case the oldest young was only 15 days old. This corresponds well with the brooding period of females in our study population (mean ± s.d. 21.1 ± 2.5 days, range 15–23 days; Korpimäki 1981). Also the brooding period of female boreal owls in a Czech study population averaged 22.2 days (± 0.9 days) in 2004, when most females deserted their brood (Zárybnická 2009b).

Females were especially likely to desert their brood and pair with a new mate during the increase phase of the vole cycle (11 of 12 polyandry cases; Fig. 7.7), when the abundance of main prey rapidly increased during the course of the summer. The same tendency also appeared to be true for the nationwide ringing data from double-brooded females (11 of 20 polyandry cases). The yearly proportion of polyandrous females of all females captured in our study population closely increased with the augmenting abundance indices of voles during 1981–2009 (Korpimäki et al. 2011).

The proportion of yearlings was 16.6% for polyandrous females and 41.7% for monogamous females that initiated egg-laying simultaneously in our study population. The corresponding proportions were 29.4% for polyandrous females and 41.0% for monogamous females in the nationwide data. Our results thus do not clearly suggest that older females are more successful in finding a new mate and in initiating a second clutch. All yearling females deserted their brood in the late nestling and fledging periods, while the frequency of brood desertion was lower for older females (Eldegard and Sonerud 2009).

There were no obvious differences in body mass, wing length and territory quality between polyandrous and monogamous females laying simultaneously with polyandrous females (Table 7.8). Primary mates of polyandrous females tended to be heavier and thus be in better body condition than mates of monogamous females laying simultaneously. A similar tendency was not found in the body mass and body condition between the secondary mates of polyandrous females and mates of monogamous females laying simultaneously (Korpimäki et al. 2011).

Polyandrous female owls produced 79% more eggs, 93% more hatched young, and 73% more fledglings within-season compared with monogamous females that laid at the same time (Table 7.8). In 11 of the 12 cases in our study area, the total number of fledglings produced by polyandrous females within a season was substantially higher than that of monogamous females that laid simultaneously with primary and secondary

Table 7.8. Mean (s.d.) laying date (−1 = 30 March, 0 = 31 March, 1 = 1 April, 2 = 2 April, etc.), clutch size, number of hatchlings and fledglings of primary and secondary nests of polyandrous females and monogamous females laying simultaneously to primary and secondary nests of polyandrous females, mean (s.d.) and total number of eggs, hatchlings and fledglings produced by polyandrous females, and body mass (g), wing length (mm) and territory quality of polyandrous and monogamous females. Pooled data from 1985, 1988, 1991, 1992 and 2008 in the Kauhava region. Number of females is 12 in each case (Korpımäkl et al. 2011).

	Polyandrous			Monogamous, simultaneous to	
	Primary	Secondary	Total	Primary	Secondary
No. of females	12	12		12	12
Laying date	−15.2 (5.6)	44.5 (7.6)		−15.1 (5.2)	38.8 (12.4)
Clutch size	6.3 (0.6)	5.5 (0.9)	11.8 (1.3)***	6.6 (0.8)	5.3 (0.9)
No. hatched	6.2 (0.6)	5.0 (1.1)	11.2 (1.1)***	5.8 (0.9)	4.8 (1.4)
No. fledged	3.3 (2.0)	3.1 (1.8)	6.4 (2.9)*	3.7 (1.9)	3.0 (1.6)
Body mass	166.8 (17.1)	168.1 (18.9)		157.9 (21.0)	160.8 (12.4)
Wing length	179.7 (4.7)	179.3 (3.2)		180.3 (2.9)	178.0 (3.4)
Territory quality	2.2 (1.6)	2.4 (1.4)		2.1 (1.5)	2.0 (1.0)

Notes. Monogamous control females laying simultaneously to primary and secondary nests of polyandrous females were selected among the monogamous females so that the laying dates of the first egg matched as closely as possible.

The significance of the difference in the total number of eggs, hatchlings and fledglings produced by polyandrous females within a season and clutch size, number of hatchlings and fledglings of monogamous females laying simultaneously to primary nests of polyandrous females was tested with a paired samples t-test ($t = 10.55$, ***$p < 0.001$ for no. of eggs; $t = 11.07$, ***$p < 0.001$ for no. of hatchlings; and $t = 2.61$, *$p < 0.05$ for no. of fledglings; df = 11 in each case).

nests of polyandrous females (Fig. 7.9). The inter-nest distance of polyandrous females correlated negatively with the laying date and the number of fledglings produced in the first breeding attempt within a season. There was, however, no obvious relationship between the inter-nest distance and the laying date of the second breeding attempt, the number of fledglings produced in the second breeding attempt, or the body mass of polyandrous females (Korpimäki et al. 2011).

We found that female boreal owls that deserted their first brood re-mated and successfully raised a second brood with a new mate. Most polyandrous females were recorded in the increase phase of the 3-year vole cycle, when the densities of voles are intermediate in the early spring but rapidly increase during the course of the summer. The frequency of polyandry was also higher in years of high vole density than in years of low vole density. Polyandrous females had nearly twice the within-season egg and offspring production as monogamous females that nested at the same time.

By using nationwide long-term ringing and re-trapping data from Finland we overcame the methodological challenge of relocating deserting females. But because the mean inter-nest distance of polyandrous females was 23 km (median 4.5 km), the frequency of double-brooding (1.1%) documented here was probably an underestimate owing to our restricted study area and lower ringing effort of females outside our main study area. Another confounding factor may be the large variation in dispersal distances:

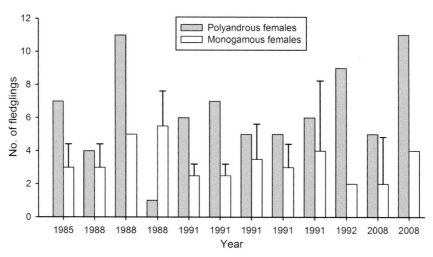

Fig. 7.9.   Total number of fledglings produced by double-brooded polyandrous females (grey bars) and mean (s.d.) number of fledglings produced by monogamous females laying simultaneously to primary nests and secondary nests of polyandrous females (white bars) during 1985, 1988, 1991, 1992 and 2008 in the Kauhava region, western Finland (Korpimäki et al. 2011).

in 4 of the 32 polyandry cases reported here, inter-nest distances of double-brooded females were as much as 95–196 km. An inter-nest distance of 194 km for a double-brooded female boreal owl has also been documented in Switzerland (Ravussin et al. 1993). However, the majority (87%) of polyandrous females in our pooled data re-mated and produced a second brood within <35 km of their first nest, which suggests that our data are not heavily biased towards shorter inter-nest distances. The future challenge will be to track deserting boreal owl females by, for example, satellite transmitters, because these females probably move very rapidly.

There appeared to be a cost for females in searching for mates over long distances, because there was a negative relationship between inter-nest distance of polyandrous females and the number of fledglings they produced in their first breeding attempt within a season. The courtship, egg-laying, incubation and nestling periods of boreal owls take 10–12 weeks, and thereafter fledglings are fed by parents for 3–7 weeks (März 1968, Eldegard and Sonerud 2009). If a female initiates her first brood as early as mid-March, she will be free from brooding duties in late May, allowing her to quickly find a new mate for a second attempt. In light of the need to re-mate quickly, the longest inter-nest distances recorded in this study appeared to be unexpectedly long, because females probably move only during the 4–6 hours of darkness that occur in midsummer at northern latitudes. In addition, some 20–25% of male boreal owls remained bachelors even if they possessed a territory and suitable nest-box, and even when these bachelors were able to feed their mates and successfully raise a brood if experimentally provided with a partner (Chapter 7.3.2). We surmise that brood-deserting females were choosy in their search for a new mate and may thus have moved over long distances in their attempt to reduce the risk of becoming a secondary female of a polygynous male. This interpretation is also supported by the negative relationship between inter-nest distance and the laying date of the first nest of polyandrous females. Those polyandrous females that were able to initiate their first nesting attempt early probably had more time to search for a second mate and thus to be more choosy in re-mating. Secondary females have substantially reduced reproductive success in terms of both quantity and quality of offspring (Chapter 7.3.5). However, despite an apparent choosy search, the first polyandrous female recorded in our study area in 1985 (Fig. 7.7) became the secondary mate of a polygynous male in her second nesting at a distance of 3.3 km (Korpimäki 1989b).

The frequency of polyandry was highest in the increase phase of the 3-year vole cycle (Fig. 7.7) and also in years of high spring vole density than in years of low vole density (Korpimäki et al. 2011). This latter result was also consistent with earlier findings that the frequency of brood desertion of boreal owls was positively correlated with the abundance of voles in spring and that this frequency can be increased with supplementary feeding (Eldegard and Sonerud 2009). We suggest that it is not only the abundance of main foods in early spring that is crucial for the adaptive advantages of brood desertion, but also that the increase in vole abundance towards autumn in the increase phase of the vole cycle is essential for the success of polyandry. Also, late owl nests have high fledgling production in the increase phase of the vole cycle (Korpimäki and Hakkarainen 1991), and the first-year survival rate of independent offspring during improving food conditions is at least twice as high as in the other phases of the vole cycle (Korpimäki and Lagerström 1988). Therefore, greater fitness benefits of brood desertion by females can be obtained during improving food conditions because double-brooding is more successful then. Temporal changes in food supply that allow females to breed again have been predicted to increase polyandry (Graul 1977). For example, the successive polyandry of black-shouldered kites *Elanus caeruleus* seems to occur in good food conditions (Mendelsohn 1983). Moreover, one of the snail kite *Rostrhamus sociabilis* parents appeared to desert the first brood during favourable food conditions (Beissinger and Snyder 1987). In addition, the skewed operational sex ratio in favour of males may contribute to re-mating opportunities of brood-deserting female owls (Emlen and Oring 1977, Pilastro et al. 2001). However, in our study, this appears to be less important than improving food conditions, because bachelor males are known to exist in both increasing and decreasing years of vole abundance (Carlsson B.-G. 1991, Korpimäki 1991b, Hakkarainen and Korpimäki 1998; see also Chapter 7.3.2).

One apparent case of simultaneous polyandry was recorded in our study area in 1986, whereby two males were trapped simultaneously at the same nest when feeding young. As the collection of blood samples for parentage analysis was not a routine method in bird population studies at that time, it remains unknown whether both males had fathered the chicks of the female, or whether one of the males was merely a visiting floater. The latter seemed quite unlikely, however, because both males took a risk and entered the box trap at the entrance hole of the nest-box to feed the female and chicks. As far as we know, the only record of simultaneous polyandry in owls in which paternity was determined by DNA fingerprinting has been reported for long-eared owls. In this case, both males also provided food for their offspring (Marks et al. 2002). In addition, an extra radio-marked eastern screech owl male provided food for the offspring of the nest of a couple, but in this case the allo-parenting male had been in captivity in a rehabilitation centre for some time before being released into the wild a few months earlier (Smith D. and Hiestand 1990). Simultaneous cooperative polyandry has been reported for two diurnal raptors, the Galapagos hawk *Buteo galapagoensis* and Harris's hawk *Parabuteo unicinctus* (Faaborg and Patterson 1981, Faaborg 1986). In the former species, DNA fingerprinting revealed that in the polyandrous groups all males (2–8, modal number of males 2) provisioned food for their young and copulated with the female. In addition, different males sired young in consecutive years in five of six groups in which male group membership was constant (Faaborg et al. 1995).

Polyandrous female owls were indeed able to produce 73% more fledglings than monogamous females. Because we compared offspring production of polyandrous females with simultaneously laying monogamous females, the well-known seasonal decline in clutch size with later laying date (see Korpimäki and Hakkarainen 1991 and Chapter 8.1.4) cannot explain our results (Table 7.8). Single-handed male owls, even if they increased their provisioning rates (Eldegard and Sonerud 2010), appeared not to be able to fully compensate, as their fledglings had an estimated 30% lower survival until independence than fledglings of non-deserting females (Eldegard and Sonerud 2009). Even if we conservatively suppose that this lowered survival rate of offspring of single-handed male owls also applies to our results, the larger offspring production of polyandrous than monogamous females indicates that polyandry confers substantial fitness benefits to females. Earlier it was shown that the number of fledglings produced by boreal owls in any one breeding attempt is closely correlated with the lifetime number of recruits to the breeding population (Korpimäki 1992b). This in turn is one of the most accurate measures of fitness in wild bird populations (Newton 1985, 1989). To our knowledge, fitness benefits for females owing to sequential polyandry have not been demonstrated for birds of prey. Double-brooded female barn owls deserting their first brood initiated their second brood 2 weeks earlier and produced significantly more eggs than non-deserting double-brooded females, but their reproductive success did not obviously differ (Roulin 2002).

To conclude, not only males but also females of altricial species with bi-parental care can increase their fitness by deserting their first brood, provided that these broods will be cared for by the males. Earlier we have shown that male boreal owls can increase their annual offspring production (Chapter 7.3.4) and also their lifetime reproductive success by simultaneous polygyny (Korpimäki 1992b). Polyandrous females desert their first brood when the chicks are 3–4 weeks old, but the deserted males substantially increase their offspring feeding rate during the 6–7 week post-fledging period (Eldegard and Sonerud 2010). Because the primary mates of polyandrous females tended to be in better body condition than mates of monogamous females laying simultaneously, these single-handed males were at least partly able to compensate for the suddenly reduced parental care by their deserting partner. High fitness costs of complete nest failures after female brood desertion probably force the single-handed males to partially compensate for the decreased care of their partners. This notion appears to be in agreement with many older models of parental care in which incomplete compensation has been suggested to be an evolutionarily stable strategy (see Wiebe 2010 with references). The peculiar mating systems of boreal owls can probably be best explained by the intersexual 'tug-of-war' over bi-parental care, where both sexes attempt to shift a majority of the parental care to their mate. In polygyny, males appeared to succeed better in the battle of the sexes, because their reproductive success was improved, whereas in polyandry females appeared to succeed better, because their reproductive success was improved. To the best of our knowledge, there are no other bird of prey species in which both polygyny and polyandry have been frequently recorded in the same population. Therefore, future studies are needed to fill the gaps in our knowledge of the mating systems and sexual conflict of owls, particularly in the early stages of the mating and reproductive season.

8 Reproduction

Source: Drawn by Marke Raatikainen

Reproduction (fecundity) and immigration affect the recruitment of new breeders to the local population, and mortality and emigration in turn determine the loss of individuals from it. Variation in these demographic variables, due to stochastic events or density-dependent factors, determines the dynamics of local population. Therefore, factors influencing reproductive success, survival and dispersal of boreal owls are the main topics of the following chapters.

8.1.   Wide variation in the start of egg-laying, clutch size and fledgling production

The importance of food supply for reproduction in birds has been stressed ever since the publication of David Lack's research findings (1947, 1948, 1954). He hypothesised that food abundance is both an ultimate factor in the timing of breeding seasons and evolution of clutch size, but also acts as a proximate factor exerting a direct effect on various aspects of breeding performance. Clutch size is considered to represent one of the most important life-history and demographic parameters (e.g. Stearns 1976, 1977). Four main hypotheses for the evolution of clutch size have been formulated, mainly on the basis of studies on small passerine birds.

1. The number of eggs in a clutch corresponds to the mean number of young which parents can rear (Lack 1947). Thus, the amount of food available during the nestling and fledging periods is the critical factor.
2. The amount of food available to the egg-laying female is the critical factor, so that fewer eggs are laid than the number of young which the parents could raise (Perrins 1970, von Haartman 1971).
3. Clutch size is related to nest predation; birds laying smaller clutches will have smaller nests and make fewer feeding visits to the nest and thus decrease the risk of nest predation (review in Martin T. 1995).
4. Clutch size is related to nest size. Several hole-nesting birds including boreal owls lay smaller clutches in holes or boxes with a small bottom area than in those with a large bottom area (review in Korpimäki 1985a).

These hypotheses and ideas on the influence of food supply on breeding performance may be studied under environmental conditions, where the abundance of the main food varies in high-amplitude and periodic population cycles (Chapter 1).

In birds of prey subsisting on voles as their staple food, breeding usually starts in early spring, although the high mortality of voles during winter and their summer reproduction cause vole numbers to peak in autumn, long after the birds' nestling phase. Furthermore, the largest clutch sizes and reproductive success are usually observed in the early-breeding pairs, rather than the late ones (e.g. Cavé 1968, Korpimäki 1987f, Pietiäinen 1989, Korpimäki and Wiehn 1998). In our boreal owl study area, during 1973–2009, the earliest breeding was initiated on 18 February and the latest on 31 May. Most females, however, start to lay eggs in late March to early April: half of the females started egg-laying before 1 April (median 31 March, mean ± s.d. 2 April ± 17 days; Table 8.1). At that

Table 8.1. Mean (s.d.) laying dates, mean (s.d.) clutch sizes and mean (s.d.) number of fledglings produced by boreal owls during 1973–2009 (n = number of nests; data from Korpimäki and Hakkarainen 1991 and unpublished).

Year	Laying date[a] Mean	s.d.	n	Clutch size Mean	s.d.	n	No. of fledglings Mean	s.d.	n	No. of nests n
1973	−10.4	5.5	8	6.0	1.1	8	3.2	2.6	10	11
1974	0.7	17.9	7	5.4	0.9	9	4.4	1.5	8	9
1975	25.5	2.1	2	4.0	0.0	2	1.0	1.4	2	2
1976	13.1	7.3	10	4.9	0.9	11	1.8	2.0	11	12
1977	−4.7	6.1	40	6.7	1.1	39	4.0	2.1	42	45
1978	5.0	5.0	6	5.0	1.8	7	2.3	2.1	7	8
1979	6.9	10.9	25	5.5	1.0	25	2.5	1.6	28	29
1980	11.0	7.4	21	5.0	1.0	20	1.0	1.1	23	23
1981	26.4	16.6	8	4.8	0.9	8	3.1	1.6	8	8
1982	0.0	9.4	25	6.4	1.4	25	4.2	2.0	26	34
1983	3.0	7.2	22	5.0	1.2	23	1.0	1.1	24	25
1984	28.1	8.9	7	5.0	1.0	7	1.1	1.6	7	7
1985	3.1	15.1	36	5.8	1.1	35	4.3	1.9	36	36
1986	1.1	9.8	73	6.1	0.9	72	3.2	2.2	72	74
1987	20.7	12.9	10	4.4	0.7	10	1.5	1.4	10	10
1988	11.1	21.1	87	6.4	1.2	87	4.1	2.1	87	87
1989	−16.8	10.1	124	5.6	1.1	121	2.8	1.8	127	125
1990	31.5	12.0	15	3.6	0.9	16	1.5	1.0	16	16
1991	−2.2	19.8	97	5.9	1.1	91	2.4	1.7	99	98
1992	−8.1	15.7	122	5.8	1.2	121	2.5	1.8	124	125
1993	32.0	12.3	9	3.7	0.9	9	1.8	1.2	9	9
1994	12.1	8.2	41	5.5	0.9	42	1.9	1.5	43	44
1995	−0.2	15.4	26	5.8	0.8	26	1.8	1.8	26	26
1996	8.5	14.2	31	5.3	1.0	34	2.3	1.8	35	35
1997	13.9	15.1	14	4.3	0.8	15	1.1	1.4	16	16
1998	14.0	8.9	3	3.8	0.5	4	1.5	1.3	4	4
1999	5.0	12.8	17	6.5	0.9	20	4.0	1.6	24	24
2000	3.4	8.9	14	4.7	0.9	22	0.5	1.0	21	22
2001	28.0	1.4	2	4.0	0.0	2	2.5	0.7	2	2
2002	−0.1	14.8	34	6.2	0.9	30	3.8	2.0	39	39
2003	−1.1	9.6	127	6.1	0.9	130	4.1	2.1	136	138
2004	18.9	8.4	9	4.4	0.9	9	2.3	1.4	10	10
2005	1.1	12.7	64	6.6	1.2	69	2.9	2.2	74	74
2006	18.8	4.1	17	5.2	0.9	19	1.4	1.5	20	20
2007	24.8	10.6	13	4.5	0.8	13	2.8	0.9	13	13
2008	−2.0	16.8	53	5.7	1.3	58	3.3	2.1	59	59
2009	−3.4	9.3	34	5.4	0.9	46	2.9	2.0	50	50

[a] 30 March = −1, 31 March = 0, 1 April = 1, 2 April = 2, etc.

time the snow layer may be 30–40 cm deep and vole populations are near their seasonal low. The boreal owl, in fact, is one of the earliest breeding bird species in our study area, as elsewhere in northern Europe (Korpimäki 1981, Mikkola 1983). This species shows large variation in the initiation of breeding, and clutch size varies greatly depending on annual changes in its main food abundance throughout its European breeding range (Korpimäki 1987f, Hörnfeldt et al. 1990). In this chapter, we focus on the wide between-year and seasonal variation in laying date and clutch size detected in boreal owls. In addition, we concentrate on parental characteristics affecting laying date and clutch size, especially from the parental age point of view.

8.1.1.   Food abundance as determinant of laying date, clutch size and fledgling production

Yearly mean laying dates in our study population varied from 13 March (in 1989) to 2 May (1993) (Table 8.1). During the 3-year cycle of voles, the start of egg-laying was earliest in the decline phase of the cycle, when most clutches were already laid in mid to late March. At that time, vole numbers had been high in the previous autumn and winter, and moderate to high during egg-laying, but declined rapidly to low numbers in the next autumn or winter. In the increase phase, most clutches were laid in late March to early April, when vole abundances were moderate at the time of egg-laying but increased to a peak in the next autumn. The latest clutches were laid in low vole years (mid to late April), when voles were scarce in spring and in the preceding winter but started to increase in late summer. Wide between-year variation was also evident in the 25 year (1985–2009) data from mean laying dates in the Jura Mountains, Switzerland (mean 11 April, range of yearly mean 18 March to 3 May; Ravussin et al. 2001b, 2010).

Our snap-trapping data in the field also revealed that abundances of voles in the field were significantly associated with the start of egg-laying. Egg-laying was initiated substantially earlier in good vole years than in poor ones. The closest negative relationship of yearly mean laying date of the owl population was found for trap indices of *Microtus* and bank voles in the preceding autumn (Fig. 8.1), although the association of laying date with vole indices for the current spring was also significant. Therefore, the timing of egg-laying appeared to be determined more by abundances of voles in the preceding winter than in the current spring, although this interpretation is constrained by the fact that our spring trap indices for voles were estimated in May, 1–2 months later than owl females began to lay eggs. In any case, these results suggest that food supply before the breeding season is the most important factor in determining the laying date, obviously via the body condition of females. This interpretation is supported by the fact that the yearly mean body mass of breeding females in the mid-nestling phase (data from Fig. 2.9) during 1980–2009 was augmented with increasing abundances of *Microtus* voles in the preceding spring ($r_s = 0.43$, $n = 30$, $p = 0.02$) but not with abundances of these voles in the current spring ($r_s = 0.16$, $n = 30$, $p = 0.39$). A similar relationship was not found for yearly mean body mass of breeding males (data from Fig. 2.9; $r_s = 0.08$, $n = 30$, $p = 0.66$ and $r_s = 0.13$, $n = 30$, $p = 0.49$, respectively) (see also Korpimäki 1990a). In addition, the laying date also advanced with the daily body mass increase of female boreal owls (Hörnfeldt and Eklund 1990).

Timing of breeding could also be influenced by weather conditions. This could be a valid explanation at northern latitudes because, for example in our study area, snow cover may prevail for 5–6 months and, even in March, temperature may drop as low as −25 to −30°C (see Fig. 3.5). However, ambient temperature and snow depth in March–April had no obvious relationships with the timing of laying in our study population (Korpimäki

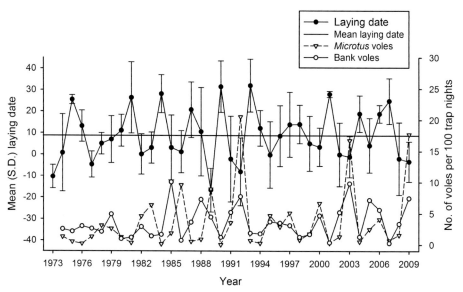

Fig. 8.1.   Yearly mean (s.d.) laying date of boreal owls (left y-axis; 0 = 31 March, 1 = 1 April, 2 = 2 April, 3 = 3 April, etc.) and trap indices (right y-axis) of *Microtus* and bank voles in the preceding autumn during 1973–2009 in our study area (Spearman rank correlation, $r_s = -0.74$, $n = 36$, two-tailed $p < 0.001$ for laying date and *Microtus* voles, and $r_s = -0.56$, $n = 36$, $p < 0.001$ for laying date and bank voles; the correlations for the trap indices of *Microtus* voles in the current spring were $r_s = -0.62$, $n = 37$, $p < 0.001$, and for trap indices of bank voles they were $r_s = -0.58$, $n = 37$, $p < 0.001$) (data from Korpimäki and Hakkarainen 1991 and unpublished).

and Hakkarainen 1991). In addition, ambient temperature and snow depth in January–April did not have any apparent influence on the timing of hatching of boreal owl broods in the Finnish nationwide ringing data, which included a total of 13 267 boreal owl broods from 1973 to 2004 (Lehikoinen et al. 2011).

Overall mean clutch size was 5.71 (± 1.24) eggs during 1973–2009. The yearly mean clutch size varied from 3.6 (1990) to 6.7 (1977) (Table 8.1). The wide between-year variation in the mean clutch size of the study population of the Jura Mountains (Switzerland) was also evident (range of yearly means 3.3–6.6 during 1985–2009; Ravussin et al. 2001b, 2010). Clutch size was largely determined by the abundance of main food animals in the field, as the yearly mean clutch size was positively associated with abundance indices of *Microtus* and bank voles in the current spring (Fig. 8.2). In good vole years approximately 5–8 eggs were laid, whereas in poor vole years there were only 3–5 eggs. This high inter-annual variation in clutch size was not due to the fact that only those females capable of laying large clutches bred in good vole years, because individual females were very flexible in laying. In good years, a female was able to lay 6–8-egg clutches, while in the next poor vole year only 3–5-egg clutches were laid, and vice versa (Korpimäki 1990b). During the whole study period (1973–2009), the smallest clutch comprised only one egg (1 nest), while in the largest clutch, ten eggs (2 nests) were recorded. The most common clutch size was six eggs (433 nests) (Fig. 8.3), followed by five-egg clutches (326), seven-egg clutches (256), four-egg clutches (154), eight-egg

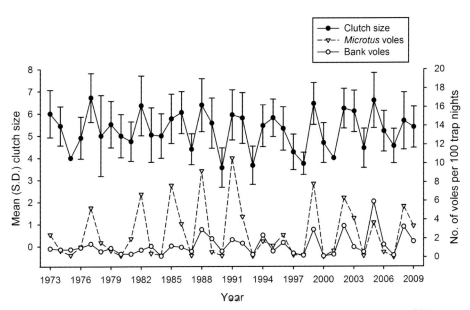

Fig. 8.2.   Annual mean (s.d.) clutch size of boreal owls (left y-axis) and trap indices (right y-axis) of *Microtus* and bank voles in the current spring during 1973–2009 in our study area (Spearman rank correlation, $r_s = 0.84$, $n = 37$, two-tailed $p < 0.001$ for clutch size and *Microtus* voles and $r_s = 0.73$, $n = 37$, $p < 0.001$ for clutch size and bank voles; the correlations for the trap indices of *Microtus* voles were $r_s = 0.49$, $n = 36$, $p = 0.002$ and for bank voles in the preceding autumn were $r_s = 0.51$, $n = 36$, $p = 0.002$) (data from Korpimäki and Hakkarainen 1991 and unpublished).

clutches (60), three-egg clutches (31), nine-egg clutches (14) and two-egg clutches (8). There was no obvious long-term trend in the yearly mean laying dates and clutch sizes during 1973–2009 (Fig. 8.1, Fig. 8.2). In addition, in the short-term data from 1980 to 1986 in northern Sweden, the yearly mean laying date correlated negatively with the vole abundance index of the preceding spring, while the annual mean clutch size was positively correlated with the vole abundance index of the current spring (Hörnfeldt et al. 1990).

The overall mean number of young hatched per nest in our study population during 1973–2009 was 4.95 (± 1.87) and that of fledglings 2.91 (± 2.08) (Table 8.1). Thus, on average 13.3% of eggs did not hatch, mainly because they remained unfertilised. An additional 41.2% of hatchlings were not able to fledge, mainly because of starvation during the nestling period (Table 8.1). The annual mean number of hatchlings per nest varied from 2.5 (1998) to 5.8 (1977 and 1982) and that of fledglings from 1.0 (1980) to 4.3 (1974) during 1973–2009. Both the annual mean number of hatchlings and fledglings per nest varied in close accordance with the abundance indices of *Microtus* and bank voles in the current spring (Fig. 8.4, Fig. 8.5). In good vole years, usually 4–6 fledglings were produced per successful nest, whereas in poor vole years the fledgling production was only half of that (Fig. 8.6). Therefore, large variation in the clutch size also resulted in a high variation in fledgling production depending on vole abundance. There were no obvious long-term trends in yearly mean hatchling and fledgling production during 1973–2009 (Fig. 8.4, Fig. 8.5).

Fig. 8.3.   The most common clutch size of boreal owls was six eggs, of which four have already hatched in this photograph. Photo: Erkki Korpimäki.

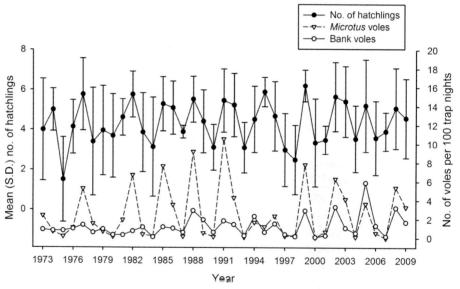

Fig. 8.4.   Annual mean (s.d.) number of hatchlings per nest (left y-axis) and trap indices (right y-axis) of *Microtus* and bank voles in the current spring during 1973–2009 in our study area (Spearman rank correlation, $r_s = 0.84$, $n = 37$, two-tailed $p < 0.001$ for number of hatchlings and *Microtus* voles and $r_s = 0.66$, $n = 37$, $p < 0.001$ for number of hatchlings and bank voles; the correlations for the trap indices of *Microtus* voles were $r_s = 0.36$, $n = 36$, $p = 0.03$ and of bank voles were $r_s = 0.43$, $n = 36$, $p = 0.01$ in the preceding autumn) (data from Korpimäki 1987a and unpublished).

8.1.2.   Vole supply in late winter determines laying date

The poor availability of voles in late winter postponed the start of egg-laying by about 1 month, but vole abundance in May and ambient temperature and snow depth in March–April had only minor effects on the timing of laying (Korpimäki and Hakkarainen 1991).

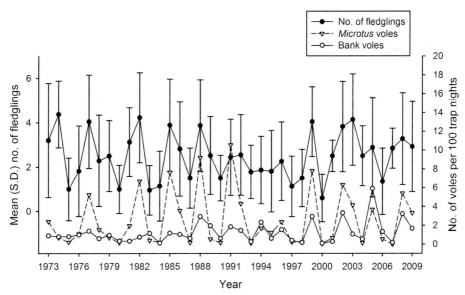

Fig. 8.5. Annual mean (s.d.) number of fledglings per nest (left y axis) and trap indices (right y-axis) of *Microtus* and bank voles in the current spring during 1973–2009 in our study area (Spearman rank correlation, $r_s = 0.72$, $n = 37$, two-tailed $p < 0.001$ for number of fledglings and *Microtus* voles and $r_s = 0.47$, $n = 37$, $p = 0.004$ for number of hatchlings and bank voles; the correlations for the trap indices of *Microtus* voles were $r_s = 0.12$, $n = 36$, $p = 0.47$ and for bank voles they were $r_s = 0.26$, $n = 36$, $p = 0.13$ in the preceding autumn) (data from Korpimäki 1987a and unpublished).

Fig. 8.6. It was very unusual that seven young were about to fledge at the age of approx. 4 weeks. Photo: Pertti Malinen.

In addition, the yearly mean body mass of breeding female boreal owls was positively correlated with the abundance of voles (Chapter 8.1.1). This indicates that food supply before the breeding season is the most important factor in determining the laying date via the body condition of females. Good correlative evidence for the causal chain of food–body condition–laying date is also available for other birds of prey (Newton et al. 1983, Hirons et al. 1984, Village 1986, Dijkstra et al. 1988).

The 'food limitation' hypothesis proposes that egg-laying begins as soon as the female accumulates enough body reserves to form eggs (Perrins 1970). We interpreted the negative relationship between yearly mean laying dates of boreal owls and abundances of voles as evidence for this hypothesis. This result suggests that food supply is a proximate limiting factor for laying females. However, our evidence in support of the food limitation hypothesis does not necessarily refute the hypothesis that delaying clutch initiation in poor vole years has an ultimate adaptive significance (see Murphy and Haukioja 1986). For example, poor vole supply in early spring may 'trigger' the start of egg-laying of owl females to coincide with the time that voles begin to reproduce and migratory songbirds arrive in the study area. Thus, young voles and birds are available when young owls demand most food.

The food limitation hypothesis predicts that the provision of surplus food early in the breeding season should advance the onset of egg-laying. A surplus feeding experiment prior to and during the egg-laying of ten boreal owl pairs was conducted in a decline vole year (1986) (Korpimäki 1989a). Food-supplemented pairs advanced their laying by an average of 6 days and laid 0.9 more eggs than did non-food-supplemented control pairs. However, it remains questionable whether supplemental feeding affected clutch size independently of laying date. Therefore we reanalysed the data by choosing non-supplemented controls from all the pairs breeding in 1986 so that laying dates were close (± 2 days) to food-supplemented pairs and that quality of nest-sites was similar. We found that fed pairs laid significantly larger clutches than unfed control pairs: 6.8 ± 0.9 vs. 5.9 ± 0.9 (laying date in March 23.7 ± 4.7 vs. 23.8 ± 4.7) (Korpimäki and Hakkarainen 1991). This was consistent with the prediction of the above-mentioned hypothesis. In addition, late-breeding female boreal owls were food-supplemented prior to and during the laying periods in 1985 in northern Sweden (Hörnfeldt and Eklund 1990). Food-supplemented females weighed more, laid eggs 8 days earlier and produced one more egg than non-supplemented controls. This experiment was, however, constrained by the fact that only late pairs were food-supplemented and that it remained uncertain whether fed females laid more eggs than unfed control females laying at the same time. Our results are consistent with a supplementary feeding experiment with Eurasian kestrels in our study area, in which food-supplemented females both advanced their start of egg-laying and laid more eggs than could be predicted on the basis of earlier laying dates (Korpimäki and Wiehn 1998). In any case, we have both observational and experimental evidence that vole supply in late winter, before the breeding season, determines the laying date of boreal owls.

8.1.3.   Spring vole supply determines clutch size independent of laying date

Food abundance could influence the amount of reserves in the body of females before the breeding season, which in turn influences both laying date and clutch size (the 'capital model'), or the rate of accumulation of body reserves in females during the laying period (the 'income model') (Drent and Daan 1980). That winter vole supply was better correlated with laying date, but spring vole supply with clutch size of boreal owls, seemed to be in agreement with the 'income' model, because this model predicts

Table 8.2. Mean (s.d.) clutch size of boreal owls in relation to laying date (1 = 1 April, 2 = 2 April, etc.) in the different phases (low, increase, decrease) of the 3-year vole cycle. Pooled data from 1980–9, n = number of nests (Korpimäki and Hakkarainen 1991). A t-test (two-tailed) was used for statistical comparisons of mean clutch size between low and increase as well as increase and decrease phases of the vole cycle.

Laying date	Low Mean	n	Increase Mean	n	Decrease Mean	n	All phases Mean	n
−37–−28					5.5 (1.4)	20	5.5 (1.4)	20
−29–−20	5.0 (0.0)	1	6.5 (0.7)	2	5.6 (1.5)	36	5.7 (1.5)	39
−19–−10			6.8 (1.2)	23***	5.7 (1.0)	65	6.0 (1.2)	88
−9–0	4.0 (1.0)	3***	6.4 (1.1)	44	6.1 (0.9)	49	6.2 (1.1)	96
1–10	5.4 (0.8)	13*	6.2 (1.2)	37*	5.6 (1.0)	44	5.8 (1.1)	94
11–20	4.9 (0.9)	16°	5.5 (1.2)	19°	4.8 (0.9)	11	5.2 (1.0)	46
21–30	4.0 (0.8)	10***	6.4 (1.0)	12**	4.3 (1.2)	3	5.2 (1.5)	25
31–40	4.7 (0.5)	6**	6.4 (0.9)	5	5.0 (0.0)	1	5.5 (1.3)	12
41–50	5.0 (0.0)	2	5.9 (0.9)	9			5.7 (0.9)	11
51–60	5.0 (0.0)	1	5.3 (1.9)	4			5.2 (1.6)	5
61–70			5.0 (0.0)	1			5.0 (0.0)	1

*** $p < 0.001$, ** $p < 0.01$, * $p < 0.05$, ° $p < 0.10$.

independent effects of food on laying date and clutch size. Far larger clutches produced in the increase phase of the vole cycle than in the decrease and low phases during the same laying periods support this prediction (Table 8.2).

In addition, the results from our feeding experiment (Korpimäki and Hakkarainen 1991) revealed that clutch size can be increased independently of laying date. We suggest that female boreal owls are too small to accumulate large internal reserves (i.e. 'capital') and thus respond to an 'incoming' amount of food during egg formation and laying. This suggestion is further supported by the fact that females lose only 10–15 g from pre-laying and laying to the early incubation period (Korpimäki 1990a; see also Hörnfeldt and Eklund 1990), although the weight of a 5–6-egg clutch is around 60–75 g (Glutz von Blotzheim and Bauer 1980). Moreover, vole supply during the egg-laying period is a better indicator of the food situation during the fledging and post-fledging periods than is vole supply prior to laying, as vole densities crash even during the breeding season in the decrease phase of the vole cycle. Thus, we suggest that it is more adaptive to adjust the clutch size to present food conditions, rather than to food conditions 1–2 months before the start of egg-laying.

8.1.4.   Seasonal trend in clutch size: food limitation or adaptation?

The clutch size of most nidicolous single-brooded bird species in temperate areas decreases continuously from the start to the end of the breeding season (Klomp 1970, Martin T. 1987). A similar pattern is usually found between years: clutch size is larger in years when laying starts early than when it starts late (see, e.g., Newton and Marquiss 1984, Korpimäki 1987f, Pietiäinen 1989 for birds of prey). But when we initiated our studies on the fecundity of boreal owls, extensive field data collected in animals living under fluctuating food conditions was mostly lacking. Under such conditions, where abundance of main food either increases, declines or remains low in the course of the breeding season, it is possible to examine the role of food abundance in determining a declining seasonal trend in clutch size.

Several hypotheses have been proposed for the seasonal decline in clutch size, and here we concentrate on the two major ones. First, differences in quality between individual birds and/or their territories may result in a seasonal decline in clutch size (the quality hypothesis). Birds may lay late and produce small clutches due to their poor nutritional condition resulting from youth, inexperience or poor territory quality, whereas pairs in better condition breed early and produce larger clutches (Perrins 1970, von Haartman 1971, Newton and Marquiss 1984, Korpimäki and Hakkarainen 1991).

Second, the seasonal decline in clutch size may be adaptive, so that the clutch size that maximises reproductive output is larger for early-breeding pairs than for late-breeding ones (e.g. Klomp 1970). Accordingly, a decreasing trend in clutch size over the breeding season could arise whenever the reproductive value of an egg declines with laying date, because the survival prospects of hatched young decrease with a later laying date. Bird parents face a territory-quality-specific gain in condition as the season advances, but territories differ in their initial food supply or the rate of gain in condition, which happens at an earlier date in high-quality territories (Rowe et al. 1994).

We found a clear relationship between the mean annual laying dates and clutch sizes of boreal owls: in early years, clutches averaged approx. three eggs more than in late years (Fig. 8.7). A similar close relationship was also found for the mean annual number of hatchlings and fledglings produced per nest during 1973–2009 ($r = -0.62$, $n = 37$, $p < 0.001$ and $r = -0.44$, $n = 37$, $p = 0.006$, respectively). Our detailed analyses on seasonal trends in clutch size in the three phases of the vole cycle (for further details, see Korpimäki and Hakkarainen 1991) showed that in the decrease phase of the vole cycle,

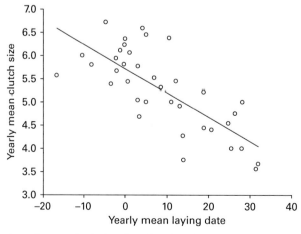

Fig. 8.7.   Annual mean clutch size plotted against the annual mean laying date during 1973–2009 (linear regression, $F = 44.3$, df $= 36$, $p < 0.001$, $r^2 = 0.56$) (data from Korpimäki and Hakkarainen 1991 and unpublished).

female body mass and clutch size declined steeply with later laying date, but only slowly (but significantly) in the increase phase of the cycle. In the low phase, such a negative relationship was not found. These trends also held when the effects of nest-site quality, female age and male age were ruled out. Nearly similar seasonal trends were also found for mean egg size of clutches in these three phases of the vole cycle (Hakkarainen and Korpimäki 1994b).

As far as we know, it has not often been documented that the seasonal trend in clutch size is independent of the age of breeding pairs and the quality of nest-sites. Seasonal declines in clutch size occur independently of female and male age in Eurasian kestrels (Korpimäki and Wiehn 1998), independently of female experience in Ural owls (Pietiäinen 1988), and independently of female age and habitat quality in sparrowhawks (Newton and Marquiss 1984). However, the breeding success of birds of prey is mostly dependent on the food collected by the male, as he is the main food provider (e.g. Newton 1979, Korpimäki 1981). Thus it was important to rule out possible effects of male age and habitat quality when studying seasonal trends in clutch size of boreal owls. Our results clearly showed that these trends remained when the effects of habitat quality, female age and male age were removed.

If some owl females always produce early and large clutches and others late and small ones, the relationship between clutch size and laying date could be due to inter-individual variation. Repeatabilities of laying dates and clutch sizes in boreal owls were very low in comparison with other birds (Korpimäki 1990b; also see Chapter 8.2), showing that the between-year variation in laying dates and clutch sizes of the owl population reflects intra-individual phenotypic plasticity, rather than changes in the genotypic composition of the population.

Our results were consistent with the idea that between-year variations in laying dates and clutch sizes were due to food supply acting as a proximate limit on egg production. When considering the seasonal trends in clutch size during different phases of the vole cycle, the simplest explanation again might be that it was due to food limitation. The abrupt decline in clutch size with later laying date in the decrease phase, the slight decline in the increase phase and the quite constant clutch size in the low phase were probably associated with the similar seasonal changes in the body mass of females. Hence, it seems that food supply determines laying date and clutch size via body condition of females. However, the following two results suggest that the seasonal trends may also have an adaptive significance.

First, clutch sizes of pairs laying at the same time were consistently larger in the increase than in the decrease phase of the vole cycle (Table 8.2), though vole supply during egg-laying and in the preceding winter was better in the decrease than in the increase phase. This suggests that the owls invest more in a clutch in the increase phase, as survival of yearlings is better at that time (Korpimäki and Lagerström 1988; and see Chapter 10.3). Though the availability of voles in the decrease phase is good prior to and during egg-laying, the survival chances of the young, especially the late-born young, are poor because voles are scarce during the independence period of juveniles (Korpimäki and Lagerström 1988). Thus, we suggest that selection for early breeding is strongest in the decrease phase, when vole abundances peak in the preceding autumn, and owls should start to lay as early as possible, even at the expense of clutch size. This results in steeply reducing clutch sizes with progressing laying dates in the decrease phase, though the vole supply is still quite good at the time of egg-laying. These ideas are consistent with the hypothesis that both laying date and food supply should influence the reproductive value of eggs under fluctuating food conditions, as the survival of yearlings markedly changes during the course of the 3-year vole cycle (Korpimäki and Lagerström 1988). Therefore, laying date along with food availability may have an important role in determining clutch size, which in turn strongly determines the number of fledglings produced.

Second, there were only minor differences in clutch size between the low and decrease phases during the same laying periods, although the vole supply during egg-laying and in the preceding winter was poor in the low phase and good in the decrease phase. We interpret this to be adaptive, so that the owls invest more in offspring in the low than in the decrease phase. The reproductive value of an egg is probably higher in the low phase, as vole numbers begin to recover in late summer, i.e. during the independence period of juveniles, which is crucial for their first-year survival. Considering the cyclic nature of vole populations, a selective advantage could be expected for those boreal owls that invest most in reproduction in the increase phase of the cycle (Chapter 12), as the contribution they can make to the future gene pool of the population is highest at that time. Although clutch size is not necessarily a good estimate of reproductive effort, our results indicate that boreal owls behave in accordance with this suggestion, investing the most in a clutch in the increase phase of the cycle.

To conclude, we suggest that the combination of laying date and clutch size maximising fitness may be complicated under the fluctuating food conditions experienced by boreal owls. The reproductive value of eggs decreases steeply within the season in the decrease phase, but remains relatively stable, or decreases only slightly, with laying date during the low and increase phases of the vole cycle. Also, in Ural owls living in fluctuating food conditions, the seasonal decline in clutch size was twice as steep in the decrease phase (0.1 eggs per day) as in the increase phase (0.05 eggs per day). Mean recruitment probability of Ural owl chicks to the local breeding population was 2–3 times higher in the increase than in the decrease phase of the cycle (Brommer et al. 2002a). Thus, in variable environments where first-year survival prospects are relatively predictable at the time of egg-laying, food supply determines both laying date and clutch size, independently of each other. The results that support this suggestion have also been obtained for Eurasian kestrels in our study area (Korpimäki and Wiehn 1998), while data from Eurasian kestrels in the Netherlands indicate that spring vole supply determines laying date, which in turn determines clutch size (Meijer et al. 1988). In temperate Europe, vole-eating birds of prey live in an area where vole populations are more or less stable between years but show predictable seasonal fluctuations, with low numbers in spring and high numbers in autumn. However, steep summer declines in vole abundance do not usually occur (Chapter 1; see also Korpimäki et al. 2005b). This means that no obvious between-year differences in the first-year survival of young birds of prey may happen, but the reproductive value of eggs decreases with laying date due to the declining probabilities of survival in the first year of life (Daan and Dijkstra 1988).

The future challenge is to study laying date–clutch size relationships and recruitment probabilities of boreal owl chicks in these mainly seasonally fluctuating food conditions in order to make comparisons with our results.

8.1.5. Prey abundance within territory is associated with breeding success

In addition to temporal variation, spatial variation in prey abundance may also influence reproductive success. Hence, animals should prefer the habitat that will confer the greatest fitness. Many approaches have been used to assess habitat quality. Most of these are based on indirect measures such as the frequency of breeding attempts and/or breeding success (review in Sergio and Newton 2003), or habitat characteristics (see Chapter 3.2). Studies with direct measurements of habitat quality, however, are scarce, probably because it is generally difficult to estimate resources accurately in the field, usually covering large areas. In the boreal owl, however, studies on habitat quality are possible because its diet composition can reliably be estimated during the breeding season (Chapter 3.6). Furthermore, its most important prey groups can be quantified quite accurately in the field by snap-trappings of voles and point counts of birds (Chapter 3.7).

We classified territories based on breeding frequency in an earlier 10-year period (with a mean value of two breeding attempts per territory) as poor if there had been 0–2 breeding attempts, or good if there had been >3 breeding attempts. In general, more voles (genera *Microtus* and *Myodes*) were snap-trapped on good than on poor territories in the increase and decrease phases of the vole cycle, but not in the low vole years (Table 8.3).

Table 8.3. Mean (± s.d.) number of small mammal prey per 100 trap-nights in poor and good home ranges of boreal owls. Pooled data from the low vole year 1990, the increase vole years 1991 and 1994, and the decrease vole years 1989 and 1992. n = number of home ranges (data from Hakkarainen et al. 1997a).

	Habitat quality				Mann–Whitney U-test	
	Poor	n	Good	n	U	p
Low vole years						
Pooled no. of voles	1.5 (2.0)	14	2.2 (2.5)	18	108.5	0.46
No. of bank voles	1.3 (2.0)	14	1.5 (2.4)	18	122.5	0.88
No. of *Microtus* voles	0.2 (0.8)	14	0.7 (1.7)	18	113.5	0.41
No. of shrews	2.1 (3.2)	14	2.2 (3.4)	18	131.0	0.83
Increase vole years						
Pooled no. of voles	8.7 (7.9)	19	14.5 (8.1)	24	129.5	0.02
No. of bank voles	2.9 (3.2)	19	5.9 (5.3)	24	149.0	0.05
No. of *Microtus* voles	5.8 (6.7)	19	8.6 (5.8)	24	154.5	0.07
No. of shrews	1.2 (2.2)	19	4.8 (5.9)	24	147.5	0.03
Decrease vole years						
Pooled no. of voles	8.8 (5.8)	23	16.7 (15.8)	29	190.0	0.008
No. of bank voles	5.7 (5.1)	23	9.1 (9.5)	29	273.0	0.26
No. of *Microtus* voles	3.1 (3.1)	23	7.6 (10.2)	29	267.5	0.21
No. of shrews	3.1 (3.5)	23	4.9 (6.9)	29	288.0	0.39

Table 8.4. Linear regression of the number of fledglings, clutch size and laying date as dependent variables on the most important prey types of boreal owls snap-trapped in individual home ranges. Pooled data from the years 1992 and 1994 (Hakkarainen et al. 1997a). Laying date and all independent variables were log-transformed because of non-normal distribution. Regression equations are based on original data.

Independent variable	Regression equation	F	R^2	df	p
Number of fledglings					
Pooled no. of voles	$y = 0.10x + 0.26$	4.73	14	29	0.04
No. of bank voles	$y = 0.17x + 0.55$	8.10	22	29	0.008
No. of *Microtus* voles	$y = 0.02x + 1.88$	0.01	0	29	0.94
No. of shrews	$y = 0.20x + 1.27$	2.71	9	29	0.11
Clutch size					
Pooled no. of voles	$y = 0.02x + 5.17$	0.40	1	29	0.53
No. of bank voles	$y = 0.01x + 5.34$	0.86	3	29	0.36
No. of *Microtus* voles	$y = 0.02x + 5.31$	1.26	4	29	0.27
No. of shrews	$y = -0.01x + 5.45$	0.02	0	29	0.89
Laying date					
Pooled no. of voles	$y = 0.00x + 4.64$	0.47	2	29	0.50
No. of bank voles	$y = 0.00x + 4.58$	0.83	3	29	0.37
No. of *Microtus* voles	$y = 0.00x + 4.63$	1.29	4	29	0.27
No. of shrews	$y = -0.01x + 4.63$	0.99	3	29	0.33

The number of fledglings produced, but neither clutch size nor laying date, was associated with the number of voles snap-trapped in individual territories (Table 8.4). This suggests that the significance of habitat quality becomes apparent especially in the late nestling phase, when the energy expenditure of hunting males seems to increase with the growth of the offspring. In accordance with this, in the decrease vole year, males delivered more prey items in good territories than in poor ones, independently of male age (see Hakkarainen et al. 1997a).

We also counted the abundance of alternative prey, small passerine birds within owl territories. The abundance of small birds was higher on good than on poor territories. In the low vole year, more small birds were censused near occupied than unoccupied nest-boxes, which suggests that in poor food conditions owls may shift to luxuriant sites with abundant alternative prey within their territories (Hakkarainen et al. 1997a).

Our results suggest that main and alternative prey may have different importance over time in determining habitat quality, which in turn affects fledgling production. Therefore, the habitat choice of boreal owls may have direct fitness consequences, as the number of fledglings produced in individual breeding attempts largely determines the lifetime reproductive success of male owls (see Chapter 8.7).

Finally, we also studied the effects of landscape composition (measured by land use and forest resource data using a geographic information system, GIS) on the breeding success of boreal owls at five different spatial scales (250–4000 m) around the nests during two consecutive 3-year population cycles of voles (Hakkarainen et al. 2003). Landscape composition had the strongest effects on owl breeding success in the decrease phase of vole cycles. Significant variation in owl breeding occurred along the productivity gradient from farmland-predominated areas to barren hinterland. Owls tended to produce earlier clutches on territories predominated by agricultural areas in increasing vole years. A similar trend was found in the decreasing phase of the vole cycle; owls breeding on the barren hinterland seemed to start egg-laying later than owls breeding near agricultural areas. Surprisingly, nestling survival and fledgling production in the decreasing phase declined steeply with increasing proportion of farmland. Clutch size was not significantly related to landscape composition. During the declining years of vole abundance, nestling survival increased from western farmland areas towards the eastern outlying district. These results indicated a sudden summer decline of vole populations on farmland-predominated habitats (Hakkarainen et al. 2003). This was probably because the number of vole-eating predators, and hence their impact on vole populations, was apparently higher in farmland areas than on forested hinterland (Korpimäki and Norrdahl 1998). This finding gives support for the 'spill-over' hypothesis, which states that predators and their exploitation tends to 'spill over' from luxuriant habitats to barren habitats (Oksanen T. 1990).

8.1.6. Geographical variation in clutch and brood sizes

An old maxim of life-history theory is that the clutch size of birds usually increases with distance from the equator (Lack 1966). This was also true for the mean clutch sizes of boreal owl populations in Europe (Table 8.5), because the mean clutch size increased with latitude ($r_s = 0.79$, $n = 16$, $p < 0.001$). Mean clutch sizes of north European populations (>60°N) were one or two eggs larger than in central and southern Europe. There was also a significant tendency that mean brood sizes at fledging increased towards the north in Europe ($r_s = 0.55$, $n = 15$, $p = 0.03$). The main reason for the lower correlation between brood size at fledging and latitude might be that these data also included failed nesting attempts, of which a considerable proportion is usually attributable to marten predation.

Lack (1966) stated that the increase in clutch size of mainly passerine birds towards the north is due to the longer days, which increase food provisioning time. This does not, however, apply to night-active boreal owls because their food provision time is markedly shorter in northern Europe than in central Europe (Klaus et al. 1975, Korpimäki 1981, Zárybnická et al. 2009). The increase in the availability of the main foods of owls (mainly rodents) towards the north is probably the main reason for an increasing latitudinal trend in clutch size. This was also assumed to be the main reason for the geographical trend in clutch size of six North American owl species, including the saw-whet owl (Murray 1976). In northern areas, voles are most readily available after snowmelt in spring before the growth of new vegetation.

Data on the clutch size and number of fledglings of boreal owls in North America has remained limited, because the owl breeding densities are in most cases low and the finding of nests in natural cavities is not easy. Egg-laying started only in late April to May in Idaho (Hayward et al. 1993). This was apparently attributable to males having difficulty with courtship feeding in harsh environmental conditions with deep snow cover, and females were slow to gain enough weight for egg production. Moreover, clutch size and number of fledglings produced over 5 years were on average extremely low (Table 8.5) and nesting

Table 8.5. Mean clutch size and number of fledglings of boreal owls in various study areas of Europe and North America. n = number of nests.

Study area	Clutch size		No. of fledglings[a]		
	Mean	n	Mean	n	Source
Finland, Kokkola region (64°N)	5.8	1799[b]	3.4	1799[b]	1
Finland, Kauhava region (63°N)	5.7	1351	2.9	1417	2
Finland, Pirkanmaa (61°30′N)	5.3	610	3.4	780	3
Finland, South Häme (61°N)	5.6	110	4.6	198	4
Sweden (64°N)	5.6	97			5
Sweden (58°N)	5.0	7	1.4	10	6
Estonia (59°N)	5.3	11	4.9	7	7
Czech Republic (51°N)	5.5	145	1.7	169	8
Germany (51°30′N)	4.5	290	2.6	290	9
Germany (51°N)	4.8	52	3.7	52	10
Germany (49°N)	4.6	38	2.4	38	11
Germany (48°N)	4.3	30	1.7	35	12
Germany (47°N)	4.7	65	2.3	65	13
Belgium (51°N)	5.0	5	2.7	5	14
Switzerland (47°N)	5.1	420	2.6	420	15
Italy (46°N)	3.7	24	2.4	24	16
USA, Alaska (65°N)	4.9	111	3.3	111	17
USA, Idaho (46°N)	3.1	11	0.9	16	18

[a] Number of fledglings per nesting attempt.
[b] Total number of nests found in the study area.
Source: 1. Vikström (2009), 2. Table 8.1 of this book, 3. Lagerström (1980), 4. Linkola and Myllymäki (1969), 5. Hörnfeldt and Eklund (1990), 6. Norberg (1964), 7. Randla (1976), Maasikamäe (1978), 8. Zárybnická (2009b) and unpublished data, 9. Franz et al. (1984), 10. Ritter et al. (1978), 11. Schäffer et al. (1991), 12. König (1969), 13. Meyer H. et al. (1998), 14. Scheuren (1970), 15. Ravussin et al. (2001a), 16. Mezzavilla et al. (1994), 17. Whitman (2010), 18. Hayward et al. (1993).

failure rate was high: 10 of 16 nesting attempts failed. All these values are only comparable to the values for poor vole years in European boreal owl populations, suggesting that the breeding habitat of boreal owls at high elevations in the Rocky Mountains was poor. Considerably higher clutch sizes and numbers of fledglings produced have been recorded at lower elevations in the interior of Alaska, where both values were at the level of European populations, although this area has deep snow cover in spring which tended to delay the start of egg-laying and thus decrease clutch size (Whitman 2010).

8.2. Repeatability and heritability in clutch size and laying date

Most studies have revealed high heritability estimates for laying date and clutch size of birds (30–40%; see Boag and van Noordwijk 1987). Hence, females in consecutive breeding attempts are prone to lay clutches of rather similar sizes and lay eggs about the same time within the breeding season. High heritability of reproductive traits may be beneficial if it is predicted that offspring who inherit parental characteristics will encounter environments similar to those that their parents have encountered. This concerns most bird species living under relatively stable environmental conditions. In the boreal owl, however, owing to large temporal variations in the quantity and quality of food, offspring often experience different environmental conditions from their parents; in such cases, even the opposite characteristics from those of their parents may be useful. Accordingly, our long-term data collected under largely varying food conditions showed that both the repeatability indices (a measure of similarity of an individual's trait between consecutive breeding attempts) and heritability estimates of laying date and clutch size were low. The repeatability estimates of laying dates and clutch sizes (range −0.21 to 0.15; Korpimäki 1990b) of individual females and males breeding in different years did not differ significantly from zero. Our data collected during 1979–93 showed that the heritability estimated between mother's and daughter's laying dates was non-significant. However, the mother–daughter regression for clutch size indicated almost significant heritability ($h^2 = 0.56$, s.e. $= 0.35$, df $= 19$, $p = 0.08$; Hakkarainen et al. 1996c). This suggests that there may be some heritable variation in the clutch size.

There seems to be a high phenotypic plasticity in these fitness-related traits of boreal owls, which means that an individual female owl is able to lay a clutch with 7–8 eggs in a good vole year but a clutch with only 3–4 eggs in a poor vole year. Boreal owls are also able to considerably modify the initiation of breeding: in good vole years they advance their breeding by as much as 1 month compared to poor vole years. This may be beneficial, especially under varying environmental conditions, when parents have to adjust reproductive output parallel to future changes in their main food abundance (see Chapters 8.1.4 and 12.2).

8.3. Incubation and hatching asynchrony

Many bird species start to incubate prior to clutch completion, which results in asynchronous hatching of the young (Stoleson and Beissinger 1995). Boreal owl females lay eggs at 48-hour intervals, and start to incubate after laying their second egg, which results in highly asynchronous hatching of eggs. Incubation of the first egg takes on average 29.2 days and that of the last egg 26.6 days, and the eggs hatch in the order in which they were laid (Korpimäki 1981).

It has been widely debated whether there is any advantage in producing offspring of different sizes rather than of similar size, but no consensus has emerged. Several hypotheses have been proposed to explain this phenomenon, most of them assuming that hatching asynchrony is an adaptive trait (e.g. Magrath 1990, Stoleson and Beissinger 1997). We tested one of the main hypotheses on this issue, which could be appropriate for the boreal owl living under varying environmental conditions. One of the main adaptive roles of hatching asynchrony is based on the idea of an optimal clutch and brood size in altricial species (Lack 1947): laying too many eggs would result in food shortage during the nestling period and, as a consequence, the nestlings would be underweight and have poor survival. Laying too few eggs, on the other hand, would result in a reduced number of offspring. Between these two extremes, there should be an optimal clutch size that is ultimately regulated by the available food conditions (Lack 1947, Magrath 1990). Lack (1954) proposed that asynchronous hatching could be an adaptation to unpredictable changes in food supply, because if food declines abruptly during the nestling period the youngest chicks would die first, without endangering the survival of the whole brood (the 'brood reduction' hypothesis). By contrast, synchronous hatching of young would make them equally competitive, which in case of food shortage could result in poor growth of the whole brood (Magrath 1990).

Birds of prey, and owls in particular, are traditionally considered classical examples of species exhibiting extreme hatching asynchrony (Mikkola 1983, Wiebe and Bortolotti 1994, Stoleson and Beissinger 1995). Yet, paradoxically, there had been virtually no detailed long-term studies on hatching asynchrony in owls when we began our study on hatching spans of boreal owls. We believe that when Lack (1947, 1954) put forward his brood reduction hypothesis, he was probably aware of extreme hatching asynchrony of owl broods. Hatching asynchrony has been comprehensively studied in diurnal raptors including, for instance, American kestrels *Falco sparverius* (Wiebe and Bortolotti 1994), Eurasian kestrels (Wiebe et al. 1998, Wiehn et al. 2000) and black kites *Milvus migrans* (Viñuela 1999).

Difference in wing chord length between the largest and the smallest nestling divided by mean wing chord length of the brood was used as an index of hatching asynchrony in boreal owls (see Wiebe et al. 1998). We measured wing lengths of nestlings on average 2–3 days after the hatching of the last egg. During the 14-year study period, full records of hatching asynchrony were obtained from 79 nests. Hatching spans averaged 6–7 days (range 0–13 days) and increased with clutch size (Valkama et al. 2002). In large broods the size difference between the oldest and youngest nestling can be very large, with a maximum age difference of 2 weeks (Fig. 8.8), although the whole nestling period covers only 4.5 weeks. In comparison, hatching spans of Eurasian kestrels in the same study area

Fig. 8.8.   In large broods, such as in this seven-chick brood, the age difference between the oldest and youngest chick can be as much as 2 weeks. Photo: Benjam Pöntinen.

averaged only 2–3 days (range 0–10 days), and those of burrowing owls *Athene cunicularia* broods in the grasslands of Saskatchewan, Canada, averaged only 4 days (range 1–7 days) (Wellicome 2005). On average, 93% of the eggs of boreal owls hatched (hatching failures being most common during low vole years) and 58% of these hatchlings survived until fledging. Chick mortality was mainly due to starvation.

We found that hatching spans of boreal owl chicks were longest in nests with the largest clutch sizes (Valkama et al. 2002). In large broods, the size difference between the oldest and youngest nestling can be approx. 2 weeks. This was quite expected, as boreal owl females usually start to incubate after laying the second egg (Korpimäki 1981), and therefore large clutches inevitably result in long hatching spans. Interestingly, food abundance (i.e. the phase of the vole cycle) did not directly influence hatching patterns, but it closely correlated with clutch size (Chapter 8.1.1). Accordingly, burrowing owl pairs given extra food during the laying period had hatching spans equal to those of non-supplemented control pairs (Wellicome 2005). Thus it seems that food abundance determines clutch size of boreal owls, which in turn is the main determinant of the degree of hatching asynchrony.

Broods of yearling boreal owl females were significantly more synchronous than those of older females (mean hatching span 5.1 vs. 6.5 days) and hatching was also most synchronous in broods in which both parents were yearlings. This may indicate that young and inexperienced individuals started incubation of the clutch later relative to laying of the first egg, and/or that their laying intervals were shorter than those of experienced females. There was also a significant interaction between male age and nest-site quality: broods of yearling males were more asynchronous on poor (mean hatching span 8.1 days) than on good (5.7 days) sites, whereas among adult males the level of asynchrony did not differ with respect to habitat quality (6.4 vs. 6.1 days). Male age may be important because males provision females prior to and during incubation and during the first 3 weeks of the chick-rearing period. Adult males breeding on poor sites may have a better knowledge of their home ranges as a result of potential earlier breeding attempts in the same area and thus have a higher hunting success. By contrast, yearling males are probably less experienced hunters and may therefore not be able to provision their mates adequately. Similarly, an interaction between female age and male age indicated that broods cared for by yearling parents were significantly less asynchronous than broods in which at least one of the parents was older (Valkama et al. 2002).

According to Lack's (1954) brood reduction hypothesis, we predicted that brood reduction (i.e. chick mortality) should be highest during the low, and especially during the decreasing vole years, when prey availability is permanently low or declines abruptly at the time when the owls are breeding. Our main results on hatching asynchrony are summarised in Fig. 8.9. During the decrease phase of the vole cycle, the mortality among broods increased with hatching span, indicating that brood reduction was most common among asynchronous broods when food became scarce during the breeding season of owls. At that time, the survival prospects of young owls are likely to be poor (Korpimäki and Lagerström 1988; and Chapter 10.2). These findings were in accordance with our prediction: when food becomes scarce during the nestling period, the youngest nestling would die first without endangering the survival of the whole brood. Interestingly, during

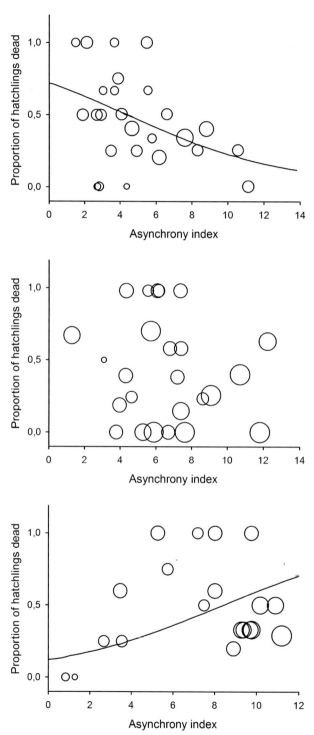

Fig. 8.9.   Relationship between chick mortality (number of dead chicks/number of hatchlings) and the level of hatching asynchrony during the low (upper panel, no. of nests = 26), increase (middle panel, n = 25) and decrease (lower panel, n = 20) phases of the vole cycle. Significance levels between mortality and asynchrony were as follows: low phase: log-likelihood χ^2 = 7.11, p = 0.01; increase phase: log-likelihood χ^2 = 1.35, p = 0.25; decrease phase: log-likelihood χ^2 = 5.70, p = 0.012 (Valkama et al. 2002).

the low phase of the vole cycle, the brood mortality appeared to decrease with hatching asynchrony. It is possible that during extremely poor food conditions, parent owls had difficulty in finding enough food for chicks that had similar energetic demands at the same time. In the increase vole years, however, with improving food conditions, brood mortality was unrelated to hatching asynchrony. At that time, owing to good food conditions and low sibling competition, all offspring probably survive relatively well independently of their body size. A proper test of Lack's (1954) hypothesis would require (1) simultaneous manipulation of food abundance and hatching patterns, as has been done, for example, in Eurasian kestrels (Wiehn et al. 2000), and (2) following juvenile survival after fledging (Magrath 1990). However, it has previously been shown that in the boreal owl there is a strong correlation between fledging success and recruitment rate of offspring (Korpimäki 1992b). It is possible that the degree of facultative manipulation of hatching asynchrony may depend on the clutch size of the species. For example, American kestrels, which have relatively small and fixed clutch sizes (4–5 eggs), can probably fine-tune their investment by manipulating hatching spans (Wiebe and Bortolotti 1994), whereas boreal owls, which have large and widely varying clutch sizes (1–10 eggs; Chapter 8.1) seem to manipulate clutch size instead. Interestingly, in Ural owls, timing of the onset of incubation, which is the reason for hatching asynchrony, varied across females and was moderately repeatable (25%), indicating that this trait may be heritable in this larger owl species (Kontiainen et al. 2010).

8.4.   Parental age and breeding performance

In long-lived animal species, breeding performance may vary considerably with age (e.g. Clutton-Brock 1988, Stearns 1992). Also, in many bird species, breeding performance improves with age in the early years of life and reaches a maximum level in middle age (e.g. Newton 1989, Saether 1990, Forslund and Pärt 1995, Martin T. 1995). Some long-term studies have also found a decline in breeding performance during the later years of the breeding span, which is often attributed to senescence (e.g. Newton and Rothery 1997, Ratcliffe et al. 1998). Two main mechanisms have been proposed to account for the changes in breeding performance of long-lived animals with age (review in Forslund and Pärt 1995).

First, the *differential mortality hypothesis* suggests the progressive disappearance or appearance of different-quality individuals within an age cohort at the population level. Individuals with poor breeding performance are more prone to die or disperse before the next breeding season (Smith J. 1981, Nol and Smith J. 1987), which increases the proportion of high-quality individuals in the breeding population. Accordingly, when proportionally more poor-quality than high-quality individuals disappear between consecutive breeding seasons, the mean reproductive performance within a cohort improves in successive years.

Second, the *constraint hypothesis* proposes that some within-individual abilities or skills may improve with age and maturation. For example, foraging efficiency may improve with experience or learning, leading to better breeding performance at an older age (e.g. Catry and Furness 1999). In addition, accumulated breeding experience with age may also lead to improved breeding success at older age (e.g. Newton et al. 1981, Ratcliffe et al. 1998). An important aspect that has been considered in only a few studies is the influence of environmental variation on the occurrence of age-related differences in breeding performance. The prediction is that age will affect breeding success less in good, and more in poor, environmental conditions (Sydeman et al. 1991, Ratcliffe et al. 1998). We used our long-term data to investigate age-related changes in laying date and clutch size of boreal owls at both the population and individual levels at different phases of the vole cycle (Korpimäki 1988c, Laaksonen et al. 2002).

In years of low vole abundance, only a few yearling boreal owl females and no males were able to breed. In years of increasing vole abundance, clutch size increased with female age (Fig. 8.10) and partners of older males initiated egg-laying earlier than those of yearling males (Fig. 8.11). In years of initially moderate to high – but decreasing – vole abundance, age-related differences in breeding performance tended to disappear (Figs. 8.10, 8.11). There were no obvious differences in clutch sizes between the partners of different-aged males. Hence, environmental variation may mask age-related differences in breeding performance at the population level, since the differences appeared to emerge more clearly in poor and intermediate food conditions. These results were in accordance with the few studies that have examined age-related breeding performance in relation to estimated environmental variation: differences become more evident in demanding conditions (e.g. Ratcliffe et al. 1998).

Inconsistent with the differential mortality hypothesis, age-related differences in breeding performance were independent of male quality, because analyses at the

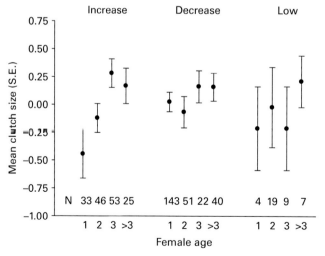

Fig. 8.10.   Standardised mean (± s.e.) clutch sizes of different-aged boreal owl females in the increase, decrease and low phases of the vole cycle. Negative values indicate smaller (and positive values larger) clutch size than yearly average of the population (N = number of clutches) (Laaksonen et al. 2002).

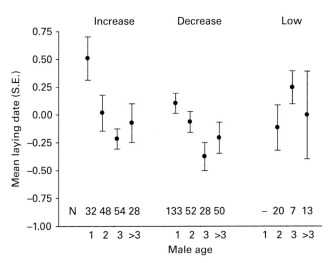

Fig. 8.11.   Standardised mean (± s.e.) laying dates of partners of different-aged boreal owl males in the increase, decrease and low phases of the vole cycle. Negative values indicate earlier (and positive values later) than yearly average laying date of the population (N = number of clutches) (Laaksonen et al. 2002).

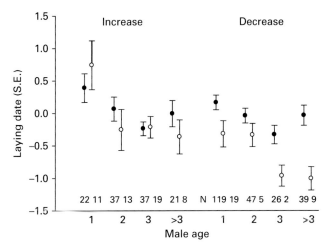

Fig. 8.12.   Standardised mean (± s.e.) laying dates of partners of males known to survive (circles) and assumed to die (dots) after their breeding in the increase and decrease phases of the vole cycle. Negative values indicate earlier (and positive values later) than yearly average laying date of the population (N = number of clutches) (Laaksonen et al. 2002).

population level did not indicate quality differences between surviving and dying males of the same age in years of increasing vole abundance (Fig. 8.12). In contrast, within-cohort analyses in years of decreasing vole abundance showed that early-breeding males were more likely to survive than late-breeding males (Fig. 8.12), supporting the idea that quality differences between individuals may induce the observed age-related differences in only some years. The relatively good conditions at the early stages of breeding in the decrease vole years may have given a breeding opportunity to most individuals, whereas in poor or intermediate conditions the lower-quality individuals were not able to breed or survive. Thus in intermediate food conditions there may not have been detectable selection on breeding performance and parental quality, because there were no quality differences between breeding individuals of the same age (Fig. 8.12). This indicates that quality differences between individuals may induce observed age-related differences only in some years, and especially that differential mortality occurs in a different year, when the difference in breeding performance is detectable. Therefore, we consider that differential mortality alone is not an adequate explanation for age-related differences in the breeding performance of boreal owls. The differential mortality hypothesis seemed to be valid only in years when the difference in breeding performance was detectable. Some earlier studies have also found support for the differential mortality hypothesis (Nol and Smith J. 1987, Gehlbach 1989, Espie et al. 2000). However, these studies have not considered the possible effect of yearly variation in food abundance on the occurrence of selection.

Some evidence, however, was found for the constraint hypothesis. The within-individual analyses showed that the yearling males beginning late in the increase phase of the vole cycle advanced their clutch initiation to the same level as others in the next breeding season, which shows that they were not consistently late breeders (Fig. 8.13). Furthermore, female age had a positive association with clutch size in the increase phase of the vole cycle, and there was a within-individual trend for clutch size to increase with female age (Laaksonen et al. 2002). The poor reproductive performance of yearling owls, as well the within-individual improvements in breeding performance could thus be explained by constraints in foraging skills. Young males may have had more difficulty in providing food for their mates during the courtship period compared with older males.

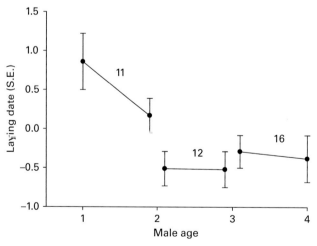

Fig. 8.13.   Within-individual changes in the standardized mean (± s.e.) laying dates of partners of individual owl males that first bred as yearlings in the increase year and then in the next decrease year of the vole cycle (on the left). The corresponding values for 2-year-old and 3-year-old males in the same phases of the vole cycle are given in the middle and on the right, respectively. The number of males in each group is shown above the lines (Laaksonen et al. 2002).

This finding is supported by the fact that young males collect smaller prey caches than older males in their nest-holes (Korpimäki 1988c). Also, in other bird of prey species where males provide food for their mate and brood, male age is expected to be more important than female age in determining breeding success (Newton 1979, Taylor I. 1994, Korpimäki and Wiehn 1998). Clutch size, however, increased with female age independent of male age (Laaksonen et al. 2002), which suggests that these age-related differences had already appeared before the courtship feeding period. Therefore, we suggest that the initial body condition of younger females has probably been low, owing to their less efficient foraging.

It has to be taken into account that although the poor performance of yearling boreal owls of both sexes (Figs. 8.10, 8.11) in the low and increase vole years could be explained by constraints in foraging and other skills, it could also be the outcome of the young males refraining from reproductive effort: young individuals may invest more in their future breeding potential rather than in their current breeding performance. To separate the roles of different mechanisms in explaining age-related differences in breeding performance, an experimental approach – for example, supplementary feeding at nests of various-aged parent owls – should be used, although it is difficult to perform such experimental studies at the appropriate spatio-temporal scale.

Within-individual analyses showed that individual boreal owl males advanced their nest initiation with age (Fig. 8.13), and there was a trend for females to lay larger clutches with increasing age (Laaksonen et al. 2002). In nests of long-lived males, however, a significant decline was found in clutch size from middle age (2–4 years) to older age (6–10 years), and this decline was not due to later clutch initiation by the partners of older males (Fig. 8.14). Possible explanations could be that the old males were restricted in their ability to feed the female for egg formation, or that the old males were no longer

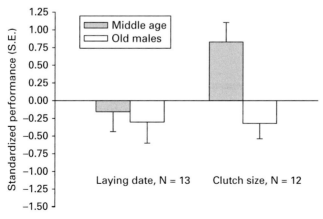

Fig. 8.14.   Mean (± s.e.) laying dates and clutch sizes of partners of owl males that were breeding at middle age (2–4 years, grey columns) and then again at an older age (6–10 years; open columns). Negative standardised laying dates mean that egg-laying began before (and positive values mean that laying began after) the annual mean laying date of the study population. For clutch size, negative values indicate lower (and positive values higher) than the annual mean performance (N = number of long-lived males) (Laaksonen et al. 2002).

attractive and could only obtain low-quality females. Some evidence for the differential mortality hypothesis was found for males, which are mainly responsible for food provision to their families during the entire breeding season. Substantial improvements in competence or increased reproductive effort from 1 to 3 years of age, and deteriorating skills due to senescence at older ages may account for within-individual changes in breeding performance of males. Future work should take an experimental approach to distinguish between different hypotheses and the mechanisms behind them. Furthermore, data on age-specific survival is also essential in order to determine whether reproductive effort and future breeding prospects change over an individual's lifetime, having the potential to improve breeding performance with age.

8.5.   Is reproduction costly?

The hypothesis for the costs of reproduction (e.g. Williams 1966) suggests that animals allocate resources between their own maintenance, growth and survival, and reproductive investment to maximise reproductive output. This is often achieved by allocating as much to current reproduction as possible without an essential decrease in future fecundity and survival. Hence, one of the major issues on which decisions must be taken, is how much time and energy should be allocated to current reproduction without endangering one's own survival and body condition, because too large an investment in offspring may decrease the future breeding potential, health and body condition of parents. Costs of reproduction may appear, for example, in terms of a lowered immuno-defence system or weakened physiological condition. A large brood size probably increases parental effort in terms of the provision of food and defence of the offspring. All this increases parental activity and apparently results in higher predation risk compared with non-breeding owls or parents that do not invest so much in breeding. In boreal owls, finding the 'right' amount of investment is probably complicated by large spatial and temporal variation in their main food resources. For example, too high an investment in reproduction in the decrease years of vole abundance apparently leads to high costs, because the crash of vole populations generally happens during the late breeding season of owls. To avoid such an excessive action, owls should adjust their breeding performance in parallel with future changes in their main food abundance and survival prospects of offspring, which will be discussed in more detail in Chapter 12. In this chapter, we concentrate on the costs of reproduction, especially from the parental point of view.

In natural conditions, fecundity can be a poor estimate of reproductive effort. For example, fecundity is generally highly correlated with individual phenotypic variation (e.g. age, experience) or differences in territory quality, which leads to problems in determining causal relationships on the costs of reproduction. For example, does high fledgling production reflect high reproductive effort of parents, or simply their high territorial or parental quality? Hence, one shortcoming of studies on the costs of reproduction has been that reproductive costs have been studied by relating current offspring production to the future survival of parents or by relating the current and future reproductive success to each other. To overcome these problems concerning causal relationships, experimental manipulations of reproductive effort are probably needed in order to control for these factors. Because most birds accept transplanted chicks and start to feed them as if they were their own, it is relatively easy to manipulate clutch or brood sizes (i.e. the workload of parents), which gives an opportunity to directly manipulate the breeding performance of parents. More than 60 studies have manipulated clutch or brood sizes in birds, but less than half of these report the effects of manipulation on survival and reproductive performance of parents in the next potential breeding season (Roff 1992, 2002). In particular, studies using long-term measures of reproductive output, such as lifetime reproductive success, to evaluate the impacts of brood manipulation experiments are rare. Hence, we need long-term studies on resident and relatively long-lived species, such as the boreal owl.

8.5.1.   Brood size manipulations

The costs of reproduction should be especially apparent in male boreal owls, because they are mostly responsible for feeding the family from courtship until the fledglings are independent. Because males are site-tenacious in this species after the first breeding attempt, it is possible to estimate their lifetime reproductive success (Korpimäki 1992h). We experimentally altered brood size of the first breeding attempt of male owls and determined their subsequent lifetime reproductive success. Furthermore, because we also have data on the recruitment probability of offspring, we also examined how brood size manipulations were associated with lifetime number of recruits produced. Both of these measures, especially the latter, are good proxies for the long-term genetic contribution of an individual to a population (Brommer et al. 2004b).

In this chapter we examine whether male boreal owls incur short- or long-term costs, or even gain benefits, from experimentally enlarged or reduced brood sizes. The costs of reproduction could arise in two ways. First, if intragenerational trade-offs exist, parental survival and/or future fecundity should be decreased. Second, if intergenerational trade-offs exist, nestling survival and recruitment rates into the breeding population should be decreased.

Brood size manipulations at 135 nests were conducted in four breeding seasons, of which two were characterised by increasing vole abundance (1985 and 1988) and two (1986 and 1989) by decreasing vole abundance (see Fig. 1.1). For brood size manipulations, nest trios with similar laying dates and clutch sizes were chosen. Within nest trios, one newly hatched young was transferred to another nest, while the third nest of the trio served as a control. Nests were randomly selected for each treatment group. Accordingly, we established two treatment groups with reduced (−1) and enlarged (+1) broods, and one control group in which we did not change the brood size (Korpimäki 1988a). This mimics the realistic situation in which a parent has to decide to lay one more or less egg. Each nest was visited in the mid-nestling phase and just before fledging to ring parents and their offspring and to measure their body mass and wing length. The number of fledglings produced was ensured by controlling for dead nestlings in the nest-box after the breeding season. Thereafter, the fate of individually marked parents and their recruits in each brood size manipulation group (reduced, increased, control) was followed until the next potential breeding season. We measured short-term reproductive success (e.g. nestling and fledgling number) and long-term fitness proxies (fledgling and recruit production during the lifetime) of male owls. The treatment groups differed in chick number for at least half of the nestling period, but the final fledgling numbers did not show differences between the treatment groups.

8.5.2.   No obvious costs of reproduction

Our correlative data showed that, although male boreal owls collect nearly all the food for their females and young in the breeding season, there was no obvious association in the brood size of males between one year and the following year. Furthermore, males rearing large broods survived no worse into the next year than those rearing small broods. The raising of large broods did not reduce the number of primaries moulted by males either. Accordingly, the proportion of males surviving and breeding in the next year was similar in the three groups that had reared reduced, control or enlarged broods in the preceding year. The enlargement of the brood did not affect breeding performance in the next year (Korpimäki 1988a and unpublished data). Broods with one experimentally added newly hatched young suffered from a higher nestling mortality in the years of declining vole abundance, but not in the years of increasing vole abundance (Korpimäki 1988a). This suggests that the amount of food available to the egg-laying female is the critical factor limiting clutch size in the increase phase of the vole cycle, but in the decline phase the food amount during the nestling period was even more limiting. However, in both phases of the vole cycle, the young of experimentally enlarged broods achieved slightly lower body mass at fledging than did young from reduced broods. This indicates that their phenotypic quality and therefore later survival, might be lowered (Korpimäki 1988a).

Males that were more than 1 year old produced significantly fewer recruits in the brood reduction than in the brood enlargement group. Among yearling males the production of recruits did not differ between the treatment groups. This suggests either that old males show higher current reproductive effort or that young males have restrained effort in the manipulation year. In this context, however, further studies are needed.

8.5.3.   Habitat quality drives costs of reproduction?

The main finding from our brood size manipulations was that fledgling condition in enlarged broods was lower when reared in spatially low-quality home ranges character- ised by a high proportion of agricultural fields (Fig. 8.15). These habitats have been found to decrease fledgling production especially in the decrease phase of the vole cycle. This probably results from a sudden summer decline in vole populations on farmland- predominated habitats. During that time, the number of vole-eating predators, and hence their impact on vole populations, is apparently higher in farmland areas than in the forested hinterland (for further details, see Hakkarainen et al. 2003). This finding suggests that a trade-off in offspring quality can be found especially for enlarged broods reared in low-quality habitats.

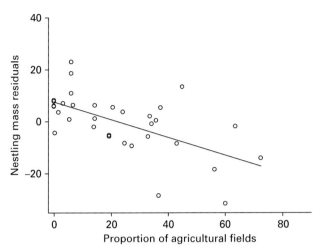

Fig. 8.15.   Residual body mass of fledglings of boreal owls against the proportion of agricultural fields in the territory of the enlarged broods (Korpimäki 1988a and unpublished data).

In conclusion, our results suggest that although huge between-year variation in vole abundance has a major effect on all phases of the boreal owl life cycle, brood size manipulations performed during both the increase and decrease phases of the vole cycle did not reveal any obvious short- or longer-term reproductive costs measured by nestling number or condition, lifetime fledgling or recruit production. This suggests that males responsible for enlarged and reduced nests did not markedly modify their parental effort as a response to brood size manipulations and/or that superior males may easily compensate for the large parental investment by increasing their food intake without risking their future survival. Only enlarged broods reared in poor-quality home ranges, which were dominated by agricultural fields, produced fledglings with poor body condition. This suggests that the minor costs of reproduction are due to survival costs of offspring and are mostly mediated by habitat quality in boreal owls.

8.6.   Detrimental effects of parasites on reproductive effort and success

Parasites living in the blood of birds may reduce reproductive success and fitness (Atkinson and Van Riper 1991, Korpimäki et al. 1993), and may even induce mortality, in conjunction with other debilitating conditions (review in Møller et al. 2009). The parasite loading of an individual may be modified by factors such as energy allocation during reproduction (Deerenberg et al. 1997), food supply and resource levels (e.g. Wiehn and Korpimäki 1998, Appleby et al. 1999), as well as investment in hunting and reproductive effort (Wiehn et al. 1999). As the maintenance of immune function seems to be energetically and nutritionally costly (e.g. Lochmiller and Deerenberg 2000), investment in immunological response may occur at the expense of reduced investment in other functions. High parental effort can increase host susceptibility to parasites, and

Table 8.6. Prevalence (percentage of infected individuals) of parasites in the blood of boreal owls in 1993–5 (50 females and 65 females, data from Ilmonen et al. 1999).

Blood parasite	Females	Males
Trypanosomes *Trypanosoma avium*	47	64
Leucocytozoids *Leucocytozoon ziemanni*	100	94
Haemoproteids *Haemoproteus noctuae/syrnii*	12	8

parasites may thus also mediate the costs of reproduction. This results mostly from the fact that both reproduction and immunodefences are energetically demanding and may hence lead to a trade-off between allocation of resources to reproduction and to immunity (e.g. Keymer and Read 1991, Toft 1991, Råberg et al. 2000). The magnitude of reproductive costs may be attributed to annual variation in environmental conditions (Schaffer 1974, Nur 1988). In fact, the costs of reproduction may emerge only under limited food resources, and not if resource levels can be elevated (Tuomi et al. 1983). Although annual variation in food resource levels has the potential to modify the costs of reproduction along with parasite effects, such studies on host–parasite effects are still scarce. In this context, boreal owls are good study objects on host–parasite interactions because they subsist on voles, whose numbers show wide between-year variation.

In the boreal owl, the most common blood parasites were intracellular haemosporidians of the genera *Haemoproteus*, *Leucocytozoon* and extracellular haemoflagellates of the genus *Trypanosoma* (Table 8.6; see also Korpimäki et al. 1993). Ornithophilic culicine mosquitoes, simulid and hippoboscid flies, and dermanyssic mites (poultry mites of the species *Dermanyssus gallinae* and related mites) are the most important vectors of these parasites.

Habitat characteristics may define the degree of exposure of birds to pathogens and parasites. Ornithophilic species of black flies (Diptera: Simuliidae) are vectors of several species of pathogenic avian leucocytozoans, the protozoans causing avian malaria (Desser and Bennett 1993). In addition, blood extraction by black flies alone, or in concert with the blood parasite, may considerably contribute to anaemia of the host (Hunter et al. 1997). The larvae and pupae of black flies live in aquatic environments requiring running water for their development (Jamnaback 1973). Adult forest-dwelling black flies mostly have a vertically stratified distribution, and many are most abundant at mid-canopy level, where the abundance of forest birds may be highest (Rohner et al. 2000). Some works on birds, including forest-dwelling owls, have indeed found an association between habitat characteristics and the prevalence of blood parasites (Rohner et al. 2000, Galeotti and Sacchi 2003, De Neve et al. 2007).

We examined intracellular haemosporidians from a drop of blood which was collected in a microcapillary tube. The drop of blood was smeared onto a clean glass slide, air-dried, and fixed in absolute ethanol some hours later (Bennet 1970). The avian haematozoans found in blood smears of boreal owl parents were *Haemoproteus noctuae* (3% of females and 2% of males infected), *H. syrnii* (prevalence 10% and 1%), *Leucocytozoon ziemanni* (97% and 94%) and *Plasmodium circumflexum* (0% and 1%) (Korpimäki et al. 1993).

Flagellated trypanosomes, which circulate in the bloodstream, were quantified by means of the haematocrit centrifuge method (Woo 1970). In a poor vole year (1993), most (>80%) of the breeding parent owls trapped in mid-nestling phase were infected with trypanosomes, while in good vole years (1994–5) the prevalence was lower (approx. 30–70%; Fig. 8.16). The prevalences of leucocytozoids and trypanosomes in the blood of forest-dwelling boreal owls were considerably higher than the corresponding prevalences in the blood of vole-eating open-country Eurasian kestrels co-existing partly in the same study sites (figure 2 in Wiehn et al. 1999). This difference is probably due to the fact that the abundance of vectors, in particular black flies, is considerably higher in coniferous forests

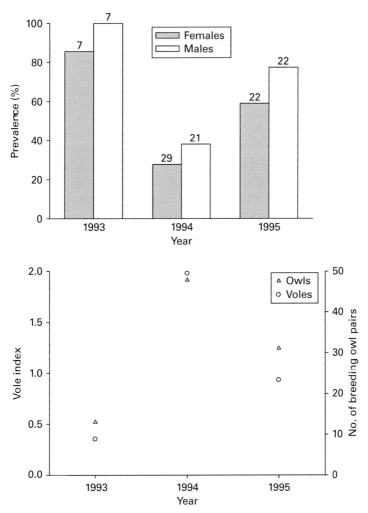

Fig. 8.16. Prevalence (% of infected individuals) of *Trypanosoma avium* blood parasites in female (grey bars) and male (white bars) parents of boreal owls during 1993–5 (upper panel). Sample sizes are shown above the bars. Vole indices (pooled number of *Microtus* and bank voles snap-trapped per 100 trap-nights) and the number of breeding boreal owl pairs recorded in the whole study area in 1993, 1994 and 1995 (lower panel) (Ilmonen et al. 1999).

Table 8.7. Prevalence (% of individuals infected) of blood parasites among parent boreal owls at food-supplemented (S) and unfed control (C) nests during the nestling period in 1996–7 (data from Ilmonen et al. 1999).

Parasite	Females		Males	
	S(11)	C(11)	S(11)	C(10)
Trypanosomes	36	82	73	80
Leucocytozoids	100	100	91	90
Haemoproteids	9	0	9	20

than in open country. This is consistent with the suggestion that habitat characteristics at least partly define the degree of exposure to pathogens and parasites. This interpretation is also supported by the finding of parasites in the blood of little owls in arid southern Portugal where *Leucocytozoon ziemanni* was the only relatively frequent (approx. 40% prevalence) haematozoan blood parasite of breeding parents (Tomé et al. 2005).

The prevalences of leucocytozoids and trypanosomes in boreal owls were much lower in years of main food abundance than in years of scarcity of main food. This result was consistent with that for vole-eating Eurasian kestrels co-existing in our study area (figure 2 in Wiehn et al. 1999). It seems that annually fluctuating food resource levels (i.e. voles) influence the susceptibility of vole-eating predators to blood parasites. To examine this experimentally, we provided extra food (rooster chicks; for further details see Ilmonen et al. 1999), from incubation onwards, in owl nests during two years of relatively low natural food availability. Food supplementation did not influence the prevalence of leucocytozoids and haemoproteids in the blood of parent owls. In accordance with our predictions, trypanosome prevalence was lower among food-supplemented than unfed control female owls, whereas in males such a relationship was not found (Table 8.7). These results showed that females invested a proportion of the food supplements in parasite clearance. Similar results have also been obtained in food-supplementation experiments on Eurasian kestrels under fluctuating food conditions in our study area (Wiehn and Korpimäki 1998).

We examined parasite effects on body condition of boreal owls under varying food conditions. In a moderate vole year, 1995, trypanosome-infected females were in poorer body condition than uninfected ones (*t*-test, $t = 2.2$, df $= 21$, $p = 0.04$), but this difference was not found in the good vole year, 1994. In males, body condition did not differ between trypanosome-infected and uninfected individuals in any year. Hence, some evidence for adverse effects of parasites on body condition was found, but only in a moderate vole year and only in females. For other parasite species, no apparent influence on body condition of either gender was found (Ilmonen et al. 1999).

Clutch size was reduced in female boreal owls infected with leucocytozoids at intermediate vole densities during the egg-laying period (in 1991), but not at peak vole densities during egg-laying (in 1992; figure 1 in Korpimäki et al. 1993). Our later analyses did not show any constant or statistically strong relationships between breeding success and parasite load. In the moderate vole year 1995, males which raised heavy fledglings had high intensities of leucocytozoids ($r_s = 0.76$, $n = 12$, $p < 0.05$). In the moderate vole year 1995, females which produced large clutches had higher intensities of trypanosomes during

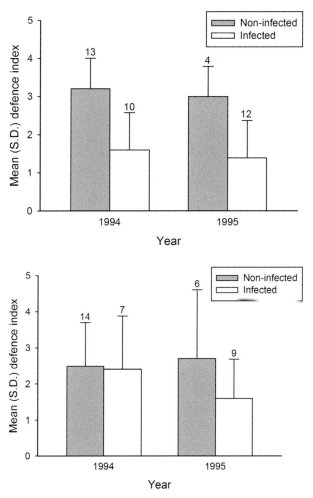

Fig. 8.17.   Responses of *Trypanosoma avium*-infected (white bars) and non-infected (grey bars) male (upper panel) and female (lower panel) boreal owls to a live American mink as measured by the nest defence index (mean ± s.d.) during 1994–5. Sample sizes are shown above the bars (Hakkarainen et al. 1998).

the nestling period ($r_s = 0.77$, $n = 10$, $p < 0.05$). No other significant relationships were found for the other parasite species (Ilmonen et al. 1999). These results are, however, based on a correlative study and a small sample size. Therefore, more studies on host–parasite interactions should be undertaken. In particular, more experimental studies are needed to examine causal relationships in host–parasite interactions in the boreal owl.

    Defending offspring against predators entails a risk of injury or even death. Therefore, the intensity of nest defence has been considered as a good measure of parental effort (for a review, see Montgomerie and Weatherhead 1988). Our nest defence experiments showed that male owls infected with trypanosomes defended their offspring against a live caged American mink *Mustela vison* less vigorously than uninfected males, whereas in females such a relationship was not found (Fig. 8.17). American mink looks like a pine marten, which is the main predator at boreal owl nests in our study area (Korpimäki 1987e).

Our studies on boreal owls showed that blood-parasite infections were observed especially in years of food scarcity, whereas in years of food abundance the prevalence of blood-parasite infections was considerably decreased in females but not in males. Parasitism was sometimes costly, as seen in reduced clutch sizes in female boreal owls infected with leucocytozoids and less vigorous defence of offspring by trypanosome-infected male owls. In general, trypanosomes appeared to have more detrimental effects on reproductive effort and success than did haemoproteids. Our results support the idea that increased parental effort may make avian hosts susceptible to blood-parasite infections, and that the level of natural food supply can considerably modify vulnerability to blood-parasite infections.

Our results strongly suggest that defending against blood parasites may be costly, and that the cost may be modified by food supply. The most noticeable blood-parasite effects in our supplementary feeding experiment were found only in female boreal owls, suggesting that the expected costs may vary between the sexes due to their different parental roles. One explanation may be that breeding female owls are dependent on the food provision of males and are also more exposed to blood-parasite vectors, such as black flies dwelling at mid-canopy level when females incubate and brood in nest-boxes. Consistent with our results on boreal owls, food supplementation in Ural owl nests resulted in better body condition in the female parents, but effects in males remained only moderate, although they delivered less food to food-supplemented nests than to control nests (Brommer et al. 2004a). In the next breeding season following food supplementation, fed Ural owl females bred 1 week earlier and produced on average 0.6 eggs more than control pairs (Brommer et al. 2004a). This difference in the responses of boreal and Ural owls to food supplementation is probably due to the fact that small boreal owl females are mainly 'income' breeders, mostly relying on incoming food of the season delivered by males in maintaining their body condition and producing eggs. In contrast, large Ural owl females are able to accumulate large body reserves and rely on 'capital' collected over as long as 1 year to clear their blood parasites and to lay eggs.

8.7. Fluctuating food abundance determines lifetime reproductive success

Lifetime reproductive success (LRS) is so far the most important determinant of an individual's fitness, which is generally defined as the contribution of an individual's genotype to subsequent generations proportional to that of other individuals in the same population (Endler 1986). In general, LRS is measured by counting the total number of offspring produced during an individual's lifetime. Therefore, studies on LRS are demanding, because the breeding history of all individuals in the breeding population needs to be recorded over several generations. Such studies, however, are advantageous for the following reasons. First, LRS studies improve our understanding of the potential conflict between reproduction and survival (i.e. costs of reproduction; see Chapter 8.5). Second, longitudinal studies provide a better estimate of between-individual variation in breeding success than do cross-sectional studies based on data collection over short periods and from individuals of unknown history (see Newton 1989). For example, if an individual produces 'too many' offspring in one breeding attempt (e.g. three), it may be unable to breed in the following breeding seasons or may even perish owing to the high costs of breeding, whereas an individual investing less by producing only two offspring in a single breeding season, but for two breeding seasons, has a better LRS (i.e. four), and hence better fitness. One advantage of LRS measures is that cross-sectional studies based on one breeding season apparently underestimate the fledgling production of multiple-mated males (i.e. polygynously mated males) and sex-related differences in reproductive output. Furthermore, a poor breeding year (e.g. an abandoned nest) may not necessarily mean low LRS (i.e. fitness) if an individual is able to compensate for the poor breeding success in later breeding seasons. Hence, one-year breeding success is not necessarily a good approximation of fitness. Although LRS can be considered an important estimate of an individual's fitness, it is only during the last two decades that measurements of LRS have become available.

The lack of LRS studies is mostly due to the fact that they need long-term follow-ups of individually marked animals at an appropriate spatial scale. Our long-term population study has fulfilled these basic conditions, allowing us to examine LRS in male boreal owls, which are site-resident after the first breeding attempt (Korpimäki 1987e). Furthermore, our study was long enough to cover the breeding history of long-lived individuals. One shortcoming of studies on LRS has been that they do not necessarily include the complete lifespans of the longest-lived individuals (Newton 1989). We did not study LRS of females because of their long breeding dispersal distances. This is a generally known problem in LRS studies; it is often difficult to keep track of a large number of individuals throughout their breeding lives if they change breeding localities during their lives (Newton 1989). Egg dumping, where females lay in nests of conspecifics, has never been documented in the boreal owl (probably because females start to incubate eggs from the first egg, which leads to high hatching asynchrony in the boreal owl; see Chapter 8.3). Furthermore, extra-pair paternity is rare in our study population (see Chapter 7.3.1). Hence, extra-pair paternity and egg dumping do not have any significant role in determining the LRS of boreal owls. Our data on LRS is based on 141 males trapped from 438 nests during 1979–90 (Korpimäki 1992b). LRS was calculated under the assumption that the male trapped at a nest was the father of all the nestlings.

In this chapter we concentrate on how LRS of male boreal owls is determined, and especially whether fluctuating food conditions (i.e. vole cycles) interact with LRS. This is important to take into account, because much of the variation in annual reproductive success and also in LRS seems to be related to annual variation in food abundance. In addition to annual variation in food abundance, however, spatial variation in LRS should also be considered. Therefore in Chapter 11.3 we also examine whether habitat character-istics are associated with LRS.

LRS was largest in males that entered the breeding population in the increase phase of the vole cycle, followed by males recruiting in the low, peak and decrease phases of the vole cycle. However, vole abundance affected breeding lifespan and brood size differ-ently. Breeding lifespan was longest in males that first bred in the low phase, followed by those recruiting in the increase, decrease and peak phases of the vole cycle (Table 8.8). Mean brood size was largest for males entering the breeding population in the increase phase, and declined through the peak to the low and decrease phases (Table 8.8).

Table 8.8. Duration of the breeding life in years and the number of fledglings produced by male boreal owls making their first breeding attempt in different phases of the vole cycle (from Korpimäki 1992b).

Phase of the vole cycle	Breeding lifespan Mean (s.d.)	Mean brood size over lifetime Mean (s.d.)	n
Decrease	1.7 (1.7)	2.6 (1.1)	13
Low	1.9 (1.1)	2.7 (1.2)	23
Increase	1.5 (0.8)	4.2 (1.7)	44
Peak	1.3 (0.8)	2.9 (2.0)	61
Kruskal–Wallis test (two-tailed)	$H = 9.39$ $p < 0.05$	$H = 16.62$ $p < 0.001$	

These results suggest that the temporal variation in habitat quality due to fluctuating vole abundance was one of the most important environmental determinants of LRS in males. It is beneficial to enter the local population during the low and increase phases of the vole cycle. At that time the number of fledglings produced during a lifetime will be maximised, along with their survival prospects, which is probably worth a high investment. Fledglings produced in the low and increase phases have a higher probability of entering the breeding population than those produced in the peak phase, owing to the crash of vole populations after the peak phase (Chapter 10.3). Accordingly, the mean recruitment probability of Ural owl chicks to the local breeding population in southern Finland was 2–3 times higher in the increase than in the decrease phase of the vole cycle (Brommer et al. 2002a). In female Ural owls, however, LRS was not greater in females which started to breed one year before a peak in voles than in those which started to breed in the peak vole year, and thus before the decline in vole populations (Saurola 1989). This is probably due to their different diet compared with the boreal owl; although Ural owl breeding is highly dependent on voles, it can also rely on alternative prey more efficiently than boreal owls do (Korpimäki and Sulkava S. 1987).

The breeding characteristics of 141 male boreal owls of known lifetime young production are summarised in Table 8.9. This shows that 25% of males entered the breeding population as yearlings, 51% at 2 years old and 24% at >3 years old. The age of first breeding did not affect LRS (mean LRS 5.5, 5.1 and 6.3, respectively). Mean brood size over the lifetime was affected by the mean clutch size of all the partners of males ($r_s = 0.31$, $p < 0.001$), and by breeding success (the percentage of eggs that produced fledglings; $r_s = 0.65$, $p < 0.001$). LRS was highly significantly correlated with the duration of the breeding lifespan ($r_s = 0.68$, $p < 0.01$). In general, offspring survival from egg to fledgling was the most important component of LRS, followed by the lifespan and clutch size. Accordingly, variation in LRS of female Ural owls breeding in southern Finland was mostly explained by breeding lifespan, nest success and clutch size (Brommer et al. 1998). Furthermore, the number of fledglings produced by boreal owl males during their lifetime was highly significantly correlated with their lifetime contribution of known recruits to the breeding population, which further stresses the importance of LRS as an estimate of fitness (Korpimäki 1992b).

Table 8.9. Breeding characteristics of 141 male boreal owls of known lifetime fledgling production (from Korpimäki 1992b).

	Mean	Median	Min.	Max.
Lifespan in years	3.5	3	2	8
Breeding lifespan in years	1.5	1	1	7
No. of breeding attempts	1.6	1	1	9
LRS (no. of eggs)[a]	9.2	6	2	51
LRS (no. of fledglings)[a]	5.2	4	0	26
Clutch size	5.7	6	2	8
Brood size	3.3	3	0	7

[a] LRS = lifetime reproductive success.

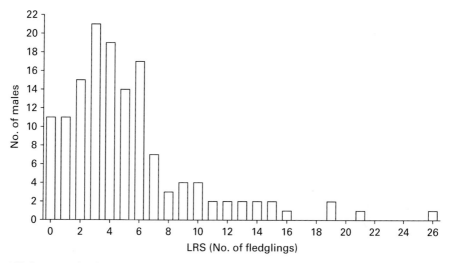

Fig. 8.18. Lifetime reproductive success (number of fledglings produced during their lifetime) of 141 males trapped from 438 nests during 1976–90 (Korpimäki 1992b).

During their lifetime, males spent on average 1.5 years (range 1–7) as breeders and reared on average 5.2 fledglings (range 0–26) (Table 8.9). Eleven out of 141 males (5%) failed to produce any young, although they attempted to breed (this was probably an underestimate, because not all the males whose nests failed in the egg stage were identified). The lifetime productivity peaked at 2–6 young, because most owls bred only once. A few males skewed the distribution of LRS because they were very productive, producing up to 26 fledglings (Fig. 8.18). Actually, 50% of all the fledglings produced in the population were fathered by one-fifth of the males (31 out of 141). This finding is in accordance with other bird studies, showing that only a small fraction of the breeding population produces the majority of gene pool for the next generation (for a review, see Newton 1989). The data based on 281 Finnish ring recoveries of boreal owls found dead made it possible to estimate age-specific survival rates and the proportion of males that are alive and not breeding, using the method of Newton (1985).

These analyses revealed that the reproductive value of male fledglings varied greatly, because most (78%) of the male fledglings died before their first breeding attempt (Korpimäki 1992b). Accordingly, 5% of fledged males in one year produced 50% of the fledglings in the next generation. These figures on male boreal owls correspond well to those for other bird of prey species: the sparrowhawk (Newton 1985, 1989), the osprey *Pandion haliaetus* (Postupalsky 1989), the screech owl *Otus asio* (Gehlbach 1989, 1994), the Ural owl (Saurola 1989, Brommer et al. 1998), the common buzzard *Buteo buteo* (Krüger and Lindström J. 2001), the barn owl (Marti 1997) and the flammulated owl *Otus flammeolus* (Linkhart and Reynolds R. 2006). As far as we know, all of these studies concern female birds of prey, and our LRS results from boreal owls were the first to examine LRS in male birds of prey. Male boreal owls do most of the hunting for their families from prior to egg-laying until the late nestling period and their costs of reproduction are thus assumed to be higher than those of female owls.

LRS was also skewed by the mating system of males. Seventeen out of 141 males (8%) were trapped at two (15) or three (2) nests in at least one year and were defined as polygynous (Chapter 7.3.3). These males produced about twice as many fledglings as monogamous ones (9.1 vs. 4.7) irrespective of territory quality, which did not obviously differ between the two mating systems. Furthermore, the breeding lifespan was longer in polygynously than in monogamously mated males (2.1 vs. 1.4 years, respectively). This finding suggests that polygynous males were probably of high quality (for further details, see Chapter 7.3.4).

In conclusion, one of our main findings was that widely varying food conditions largely determined the LRS of male boreal owls: LRS was largest in males that entered the breeding population in the increasing phase of the vole cycle, whereas the population crash of voles after the peak vole years decreased LRS considerably. Accordingly, the Ural owl females that started breeding in decrease vole years had only half of the LRS of females that started to breed during increasing vole years (Brommer et al. 1998). In tawny owls breeding in an area of low-amplitude vole cycles in northern England, most of the recruits (>75%) were reared in the increase phase of the vole cycle and LRS was highest in individuals beginning their breeding career at that phase (Millon et al. 2010). In accordance with other studies on LRS in birds of prey, our data also revealed high inter-individual variation in LRS. A small proportion of males in the breeding population produced the majority of fledglings for the next generation. Furthermore, polygynously mated males were found to have a considerably higher LRS than monogamously mated males. This is quite evident, because the number of breeding attempts increases LRS.

9   Dispersal and autumn movements

The dispersal behaviour of animals is one of the perennial problems in ecology, as it has consequences not only for individual fitness, but also for population dynamics and the genetic composition of the populations (e.g. Johnson and Gaines 1990, Bowler and Benton 2005, Ronce 2007 with references). Dispersal is usually divided into two categories: natal dispersal, which is the movement between the birth site and first breeding site, and breeding dispersal, which is the movement between successive breeding attempts (e.g. Greenwood 1980, Greenwood and Harvey 1982). Theoretical models on the evolution of dispersal indicate that temporal and spatial variations in environmental quality play a major role in determining dispersal rate and distance. In general, temporal variation tends to increase dispersal, whereas spatial variation seems to act in the opposite way (Johnson and Gaines 1990).

In the following four sections, we first look at sex-related and inter-areal variation in dispersal distances of boreal owls. Next we examine the most important factors, such as food supply and predation risk, that may affect breeding dispersal distances of boreal owls in our study population and elsewhere. Finally we describe the autumn movements of boreal owls that result from long-distance natal and breeding dispersal.

9.1.   Sex-related and interregional variation in dispersal

Finnish ringing data included 83 male and 211 female boreal owls that were ringed as nestlings and later recaptured as breeders elsewhere. As in most other birds, including birds of prey, there were marked intersexual differences in the natal dispersal distances. Females dispersed over longer distances than males between the birth site and first breeding site (Fig. 9.1; median 14 and 62 km, maximum 409 km and 1099 km, respectively). Correspondingly, the median natal dispersal distances of boreal owls ringed in Finland (4443 recoveries of 89 829 owls ringed up to 1999) were 19 km for males (max. 382 km) and 110 km for females (max. 588 km) (Korpimäki et al. 1987, Saurola 2002). The more extensive dispersal in the latter data set is attributable to the fact that dispersal distances were calculated from recaptures at the nest of breeding parents and also from recoveries of birds found dead during the breeding season. These latter recoveries do not reveal whether specimens were breeding or not. In Norwegian data, median natal dispersal distance of males was 6 km (range 5–11 km, $n = 3$) and that of females 10 km (3–239, $n = 9$; Sonerud et al. 1988). The corresponding distances were 4 km for males

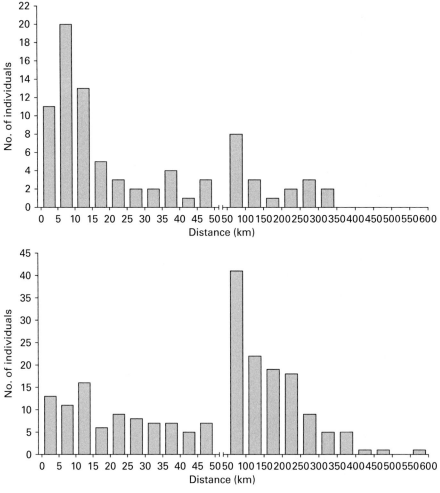

Fig. 9.1.   Natal dispersal distances of 83 males (mean ± s.d. 49.1 ± 78.7 km, median 14 km; upper panel) and 211 females (104.1 ± 103.8 km, median 62 km; lower panel) ringed as nestlings and later recaptured as breeders in Finland during 1973–2009.

(range 0–90 km, $n = 8$) and 30 km for females (4–110 km, $n = 10$) in the data from northern Sweden (Löfgren et al. 1986), and 15 km (0–45 km, $n = 7$) for males and 13 km (0–451 km, $n = 50$) in the German data (Meyer W. 2010). The small sample sizes for the Norwegian, Swedish and German data probably bias the distributions towards shorter distances of natal dispersal.

There were a total of 131 males and 398 females that were ringed as breeding parents at their nests and recaptured as breeders 1 year later in the Finnish data collected in 1973–2008. The mean breeding dispersal distance of males was only 2.7 km (± s.d. 5.9 km, median 1.3 km, max. 28 km) and that of females was 60.9 km (± 102.7 km, median 2.5 km, max. 580 km) (Fig. 9.2). In addition, several cases of long-distance (500–630 km) breeding dispersal are known among female owls in northern Europe (Löfgren et al. 1986,

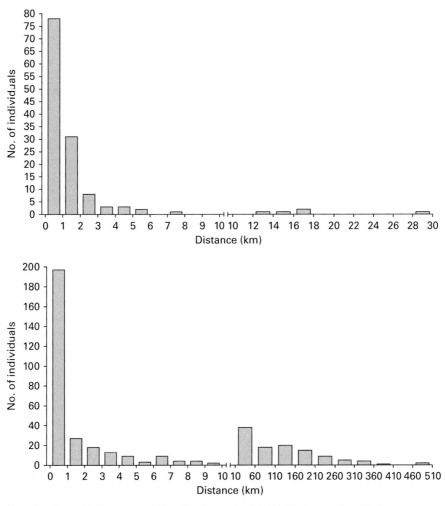

Fig. 9.2. Breeding dispersal distances of 131 males (mean ± s.d. 2.7 ± 5.9 km, median 1.3 km; upper panel) and 398 females (mean ± s.d. 60.9 ± 102.7 km, median 2.5 km; lower panel) ringed and re-trapped as breeding parents in Finland during 1973–2009.

Korpimäki et al. 1987, Sonerud et al. 1988) and even in central Europe (120–200 km; Franz et al. 1984, Schwerdtfeger 1984, Meyer W. 2010). There were no obvious differences in the median breeding dispersal distances between Finnish and Norwegian females (median 2 km, $n = 28$; Sonerud et al. 1988). However, females breeding in our study area tended to disperse more between successive breeding attempts than conspecific females breeding on Harz Mountain, Germany (median 4 km, $n = 72$ vs. 2 km, $n = 140$), while there was no obvious inter-areal difference in the breeding dispersal distances of males (median 1 km, $n = 83$ vs. 1 km, $n = 180$) (data from Schwerdtfeger 1997).

Our finding that female owls dispersed more between successive breeding attempts than males did was consistent with results found for other birds. Reviews have shown that female birds often disperse more often and for longer distances between successive breeding attempts than males do (Greenwood and Harvey 1982, Paradis et al. 1998). One popular explanation for female-biased dispersal is that males are expected to benefit from familiarity with their home range through more efficient use of resources and better defence of the nest-sites (Korpimäki 1987e), or through better territories and higher mating success, resulting in better breeding performance (Newton and Marquiss 1982). Females, in turn, disperse more to acquire better resources, such as mates, nest-holes and territories.

It was evident that both genders of owls, particularly juvenile females, were capable of dispersing over several hundreds of kilometres between their birth and first breeding sites (Fig. 9.1). In addition, a considerable proportion (14%) of adult females dispersed more than 100 km between successive breeding attempts (Fig. 9.2), while the great majority of males were site-tenacious, occupying their home ranges continuously after their first breeding attempt. Extensive dispersal of juveniles and adult females has probably led to a great deal of gene flow between local populations of boreal owls in northern Europe and elsewhere at northern latitudes. This conclusion is supported by the fact that genetic analyses have found no obvious evidence of metapopulation structure of boreal owl populations within North America and Europe (Koopman et al. 2005, 2007a). In future, it could be interesting to study possible genetic differences between local populations in Eurasia.

9.2.   Natal dispersal and food supply

In the pooled data from both sexes, there were marked differences in relation to food supply in the year of birth. In the low phase of the vole cycle, median natal dispersal distance was only 46 km (max. 499 km); in the increase phase it was 118 km (489 km); and in the decline phase 143 km (579 km) (Saurola 2002). However, these analyses were flawed because both sexes were pooled, and because the data set covered the whole of Finland and the classification of the vole cycle phases was not based on the monitoring network of small rodent populations in the field.

Because regional synchrony of population fluctuations of boreal owls, and probably also of voles, covered >250 km (Chapter 13.3), we used only the data from a distance of ≤250 km from our study area to analyse natal dispersal distances in relation to the phases of the vole cycle. This enabled the classification of vole cycle phases based on the bi-annual snap-trapping data from our study area and from the Seinäjoki, Ähtäri and Keuruu regions. We also found that females tended to disperse over longer distances between the birth site and first breeding site in the decrease than in the increase phase of the vole cycle (163.4 ± 175.5 km vs. 86.2 ± 80.0 km, median 117.5 vs. 56.5 km, respectively). The dispersal distances were intermediate in the low vole phase (114.9 ± 111.1 km, 77.5 km). However, the differences between vole cycle phases were not statistically significant. Cycle-phase-related differences seemed even less obvious for natal dispersal of males, although they also tended to disperse more in the decrease than in the increase phase of the vole cycle (53.5 ± 50.6 vs. 41.6 ± 75.0 km, median 33.5 vs. 13 km, respectively). In addition, their dispersal was intermediate in the low phase (48.6 ± 73.2 km, 16.5 km).

Natal dispersal of female boreal owls was twice as extensive in the decrease than in the increase phase of the vole cycle. Therefore, we conclude that juvenile females really do disperse over longer distances in declining than increasing abundances of main foods to search for better resources for over-winter survival and future breeding. Finnish tawny and Ural owls also moved longer distances between birth site and first breeding site when food supply was poor (Saurola 2002), whereas food-abundance related differences in the natal dispersal distances of Scottish barn owls remained undetected (Taylor I. 1994).

9.3. Annual variation in food supply drives breeding dispersal

Animals living in stable environments typically show site fidelity or restricted movements between successive reproductive seasons (Greenwood and Harvey 1982). Most studies on dispersal in birds are consistent with the *resource competition hypothesis* (Greenwood 1980): males tend to be philopatric in order to increase their chances of territory acquisition, while females tend to disperse in order to find the mates and gain the resources necessary to breed successfully (Johnson and Gaines 1990). Among birds nesting in holes or boxes, the nest-hole quality is an important resource that may determine the mating and breeding success of males (e.g. Alatalo et al. 1986, Slagsvold 1986). If the high-quality nest-hole is in a territory with good food resources, the habitat quality may also promote philopatry of males (e.g. Korpimäki 1988d).

Poor breeding performance or breeding failure are usually causes of increased dispersal (e.g. Newton and Marquiss 1982, Pärt and Gustafsson 1989). Low breeding success is usually associated with poor territory or mate quality (Newton et al. 1981, Korpimäki 1988c, 1988d). Nest predation is one of the main factors causing nesting failures. Although hole-nests are relatively safe from predators, high predation rates of eggs, nestlings and incubating adults may still occur, even in large, hole-nesting species (e.g. König 1969, Eriksson 1979, Zang and Kunze 1985, Pöysä and Pöysä 2002). Therefore, a high risk of nest predation may select for breeding dispersal (e.g. Dow and Fredga 1983, Sonerud 1985a).

Several resource competition-related hypotheses (Greenwood 1980) have been put forward to explain the breeding dispersal of boreal owls. The *nest-hole quality* and *predation risk hypotheses* are based on observations that the breeding frequency in nest-boxes decreases with the age of the box. An experimental test of these hypotheses was conducted in our study area by relocating and/or renewing nest-boxes, but we did not find any obvious support for these hypotheses (see Chapter 4.3.3).

A third hypothesis (the *reproductive success hypothesis*) is based on the nesting attempt of the previous year: the owls disperse further after poor breeding success than after good success. This hypothesis predicts that dispersal distances should be negatively related to the reproductive success in the preceding breeding season.

A fourth hypothesis is based on the amount of food in the territory. Data on the vole predation rate of breeding boreal owls indicate that they consume a larger proportion of the available voles in good vole years than in poor ones (Korpimäki and Norrdahl 1989, Chapter 5.9.2). A breeding owl pair may thus substantially reduce the density of voles near the nest-box, so that food abundance is also poor in the next breeding season.

Therefore, it is adaptive to shift nest-sites between successive breeding seasons (Korpimäki 1987e). This *food depletion hypothesis* states that the owls should shift nest-holes less often and over shorter distances in the increase phase than in the decrease phase of the vole cycle. In the former phase, vole densities rapidly recover after the breeding season of owls, whereas in the latter phase they crash abruptly.

We used data from male boreal owls ringed or re-trapped at 529 nests (in 1979–91) and from females at 655 nests (in 1976–91) early in the nestling period to test the food depletion hypothesis (Table 1). These data included a total of 74 males and 24 females that after 1 year were re-trapped as breeders within the study area, and 7 females that were re-trapped as breeders outside the study area (no males were recovered as breeders outside our study area). Dispersal distances of re-trapped males and females were estimated as the linear distance between the nest-boxes that they used in successive breeding seasons. The dispersal of polygynous males was measured as the distance between their primary nests (Chapter 7.3.3). Breeding attempts with two or more years in between were excluded. When comparing the dispersal distances in the different phases of the vole cycle, the pooled data including more than one observation from individual owls were used, though the observations from one owl may not be independent of each other. However, this source of error does not bias the results because the aim was to analyse dispersal distances under different food conditions, rather than inter-individual variation in dispersal behaviour.

Using the pooled data, the median dispersal distance of males was 1.3 km and that of females 5.5 km between successive breeding seasons. Dispersal distances were not obviously correlated with the number of eggs or fledglings (clutch and brood sizes) produced in the preceding breeding season (see table 3 in Korpimäki 1993a), and nesting failure did not appear to increase dispersal. Because we were able to trap breeding owls only during the early nestling period, the identity of owls that failed early in their nesting attempts remained unknown. No significant relationship was apparent between breeding success (proportion of eggs producing fledglings) and dispersal distances of male and female parents.

The median dispersal distance of males was 0.8 km in the vole increase phase, 1.5 km in the decrease phase, and 1.3 km in the low phase. In the increase phase, a majority of males (51%, 22 out of 43) stayed to breed in the same box, while the corresponding figures tended to be smaller in the low phase (31%, 5 out of 16), and significantly smaller in the decrease phase (13%, 2 out of 15) (Fig. 9.3). Differences in dispersal distances of males between the three phases of the cycle were almost significant. This difference was mainly because male dispersal was more extensive in the decrease than in the increase phase of the vole cycle.

Dispersal distances of females differed significantly between the three phases of the vole cycle (Fig. 9.3). In the decrease and low phases, most females moved outside the study area, where the likelihood of recovery was lower than within our study area. Therefore, the data from these phases are scarce. Despite this, the dispersal of recorded females was more extensive in the decrease than in the two other phases of the vole cycle, assuming that the recovery probability outside our study area did not vary with the phase of the vole cycle.

Our findings gave support to the prediction of the food depletion hypothesis (Korpimäki 1987e): male and female owls dispersed over shorter distances in the increase than in the

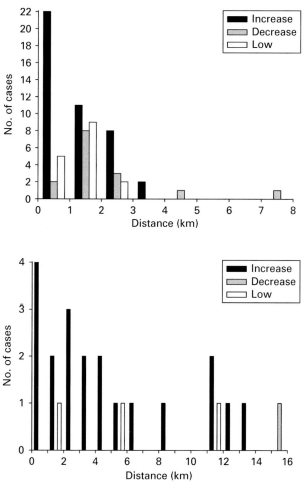

Fig. 9.3.   Breeding dispersal distances of male owls breeding in our study area in the increase (black columns, median 0.8 km), decrease (grey columns, 1.5 km) and low (white columns, 1.3 km) phases of the vole cycle (upper panel). The same but for females in the increase (black columns, median 4.7 km), decrease (grey columns, 125 km), and low (white columns, 5 km) phases of the vole cycle (lower panel). The dispersal distances of females not included in the figure are as follows: increase phase 29, 44, 70, 89 and 176 km, decrease phase 125 and 216 km. Differences in dispersal distances between the three phases of the cycle were almost significant for males (Kruskal–Wallis test, $H_{2,71} = 5.06$, $p = 0.08$) and significant for females ($H_{2,28} = 5.82$, $p = 0.05$) (Korpimäki 1993a).

decrease phase of the vole cycle. In addition, a larger proportion of males bred in the same box in the increase than in the decrease phase of the cycle. These results agree with the theoretical models on the evolution of dispersal, which predict that temporal variation in environmental quality increases dispersal (Johnson and Gaines 1990). Food abundance in the territory was apparently the most important determinant of the dispersal behaviour of male and female boreal owls. When vole populations increased during and after the breeding season of owls, most males bred in the same or nearby box in the next spring.

This was probably because the depleted food supply caused by the hunting male in the vicinity of the nest improved rapidly, and food supply was again good in the next spring. In the decrease phase of the vole cycle, food depletion caused by hunting males probably contributed to the crash of vole populations, reducing further the food supply near the box. Therefore, it was adaptive to shift to breeding in another box. In the low phase, vole populations start to recover slowly in late summer, after the owls' breeding season, so that there is probably no need to disperse as far as in the decrease phase.

Boreal owls are the main avian predators of voles in north European coniferous forests. Male owls mostly hunt within 1 km of the nest (Chapter 4.1.3) and their predation rate of voles is density-dependent (Chapter 5.9.2). Accordingly, mortality due to owls may substantially reduce vole densities and thus select for increased breeding dispersal of the owls between successive breeding attempts. A radio-tracking study on sparrowhawks in Scotland has shown that the better the food supply, the more sedentary hawks become, and the worse the food supply, the more wide-ranging they become (Marquiss and Newton 1982).

The yearly breeding dispersal distances of male owls were associated with the food abundance during autumn and winter. Although males dispersed more in years of food scarcity than in years of food abundance, most of them were still philopatric. The maximum recorded dispersal distance was only 7.5 km, although they could have dispersed up to 30 km within our study area. Accordingly, the study area was probably large enough that male dispersal was not much biased towards short distances owing to a boundary effect (e.g. van Noordwijk 1984). As males can make long trips from their nests to hunt and to attract additional mates (the maximum distance between the nests of polygynous males is >4 km; Chapter 7.3.3), most males would be expected to know the area over which they disperse. If low food supply was likely to stimulate male dispersal regardless of food depletion, one could argue that males, like females, would mostly disperse outside our study area and over longer distances within our study area.

We conclude that philopatry of male boreal owls is probably adaptive, because the costs of travelling and acquiring familiarity with new areas will be avoided. In addition, scarcity of high-quality territories and nest-holes may select for philopatry of male owls, even in the low phase of the vole cycle. Our results suggest that temporal variation in environmental quality is the most important determinant of breeding dispersal in boreal owls with predictable multi-annual fluctuations in food abundance at northern latitudes. In extreme conditions, however, when vole abundances have crashed to very low levels and deep snow cover protects small mammals against avian predators, the only option to survive over winter might be long-distance breeding dispersal. It is likely that yearling males are more inclined to leave their home ranges and nest-sites than older males are.

Our results provide evidence for the suggestion that female dispersal is driven by declining food abundance: on a local scale, females moved over longer distances in the decrease phase than in the other phases of the vole cycle. The breeding dispersal of female owls was also more extensive in autumns of vole scarcity than in those of vole abundance (Korpimäki 1987e). In contrast, the tendency of females to disperse more after a breeding failure or poor breeding success was weak, although this possibility cannot be totally excluded, as many females at nests that failed at the early stages of the season were not trapped and identified. Studies in Sweden and Norway also demonstrated that females dispersed farther between the two vole peaks separated by 3–4 years than between the two successive good vole years (Mysterud 1970, Wallin and Andersson 1981), but this may be due to the positive correlation between the time elapsed from ringing and dispersal distance. Because local vole populations fluctuate in a predictable manner in time and asynchronously between areas, long-distance movements of female boreal owls between successive breeding seasons are probably a response to changes in vole abundance (e.g. Mysterud 1970, Lundberg 1979, Korpimäki 1981, Wallin and Andersson 1981). Many field studies supporting this suggestion have been published both in northern Europe (Löfgren et al. 1986, Korpimäki 1987e, Korpimäki et al. 1987, Sonerud et al. 1988) and in central Europe (Franz et al. 1984, Schwerdtfeger 1984, 1997, 2008, Meyer W. 2010).

9.4.　　Predation risk induces breeding dispersal

Nest predation and its avoidance are critical components of an individual's fitness and play an important role in life-history evolution (Martin T. 1995). Almost all studies on this topic and on other factors possibly affecting breeding dispersal have been observational, and thus have not been able to separate the effects of individual quality, habitat selection and predation risk of given nest-sites from each other. For example, nest-site shift due to predation risk may be difficult to distinguish from nest-site shift caused by food depletion (Chapter 9.3). Nest predation risk may also be associated with other factors, confusing causal interpretations such as individual and habitat quality (Andrén 1995). One problem in dealing with observational data on the presence or absence of predators is that their number and influence may be difficult to estimate in the field.

We simulated a nest predation attempt by a pine marten on the nests of boreal owls by placing a caged American mink *Mustela vison* on the roof of 106 occupied nest-boxes over a period of 6 years (1990–2, 1994–6) at night. Nests without exposure to a mink served as controls. We used a live mink because owls responded to a live mink more actively than to a stuffed pine marten (Chapter 12.3). An American mink also resembles the pine marten in size, colour and movement. In most cases male owls responded to the mink with several warning calls, beak-snaps and pseudo-attacks. Some males showed extremely high activity in the defence of the nest by direct strikes in which the male touched the cage of the mink (Hakkarainen and Korpimäki 1994c). Therefore, we were able to simulate a 'true-to-life' predation danger at the nests, because all males responded to the mink.

Our prediction was that if male owls try to minimise nest predation risk in the following breeding season, the probability of a nest-hole shift should be higher and breeding dispersal distances longer in the treatment group than in the males from the control nests. To exclude the effects of other possible factors which could obscure the interpretation of our results, all the known important background variables, such as parental age, home range and nest-box quality, food abundance (as estimated by the number of prey items stored in nest-boxes), laying date, brood size and nest-box density were similar between the treatment and control groups (Hakkarainen et al. 2001). Male boreal owls are suitable for this study because the choice of nest-site is mostly dependent on the decision of the male. At the beginning of the breeding season, a male occupies a certain nest-hole and tries to attract a female to breed in it by hooting and delivering courtship prey to a suitable nest-hole (Hakkarainen and Korpimäki 1998).

As annual sample sizes for male owls that were recaptured in the following breeding season were small (14 males in 1991 and 9 males in 1994), we pooled the data from both years in our analyses. Territory quality, the number of voles stored in the nest-box, male age, breeding dispersal distances and breeding characteristics did not obviously differ between 1991 and 1994 (these years represented the increase phase of the 3-year vole cycle), which suggests that dispersal behaviour and breeding performance of male owls were not greatly affected by between-year variation (Hakkarainen et al. 2001). Nesting failure seemed to be more common among some males than others, suggesting that individual or nest-site quality may differ in the breeding population. Among six males which failed in their nesting attempts, four (67%) were also unsuccessful in the following breeding season, whereas only three (18%) nesting failures were observed among the males that bred successfully in the previous year. This probably reflects variation in home range quality or differences in feeding efficiency of males, which are mostly responsible for feeding the whole family during the breeding season. The proportion of males producing no fledglings, however, was the same between the control and treatment groups (2 out of 8 in the control group and 4 out of 15 in the treatment group), suggesting that either male or habitat quality did not differ between the control and treatment groups. In addition, habitat quality was not related to the probability of nesting failure at the nestling stage (Hakkarainen et al. 2001).

A majority (80%) of the males exposed to the mink visit shifted to breeding in a different nest-box in the following year, while only a minority (25%) of control males shifted nest-box. Likewise, breeding dispersal was longer in the treatment than the control group (median 1.25 km and 0 km, respectively). In fact, the longest breeding dispersal distance of males was recorded as a response to mink treatment (13 km). Despite the relatively high efficiency of owl trapping around our study area (Hakkarainen et al. 1996c), no males belonging to the experimental or control groups were later found outside our study area. Moreover, we did not find any costs of nest-hole shift, because laying date, clutch size and fledgling production did not show obvious differences between males that reoccupied the same nest-box or changed nest-box in consecutive breeding seasons. In addition, the distance that males dispersed between consecutive breeding seasons was not related to the next year's laying date, clutch size, or number of fledglings produced. Pine martens or other predators destroyed none of the 23 nests examined (Hakkarainen et al. 2001).

Despite the large body of literature published on the various consequences of dispersal, the question of why particular dispersal strategies evolve has received much less attention (Dieckmann et al. 1999, Ronce 2007). We were able to examine the causal relationship between nest predation risk and breeding dispersal of boreal owls. Our experiment showed that even a short-term nest predation risk due to a 10-minute presentation of a mink increased nest-hole shift and breeding dispersal distances of male boreal owls. The distance between nests in consecutive years was at least three times larger in the treatment than in the control group, although the distance to the nearest unoccupied nest-box did not differ between the two groups. After predator simulation, males in the following year apparently attracted females to breed in a nest-box where the nest predation risk was expected to be lower. Nest-hole shift and long breeding dispersal distances probably decrease the risk of nest predation, because pine martens seem to revisit nest-holes they have found (Dow and Fredga 1983, Sonerud 1985a). Burrowing owls also responded to experimental egg removal with increased dispersal probability, nesting attempts and egg production (Catlin and Rosenberg 2008). These results were consistent with earlier observational studies. For example, female boreal owls dispersed farther after nest pre-dation than after successful nesting (Sonerud et al. 1988), and nest predation resulted in a lower return rate of female merlins *Falco columbarius* to the breeding population (Wiklund 1996). Results of observational studies, however, may be affected by factors other than predation risk, such as individual or habitat quality. In boreal owls, males that failed in their nesting attempts had a high probability of failure in the following breeding season, whereas among successful breeders the probability of nesting failure was low. Similarly, habitat quality may also be associated with predation risk, as nest predation risk may be higher at the edges of forests (Andrén 1995). These sites may also be poor habitats for boreal owls and may be occupied by subdominant and low-quality individuals.

9.5.   Autumn movements

Spatial synchrony of population fluctuations of voles and boreal owls extends up to 250–300 km (Chapter 13.3). More than 40 years ago, Mysterud (1970) put forward a hypothesis that boreal owls are highly mobile and that both sexes and all age classes settle to breed in areas where voles are temporarily abundant and move on when vole populations decline. These nomadic movements cover the huge boreal forest belt, for example between the Scandinavian Peninsula and the Ural Mountains. In accordance with this hypothesis, large numbers of irruptive boreal owls have been captured and ringed south of the boreal forest region in bird observatories of southern Sweden in the autumns of 1964, 1967, 1974 and 1982 (Källander 1964, Wirdheim 2008), as well as in bird observatories of the west coast and south-western archipelagos of Finland in the autumns of 1967, 1971, 1975 and 1978 (Saurola 1979, Korpimäki and Hongell 1986, Pakkala et al. 1993). Large-scale autumn irruptions of boreal owls south of the boreal forest have also been documented in North America a long time ago (Catling 1972).

    According to Andersson (1980), the boreal owl is a candidate for nomadism. It subsists on a cyclically fluctuating food source (voles) with 3–4 years between population peaks, its lifespan is relatively short (Chapter 8.7), and it has a large clutch size – up to ten eggs. On the other hand, boreal owls originally bred in natural cavities, which, in contrast, should favour the evolution of resident habits (von Haartman 1968). One could therefore expect a serious conflict between these opposing selective pressures. It has been sug-gested that, as a response to the conflicting selective forces, the strategy of partial migration has evolved for boreal owls; i.e. males are resident while females and young owls are migratory or irruptive (Lundberg 1979, Korpimäki and Hongell 1986). In this section, we first examine between-year variation and inter-areal synchrony in the number and age composition of boreal owls captured in autumn along the Finnish coast of the

Baltic Sea. Second, we relate the yearly numbers of different-aged owls captured in autumn to breeding density estimates of owls in different localities of Finland. Third, we compare our results with our findings of intersexual differences in natal and breeding dispersal (Chapter 9.1) and with results from other studies in northern Europe and North America.

There are two long-term ringing sites (Kokkola and Kemi) and three bird observatories (Hanko, Valassaaret and Tauvo) on the southern and western coasts of Finland (see map, Fig. 3.1) where boreal owls have been permanently captured with similar methods and relatively constant capturing effort from late 1970s until 2008 (Fig. 9.4). Four of these capture sites are on the coastline, while Valassaaret is the small island in Quarken between Vaasa, Finland and Umeå, Sweden (see Fig. 3.1, Fig. 9.5). At least six additional bird observatories were operating on the southern and western coasts and in the south-western archipelago of Finland in the 1970s and 1980s (Saurola 1979) but recently their efforts to capture boreal owls have considerably declined. Therefore, we use only the data from these five sites in the following analyses of autumn movements of boreal owls. Most boreal owls seem to move along the coastline rather than venturing inland, because 50% of the boreal owls were captured on the coastline in Kokkola, where there was a line of five capture sites with similar trapping efforts extending up to 17 km from the coastline towards the inland areas (Sykkö and Vikström 1987).

Autumn movements of boreal owls usually started in late July to early August. At the age of 6–8 weeks fledglings have already become independent and have dispersed several kilometres from the hatching site (mean 7.8 km, range 5–21 km; Sykkö and Vikström 1987, Vikström 1988). The peak of captured and ringed owls was on 10–20 September at the two northernmost sites (Kemi and Tauvo; Pakkala et al. 1993, Suopajärvi M. 2001), while the corresponding peak was on 1–15 October at the south-ernmost capture site (Hanko; Lehikoinen et al. 2000). The last owls were usually captured in early October in the north (Kemi and Tauvo) but in early November in the south (Hanko). Most owls in Tauvo were captured during the late evening (between 22:00 hours and midnight; Pakkala et al. 1993).

The majority of the boreal owls caught at these five permanent sites were hatch-year specimens (67.3%, 13 605 out of 20 223; Fig. 9.4). The proportion of hatch-year owls was highest at Kokkola (74.7%), followed by Hanko (73.8%), Tauvo (70.2%), Valassaaret (57.4%) and Kemi (47.5%). The considerably lower proportion of hatch-year owls at Valassaaret may be because this site is situated approx. 30 km from the coastline. The study period at Kemi was shorter than at the other four sites, which may have been the reason why this data set included more poor fledgling production years of boreal owls in the late 1990s and 2000s (see Chapter 13.4).

The number of boreal owls captured at Hanko during 1979–2008 and at Tauvo during 1973–2008 decreased significantly with year (Fig. 9.4; $r_s = -0.66$, $n = 30$, $p < 0.001$ and $r_s = -0.47$, $n = 36$, $p = 0.004$, respectively), while no obvious long-term trend was found for the other three capture sites (Valassaaret, Kokkola and Kemi). The proportion of hatch-year birds varied widely, in particular in the data set from Valassaaret: the propor-tion of 1+-year-old owls tended to be higher in the decline years of the 3-year vole cycle (for example, in 1986, 1989 and 2006). There was also a large-scale synchrony in the

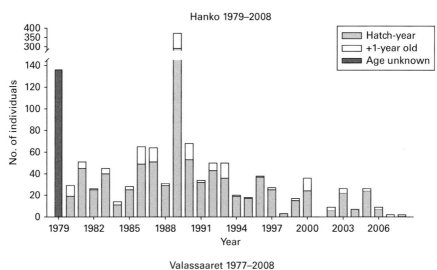

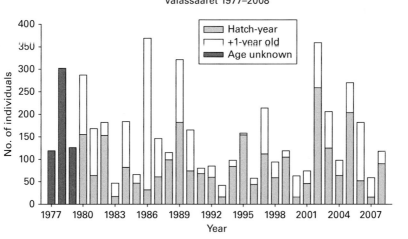

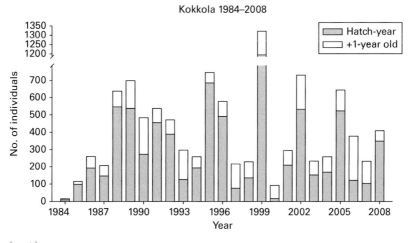

Fig. 9.4. (cont.)

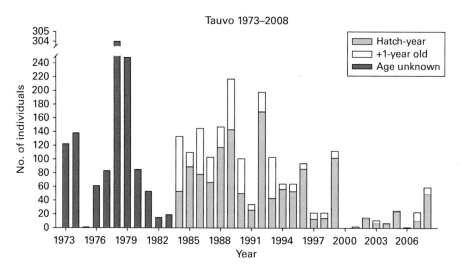

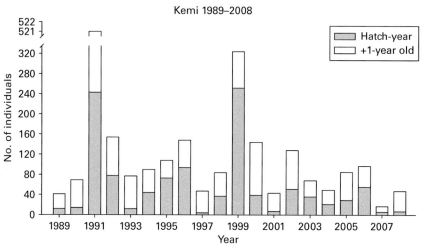

Fig. 9.4.   Yearly numbers of hatch-year (grey column) and older (white column) boreal owls captured in autumn in Hanko during 1979–2008 (total number of owls captured 1300, 1 year old 959, 1+ years old 341; Lehikoinen et al. 2008, Lehikoinen 2009 and unpublished data), Valassaaret during 1977–2008 (total 4974, 1 year old 2540, 1+ years old 1887, age unknown 547), Kokkola during 1984–2008 (total 10 341, 1 year old 7720, 1+ years old 2621; Sykkö and Vikström 1987, Vikström 1988 and unpublished data), Tauvo during 1973–2008 (total 2941, 1 year old 1272, 1+ years old 540, age unknown 1129; Pakkala et al. 1993 and unpublished data), and in Kemi during 1989–2008 (total 2343, 1 year old 1114, 1+ years old 1229; Suopajärvi M. 2001, 2006 and unpublished data). The panels showing the different trapping locations have been ordered from south to north (see locations in map, Fig. 3.1).

yearly numbers of owls captured at the five permanent ringing sites. The annual number of owls captured at Hanko was positively correlated with the corresponding figures for Tauvo and Valassaaret, and the number of owls at Valassaaret with those from Kokkola, Hanko and Kemi (Fig. 9.4, Table 9.1). The yearly number of owls captured at Kokkola

Fig. 9.5.   Capturing irruptive boreal owls on the small island of Valassaaret in the Quarken area between Vaasa, Finland and Umeå, Sweden. Photo: Pertti Malinen.

was closely positively related to those from Valassaaaret, Tauvo and Kemi, the number of owls at Tauvo with those from Hanko and Kokkola, as well as the number of owls captured at Kemi with that from Kokkola. In addition, the yearly number of owls captured at Kemi decreased with increasing numbers caught at Valassaaret.

We also investigated whether the yearly numbers of boreal owls captured at the five permanent ringing sites fluctuated in synchrony with the numbers of owl nests found at 23 localities or provinces in Finland (Table 9.2). The annual number of owls captured on the southern coast of Finland (Hanko) was closely positively correlated with the owl breeding density estimates at distances of 100–500 km (Quarken, South Karelia, Varsinais-Suomi, North Karelia and Central Ostrobothnia) but not with the breeding density estimate in the province nearby (Uusimaa). The yearly number of owls caught at Valassaaret was positively related to the breeding density estimates for the neighbouring study area (Quarken), but also to the corresponding figures in four provinces (South Karelia, Kanta-Häme, South Savonia and Satakunta) 200–600 km away. The

Table 9.1. Spearman rank correlation coefficients (number of years) between the yearly numbers of hatch-year (first calendar year, 1-year-old) and older (1+-year-old) boreal owls trapped in autumn at five permanent ringing sites (Hanko = Ha during 1979–2008; Valassaaret = Va during 1977–2008; Kokkola = Ko during 1984–2008; Tauvo = Ta during 1973–2008; and Kemi = Ke during 1989–2008) (see Fig. 3.1 for the location of ringing sites and Fig. 9.4 for the data). Only significant correlations are shown. The different trapping locations have been ordered from south to north.

Ringing site	Permanent ringing site		
	Tauvo 1-year	Tauvo 1+-year	Valassaaret 1+-year
Ha, 1-year	0.58** (25)	0.55** (25)	
Ha, 1+-year		0.57** (25)	0.53** (29)
	Kokkola 1-yr	Hanko 1+-yr	Kemi 1-yr
Va, 1-year	0.55** (25)		
Va, 1+-year		0.53** (29)	0.49* (20)
	Valassaaret 1-yr	Tauvo 1-yr	Kemi 1-yr
Ko, 1-year	0.55** (25)	0.46* (25)	0.48* (20)
	Hanko 1-yr	Kokkola 1-yr	
Ta, 1-year	0.58** (25)	0.46* (20)	
Ta, 1+-year	0.58** (25)		
	Kokkola 1-yr	Valassaaret 1+-yr	
Ke, 1-year	0.46* (25)	−0.49* (20)	

Two-tailed *$p < 0.05$, **$p < 0.01$.

annual number of owls captured at Kokkola was not associated with the breeding density estimates of owls in the area nearby (Central Ostrobothnia), but was positively associated with the corresponding figures for two areas approx. 100 km (Kauhava) and 600 km (South Karelia) away. The yearly number of owls caught at Tauvo was closely positively correlated with the breeding density estimates of owl populations in areas 300–600 km away (North and South Karelia, North Savonia, Seinäjoki, Kauhava, Kainuu, Uusimaa and Kymenlaakso). The annual numbers of owls captured at the northernmost site (Kemi) were closely related to owl breeding density estimates for the area nearby (Kemi-Tornio) and at distances of 500 km (North Savonia) and 600 km (North Karelia) away.

Based on the preliminary results presented in the two previous paragraphs, we can conclude that regular autumn movements of boreal owls extend at least 500–600 km. This was supported by our finding that the numbers of owls captured on the south and west coasts of Finland were usually closely related to the numbers of nests per 100 boxes found in south-eastern and eastern Finland, although there are many large lakes possibly slowing down autumn movements of owls between eastern areas and the west coast of Finland. This interpretation is further supported by the fact that there have been some recoveries of boreal owls ringed in Finland from the northern coast of Germany and southern Russia, more than 1000 km away, where they have been found dead (Saurola 1979). In addition, owls tended to arrive at the northernmost capture site (Kemi) from somewhat more northern areas than at the other four more southern ringing sites. Further analyses using the numbers of fledglings produced in different areas and numbers of

Table 9.2. Spearman rank correlation coefficients (number of years) between the yearly numbers of hatch-year (first calendar year, 1-year-old) and older (1+-year-old) boreal owls trapped in autumn at five permanent ringing sites (Hanko = Ha during 1979–2008; Valassaaret = Va during 1977–2008; Kokkola = Ko during 1984–2008; Tauvo = Ta during 1973–2008; and Kemi = Ke during 1989–2008) and the number of boreal owl nests per 100 boxes at 23 different localities and provinces in Finland (see Fig. 3.1 for the map of Finland and Fig. 9.4, Fig. 13.2, Fig. 13.3, Fig. 13.4 and Table 3.4 for data). The different trapping locations have been ordered from south to north.

Ringing site	Locality or province							
Ha, 1-yr		S. Karelia 0.48* (23)	V-Suomi 0.37* (29)					
Ha, 1+-yr	Quarken 0.46* (29)	S. Karelia 0.45* (23)		N. Karelia 0.39* (29)	C. Ostrobothnia 0.37* (28)			
Ha, all		S. Karelia 0.46* (23)	V-Suomi 0.42* (30)					
Va, 1-yr	Quarken 0.40* (29)							
Va, 1+-yr	Quarken 0.44* (29)							
Va, all	Quarken 0.61** (30)	S. Karelia 0.51* (23)	K-Häme 0.51* (26)	S. Savonia 0.47* (23)	Satakunta 0.42* (27)			
Ko, 1-yr	Kauhava 0.47* (25)							
Ko, all	Kauhava 0.42* (25)	S. Karelia 0.42* (23)						
Ta, 1-yr	N. Karelia 0.56** (25)	S. Karelia 0.49* (23)	N. Savonia 0.53** (25)	Seinäjoki 0.49* (25)	Kauhava 0.46* (25)	Kainuu 0.43* (25)	Uusimaa 0.41* (25)	Kymenlaakso 0.40* (25)
Ta, 1+-yr	N. Karelia 0.43* (25)							
Ta, all	N. Karelia 0.57** (30)	Kemi-Tornio 0.51* (23)	N. Savonia 0.46* (30)	Seinäjoki 0.45* (36)	Kauhava 0.37* (36)			
Ke, 1-yr	N. Karelia 0.62** (20)	Kemi-Tornio 0.53* (20)	N. Savonia 0.45* (20)					

Two-tailed *p < 0.05, **p < 0.01.

irruptive owls caught at all the bird observatories of northern Europe will probably cast new light on the autumn movements of boreal owls.

In any case, our preliminary results lend some support to the old model of Saurola (1979) on autumn movements of boreal owls in northern Europe. It is now also possible to draw up some definitions for this preliminary model. When the production of young is at a maximum in southern and central Finland (for example in 1973, 1977, 1982, 1985 and 1988), a considerable proportion of the young individuals and 1+-year-old females move northwards, even as far as northern Finland. This was shown by the directions of recoveries and controls (Saurola 1979), as well as by the low numbers of captured owls at the bird observatories of the south-western Finnish archipelago, southern Sweden and western Estonia (Lohmus 1999). In the next year (i.e. 1974, 1978, 1983, 1986 and 1989) the production of young peaks in northern Finland, and the owls make their way towards the south and south-west, even reaching the west coast of Norway. One year later (i.e. 1975, 1979, 1984, 1987 and 1990) vole populations, and thus the production of young owls, have declined to the low phase in most parts of the country. The numbers of irruptive owls, consisting more of adult individuals, peak at least at the bird observatories of south-western Finland, southern Sweden and West Estonia (Lohmus 1999). These long-distance autumn movements can be considered as responses to the lack of food in the previous breeding areas. However, it is not precisely known how many of these owls have spent the breeding season in Finland but have not been able to breed because of food shortage, and how many of them are of more eastern origin, because only small numbers of owls are ringed in Russia. In addition, only future studies will show how this 'model' of nomadic movement patterns will change with possible long-term alterations in spatial synchrony and amplitude of population cycles of northern voles, and with decreasing population sizes of boreal owls in northern Europe (see Chapter 13.4).

Large-scale irruptive movements of boreal owls probably also happen in central Europe. The year 1984 was extraordinary at the Harz Mountain (Germany) study area, because the number of broods (67) was four times as high as in the previous year and twice as high as in other good years of the 13-year study period. A total of 241 fledglings were raised; five times more than the pooled number for the years 1983 and 1985 (Schwerdtfeger 1993). The rapid increase in owl breeding density was mainly due to the immigration of juvenile owls from a distance of up to 198 km. The breeding season of 2000 was exceptional for boreal owls in the Swiss Jura Mountains. The number of breeding pairs (52), the clutch sizes (mean 6.6) and numbers of fledglings produced (5.5) were the highest recorded during the 15-year study period (Ravussin et al. 2001a). The majority of female owls were yearlings in spite of the fact that breeding success in the Jura had been poor in the preceding spring. Ring recoveries confirmed that at least three juveniles had moved from northern Germany to the Jura (distance close to 800 km; Patthey et al. 2001). In central Europe, autumn movements appeared to be mainly governed by more or less irregular outbreaks of wood mice populations, which strongly respond to good years of seed production of beech and oak. This in turn induced a major influx of young owls to the high-density mice areas, in this case to the Harz and Jura mountains. In the future, there is a need to study autumn movements of boreal owls in central Europe as well.

In the eastern Canadian coniferous forests, boreal owls are considered to be mainly residents, with irruptive movements to the south thought to occur during periods of low abundance of small rodent prey (Hayward and Hayward 1993), whereas saw-whet owls are mostly migratory (Cannings 1993). Data on the numbers and ages of irruptive and migratory boreal and saw-whet owls were collected during 1996–2004 in mid-September to late October, using audio lures and mist nets, in Quebec, Canada (Cote et al. 2007). These authors found a negative relationship between the annual abundance of small rodents in late summer and the number of boreal owls captured, while the corresponding relationship was positive for migrating juveniles and second-year saw-whet owls. The authors suggested that the breeding success of saw-whet owls, which were at the northern limit of their breeding range, was largely determined by small rodent abundance. There were more migrating juvenile saw-whet owls after good years of offspring production. Autumn movements of boreal owls were more intense in years of low rodent abundance, while no boreal owls were captured in years when rodent abundance was high. These results support the conventional belief that boreal owls mostly over-winter in the boreal forest zone of Canada but become nomadic when rodent prey is less abundant (Cheveau et al. 2004, Cote et al. 2007). Thus, boreal owls seemed less inclined to disperse when they were able to find suitable food-rich over-wintering and subsequent breeding areas in boreal forests.

Most (67%) boreal owls captured at Finnish permanent ringing sites during autumn were hatch-year birds. Similarly, a great majority (78%) of the irruptive boreal owls caught at Tankar bird observatory (an island situated 15 km west of the permanent capture site of Kokkola) were hatch-year birds ($n = 88$, pooled data from 1976–83; Korpimäki and Hongell 1986). The corresponding figure was much less (55%) for migratory saw-whet owls mist-netted in south-central Indiana, USA (Brittain et al. 2009). Based on wing lengths (Korpimäki and Hongell 1986) and molecular sexing (Hipkiss et al. 2002b), the majority (65%) of irruptive hatch-year birds were found to be females, corresponding well with the longer natal dispersal distances of females compared with males (Chapter 9.1). The probable reason for autumn movements of juveniles is that they are attempting to find areas of vole abundance in order to survive their first winter, which is usually risky for inexperienced hunters, and then to breed for the first time.

Similarly, the great majority of ±1-year-old boreal owls trapped during autumn movements are females, based on wing lengths (Korpimäki and Hongell 1986) or molecular sexing (Hipkiss et al. 2002b). This is in accordance with the much more extensive breeding dispersal of females than males (Chapter 9.1). Nomadic movements in autumn are thus a way for adult females to find an area with vole abundance so as to survive over winter, and to find a high-quality male with a suitable nest-hole to breed successfully again. When vole populations decline, females are more inclined to disperse between successive years than when vole abundance is increasing (Chapter 9.3). An additional, but possibly weaker, factor inducing long-distance breeding dispersal of females is a failed nesting attempt due to predation (Chapter 9.4). Most old males, on the other hand, stay on their home ranges after their first breeding attempt, particularly if it is successful. This was shown by the fact that no males breeding in our study area have been re-trapped in autumn movements at permanent ringing sites in Finland, although the long-term ringing sites at Kokkola and

Valassaaret are nearby. On the island of Stora Fjäderegg, 15 km off the east coast of Sweden (in the vicinity of Umeå), a few adult boreal owl males were captured during autumn movements in a decline vole year 1999, but 73% of them were 1-year-old individuals (Hipkiss et al. 2002b) that had probably not yet bred. Thus, these males were probably searching for an area of abundant food resources to survive over winter. In addition, one male that had bred previously on the mainland 50 km away was re-trapped on the island (Hipkiss et al. 2002b), indicating that this male might have been nomadic. Although males are probably under selective pressure to remain on their home ranges after a successful breeding attempt, the strength of this pressure may vary in relation to habitat quality. A male in a poor nest-site might benefit from searching for a better nest-site when vole populations decrease (Korpimäki 1988d). Selection for adult nomadism may also increase with latitude because of more high-amplitude vole cycles and deeper snow cover in winter, making site-tenacity of males less favourable at the northernmost latitudes of the boreal forest zone (Korpimäki 1986b). Clear-cutting of boreal forests probably also reduces the habitat quality of wintering resident males (Chapter 10.1), which might increase the tendency towards nomadism of males in the future.

Plate 2.1.   The greyish-white facial disc of the boreal owl is framed by a brown-black border and highlighted by raised white eyebrows. The upper part is dark brown with white spotting, and the under-part is creamy white, broadly streaked with dark brown or russet. Photo: Pertti Malinen.

Plate 2.2. Fledglings of boreal owls are dark brown with very distinct white 'eyebrow' markings and some white spots on the scapulars and wing coverts. Photo: Benjam Pöntinen.

Plate 2.3. The wings of yearling (2nd calendar year) boreal owls have only dark brown and unworn primary and secondary feathers. Photo: Rauno Varjonen.

Plate 2.4. Two-year-old (3rd calendar year) boreal owls have the outermost dark-brown unworn 1–6 (in this case 5) primary feathers and the light-brown worn innermost 4–9 (in this case 5) primaries. Photo: Rauno Varjonen.

Plate 2.5. Three-year-old (4th calendar year) boreal owls have three age classes of primaries on their wings. This male has moulted the fourth to sixth primary feathers in the previous autumn, the first three primaries in the autumn before that, and the four innermost primaries are still unmoulted. Photo: Rauno Varjonen.

Plate 2.6. At the age of 4 years, the wave of primary moulting proceeds towards the innermost primaries; the 9th and 10th innermost primaries are first moulted only at the age of 5 or more years. This >7-year-old male has not yet moulted the 10th innermost primary. Photo: Rauno Varjonen.

Plate 2.13. The North American subspecies of the boreal owl, *Aegolius funereus richardsoni*. Photo: Jackson S. Whitman.

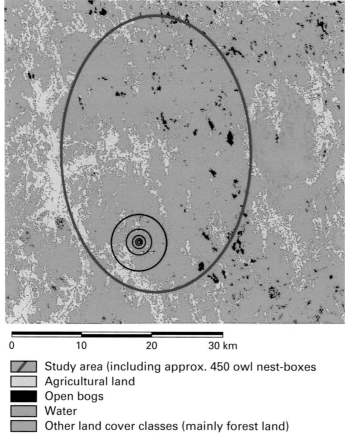

	Study area (including approx. 450 owl nest-boxes
	Agricultural land
	Open bogs
	Water
	Other land cover classes (mainly forest land)

Plate 3.2.   Landscape composition of the study area in the Kauhava region, western Finland. Lappajärvi is the large lake on the eastern border of the study area. Concentric black circles show the five landscape scales (250 m, 500 m, 1000 m, 2000 m, 4000 m) around a nest-box (see Hakkarainen et al. 2003).

Plate 3.6.   Old-growth forests nowadays comprise less than 1% of the study area. The forest nature reserve at Passinmäki, Kauhava covers some 40 ha and is restricted to clear-cut areas on the eastern side. Photo: Erkki Korpimäki.

Plate 3.7.   Main crops cultivated or agricultural fields are oats, barley, potatoes and hay. Hay fields are good habitats for field and sibling voles.
Photo: Erkki Korpimäki.

Plate 3.8. Clear-cut and sapling areas cover 6% of the study area. Recent clear-cut area in the picture. Photo: Pertti Malinen.

Plate 3.9. Wet peatland bog with small pine trees is called '*räme*', and treeless even wetter peatland bog is called '*neva*' in Finnish. Ympyriäisneva, Kauhava. Photo: Erkki Korpimäki.

Plate 3.10. Lakes, rivers and creeks cover 2% of the study area. The creek Mustalamminluoma, Kauhava is flooding in April. Photo: Erkki Korpimäki.

Plate 3.11. Pine (*Pinus silvestris*) forests cover approx. 65% of our study area. These pine forests mainly grow on dry, less productive acid soils.
Photo: Erkki Korpimäki

Plate 3.12. Spruce (*Picea abies*) forest with some small deciduous trees, birch (*Betula* spp.) and aspen (*Populus tremula*), scattered in between the spruces. Photo: Erkki Korpimäki.

Plate 3.13. Bank voles and common shrews mainly occupy woodland in our study area. Male boreal owl about to deliver a bank vole (upper panel) and a common shrew (lower panel) to the nest-box. Photo: Benjam Pöntinen.

Plate 3.19. The study area of boreal owls in the forest near Fairbanks, interior of Alaska, USA, where snow covers the landscape in late spring (photo taken on 15 April by Jackson S. Whitman).

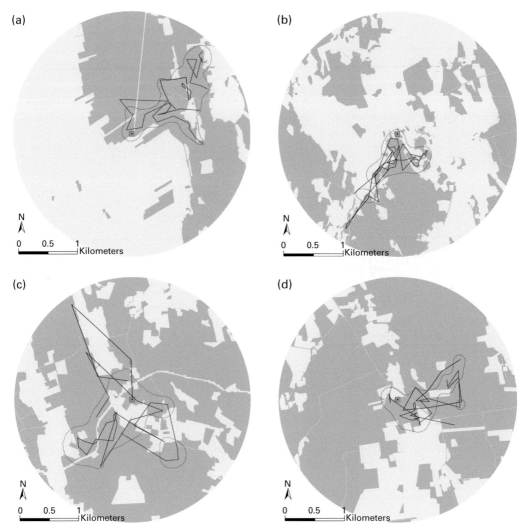

Plate 4.2.   Coverage of open terrain (agricultural land and clear-cut areas, yellow) and forests (green) in four home ranges of male owls in 2009 in the 2-km radius around the nest-box (brown square). Brown line is the 95% kernel home range. Blue, dark red and violet lines represent the routes of male owls in radio-tracking and hunting bouts lasting for at least 2 hours on three different nights at the end of the nestling to post-fledging periods. Panels (a) and (b) represent home ranges where the proportion of open terrain is large (nest-boxes 278 and 351, respectively) and panels (c) and (d) represent home ranges where the proportion of forests is relatively large (nest-boxes 819 and 313, respectively).

Plate 4.3.   In Finland, most natural cavities excavated by black woodpeckers and later used by boreal owls are in aspen trees. Photo: Benjam Pöntinen.

Plate 5.2.   Female owls pile up the prey items at the edges of the nest-box. In the case of big food stores, the bottom of the box is full of *Microtus* and bank voles. Photo. Jorma Nurmi.

Plate 5.7.   Boreal owl male about to deliver a decapitated bank vole to the nest. Photo: Berjam Pöntinen.

Plate 5.8. Male delivers a common shrew to the nest. Photo: Benjam Pöntinen.

Plate 5.9. Boreal owl male about to deliver a wood mouse to the nest in Belgium. Photo: Serge Sorbi.

Plate 5.10.   Male owl delivers a male chaffinch to the nest. Photo: Benjam Pöntinen.

Plate 5.11. Male delivers a young water vole to the nest. Photo: Benjam Pöntinen.

Plate 5.12.   Male provides a harvest mouse to the nest. Photo: Benjam Pöntinen.

Plate 6.5.   Male boreal owl delivers a bank vole to the nest-box where the female is incubating or brooding. Photo: Benjam Pöntinen.

Plate 7.1. Female boreal owls lay eggs every second day. The incubation of eggs and brooding of young until they are approx. 2–3 weeks old is the duty of females. Photo: Benjam Pöntinen.

Plate 7.2. Male parents provide for their families from before the egg-laying until the young fledge at the age of 30–33 days. In this case the young were about to fledge. Photos: Pertti Malinen.

Plate 10.1. The mortality of a boreal owl due to starvation in Seurasaari, Helsinki, Finland on 30 January 2010, when the snow layer had been unusually deep and ambient temperatures below −20°C for more than 2 weeks. Photos: Tomi Muukkonen.

Plate 14.2. A boreal owl with a freshly killed pygmy owl on 17 April 2009 in Tornio, southern Lapland of Finland. Photo: Matti Suopajärvi.

Plate 14.4. Ural owls proved to be the worst avian enemies of boreal owls in north European coniferous forests. Photo: Aku Kankaanpää.

10 Survival and mortality under temporally varying food conditions

10.1. Survival: an important determinant of lifetime reproductive success

Survival is a fundamental consideration in an animal's life. In annually reproducing species, such as owls, longevity is closely related to the number of breeding attempts, and hence survival is an important determinant of lifetime reproductive success (Clutton-Brock 1988, Newton 1989). Earlier analyses on the survival of birds have mainly concentrated on species that live under relatively stable environmental conditions in temperate areas, and the main question has been about age-related changes in annual survival (but see Taylor I. 1994 for excellent survival data on barn owls in Scotland). Although survival and lifetime reproductive success are closely connected from a reproductive output point of view, in our approach we separate these issues from each other. In this chapter we concentrate on the survival of boreal owls under spatially and temporally varying food conditions. The vole-eating boreal owl is a good model species in this context, as the maximum lifespan of free-living males is 11 years in Finland (Korpimäki 1992b), and even longer lifespans have been documented for two females in Germany (13 years; Franz et al. 1984, Meyer W. 2010). Hence, boreal owls probably face widely varying vole abundances during their lifetimes. On the basis of 281 Finnish ring recoveries of boreal owls found dead, the estimates of mean annual survival were 50% (95% confidence limit 43–57%) during the first year of life and 67% (61–75%) thereafter (Korpimäki 1991a).

Adult survival to the next potential breeding season is an essential component of fitness, which is generally estimated by the number of offspring produced during an individual's lifetime. The existing literature on fitness-related traits, however, is biased towards estimating the reproductive success of individuals, as reproductive traits are easier to measure than survival. This is apparently due to the fact that rigorous survival estimates generally need data for follow-ups of marked individuals collected at appropriate temporal and spatial scales. Furthermore, measuring survival is troublesome, as the time of death is often unknown. For example, it is generally known that many vole-eating boreal owl species, especially after the crash of vole populations, die during the winter or early spring, owing to harsh weather conditions (Mikkola 1983, see also Chapter 10.4). At that time, a deep snow layer protects voles against hunting owls and low temperatures probably increase thermoregulatory costs and thus decrease physiological condition. Direct field observations on the decease of individual owls, however, are extremely rare. The pictures in this book illustrate one such case (Fig. 10.1).

Fig. 10.1.   The mortality of a boreal owl due to starvation in Seurasaari, Helsinki, Finland on 30 January 2010, when the snow layer had been unusually deep and ambient temperatures below −20°C for more than 2 weeks. Photos: Tomi Muukkonen. (For colour version, see colour plate.)

Because it is difficult to study deaths in the field, survival or mortality has to be estimated by modelling the probability of an individual owl surviving to the next breeding season. For this purpose, the mark-and-recapture method is commonly used in animal ecology. This method is most valuable when a researcher fails to detect all the individuals present in the study population every time. For example, in our study population, owls may hide in several ways; most males do not breed in low vole years and a few may breed in unknown places (e.g. in undiscovered natural cavities). In these cases, we can model expected survival estimates for successive breeding seasons by taking into account the recapture probability for the corresponding time interval. Hence, survival estimates for low vole years tend to be higher owing to misses in the trapping process (most owls simply do not breed at that time). In contrast, in increasing vole years, survival and recapture probabilities apparently correspond to each other quite well, because at that time the majority of owls are able to breed and therefore can easily be trapped. In this context, the fundamental idea is that survival and recapture probabilities should be considered as separate factors. In general, however, recapture probability has been taken as an estimate of survival, which can be confusing because the two are not necessarily related. This is a serious shortcoming; only a few studies have actually measured rigorous survival estimates on adult birds despite there having been a methodological advance in survival analyses using mark–recapture models. Therefore, we used a sophisticated survival program, MARK, to model the survival of boreal owls (Hakkarainen et al. 2002, 2008). We examined the survival of males only, because males are sedentary after the first breeding attempt, whereas females usually disperse widely up to 500 km between successive breeding seasons. Therefore, our mark–recapture data are sufficient and comprehensive enough when dealing with males only.

10.2. Vole supply governs survival

As shown in Chapter 1, annual numbers of the main prey of boreal owls vary considerably over time. In increasing vole years, prey numbers increase between successive breeding seasons, whereas in the declining years of vole populations sedentary male owls may face as much as a 100-fold decrease in their main prey numbers. This may happen over a period of a few months, and may greatly increase the mortality of owls. Although this is quite an obvious prediction, no-one had previously examined it using representative and long-term follow-ups of marked individuals. Therefore we investigated, using 459 males that recruited to our breeding population in 1981–95, whether annual survival estimates of boreal owls vary according to changes in the abundance of their main prey, *Microtus* voles.

   We predicted that owl survival should be high in the increase phase of the vole cycle, and should decline during the crash of vole populations. In accordance with our predictions, the survival of male boreal owls was highly dependent on annual changes in *Microtus* vole numbers. In increasing vole years, survival of male owls increased along with increasing vole numbers in the field, whereas the opposite was true in the decrease vole years; survival decreased steeply at the moment of the crash of vole populations (Fig. 10.2). Our study is among the few documenting a large annual variation in the over-winter survival of adult birds: in low vole years only about 25% of males survived, while the corresponding proportion was three times higher (approx. 75%) in good vole winters. Vole cycle phase-related differences in over-winter survival of female Ural owls were much less: survival decreased to 62% after the crash of vole populations compared with 85–90% observed in the increase phase of the vole cycle (Brommer et al. 2002b). In tawny owls, over-winter survival of experienced breeders did not show any obvious variation associated with the vole cycle, but over-winter survival of first-time breeders was low after the crash of vole populations (25%), whereas it was much higher in the low (60%) and increase phases of the vole cycle (Karell et al. 2009). Our survival models show that temporal variation in the abundance of their main prey largely governs the survival of male boreal owls but also,

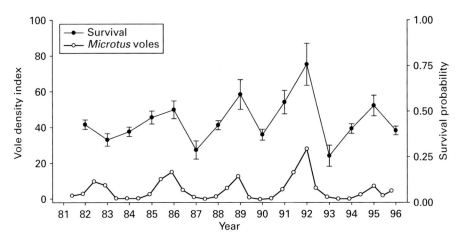

Fig. 10.2.   Annual over-winter survival probabilities (± s.e.; right y-axis) of male boreal owls, and density indices of *Microtus* voles (left y-axis) in the study area during 1981–95 (Hakkarainen et al. 2002). Annual vole density index describes the number of *Microtus* voles snap-trapped per 100 trap-nights (redrawn from Hakkarainen et al. 2002).

quite unexpectedly, that the severity of the winter did not play a crucial role in over-winter survival. However, there could be interactive effects of vole scarcity and severe winter conditions on over-winter survival. We suggest that this kind of combination is clearly fatal for boreal owls attempting to over-winter in northern boreal forests. The high mortality induced by severe winters has been well documented for barn owls in Scotland (Taylor I. 1994) and Switzerland (Altwegg et al. 2006), but this species is probably more sensitive to the severity of winters than boreal owls are. In addition to survival, recapture rates of male boreal owls also varied greatly due to the fact that in poor vole years the majority of males failed to breed. The large between-cycle-phase variation in survival of boreal owls probably creates selection for phenotypic plasticity in life-history traits related to survival and reproduction (Hakkarainen et al. 2002).

10.3.   Fluctuating vole abundance controls recruitment rate

An old dogma of life-history theory is that animals are probably selected to maximise the number of surviving offspring produced in their lifetime (Williams 1966). Therefore, the reproductive success of birds should not be measured by the number of fledglings per breeding attempt or season, but by the number of surviving offspring, and especially by the number entering the breeding population. This is a particularly straightforward idea when applied to boreal owls, in which even adult survival of males varies profoundly between the different phases of the vole cycle (Chapter 10.2). In addition, because boreal owls may disperse over long distances between their birth site and the first breeding site (Chapter 9.2), one cannot just use the number of recruits produced to the local population in analyses of recruitment rate. The most risky time for juvenile birds of prey outside the

Table 10.1. Number of nests and number of nests producing one, two and three recruits to the nationwide boreal owl population in the three different phases of the vole cycle. Pooled data from the Kauhava region during 1977–2009.

| Phase of the cycle | No. of nests | No. of recruits producing nests | | | Proportion of recruits producing nests |
		1 recruit	2 recruits	3 recruits	
Low	167	13	0	1	8.4
Increase	534	75	7	3	15.9
Decrease	714	32	3	0	4.9
Total	1415	120	10	4	9.5

nest is the period when parents stop feeding and they have to begin to hunt for themselves (i.e. the independence period; see e.g. Newton et al. 1982 for sparrowhawks, and Taylor I. 1994 for barn owls). Therefore, we chose to use the pooled data from recruits entering both the autumn and breeding populations in our analyses of recruitment rate. This is justified because the hatch-year birds recovered during their autumn movements at bird observatories had already survived over their riskiest time and travelled at least 100 km (Chapter 9.5).

Of the 1415 nests recorded in our study area during 1977–2009, it was found that 99 nests produced at least one recruit to the nationwide breeding population and an additional 35 nests produced at least one recruit to the autumn population (Table 10.1). The proportion of offspring later recruited to the nationwide autumn or breeding population in the increase phase of the vole cycle was three times as high as in the decrease phase and twice as high as in the low phase of the vole cycle. Furthermore, in the increase phase of the vole cycle, many nests tended to produce two or three recruits (1.9%), whereas in the low and decrease phases this proportion appeared to remain low (0.6% and 0.4%, respectively). These results were consistent with our earlier results on recruitment rate of boreal owls to the breeding population, although they were based on a smaller sample (Korpimäki and Lagerström 1988). Hatching date within the season did not seem to have any obvious relationship with recruitment probability, although there was a tendency that in the increase phase of the cycle late nests also produced many recruits, whereas in the decrease phase early nests tended to perform better. The higher recruitment rate of offspring hatched in the increase phase rather than in the decrease phase of the cycle seems to be a rule under periodically fluctuating food conditions, because it has also been documented in the nationwide data for Eurasian kestrels (Korpimäki and Wiehn 1998) and in the local recruitment data for Ural owls (Brommer et al. 2002a) and tawny owls (Karell et al. 2009) in Finland.

The yearly proportion of nests producing recruits during 1977–2009 varied widely (from 0 to 30%) and was closely positively related to the abundance index of *Microtus* voles in the autumn of the hatching year (Fig. 10.3). In addition, there was also a tendency for the recruitment rate of owlets to increase with augmenting abundances of bank voles in the autumn of the hatching year, whereas the correlations for the vole indices in the spring of the hatching year were substantially less. These results strongly suggest that the abundance of main foods during the post-fledging and independence

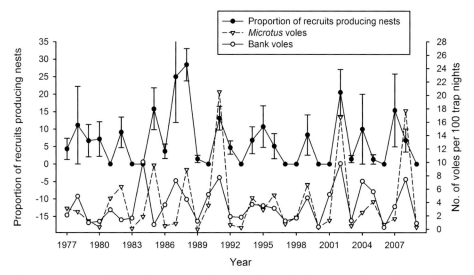

Fig. 10.3.   Yearly proportion (95% confidence intervals) of recruits producing nests and abundance indices of *Microtus* and bank voles in the autumn of the hatching year during 1977–2009 ($r_s = 0.53$, $n = 33$, $p = 0.001$ for *Microtus* spp., $r_s = 0.34$, $n = 33$, $p = 0.05$ for bank voles). Correlations for the vole indices in the spring of the hatching year were: $r_s = 0.39$, $n = 33$, $p = 0.02$ for *Microtus* spp. and $r_s = 0.34$, $n = 33$, $p = 0.05$ for bank voles.

periods of owlets is crucial for their survival over the critical independence period and first winter. In particular, the owl offspring hatched in the increase phase of the vole cycle experience improving food conditions, while those hatched in the decrease phase experience deteriorating food conditions during their first year of life. The few owl offspring hatched in the low phase, in turn, face poor food conditions during the independence period, because vole populations usually start to recover only in late July to August in the low phase of the vole cycle. Therefore, one can say that owlets hatched in the increase phase of the vole cycle are 'born with a silver spoon in their beak' (see, e.g., Lindström J. 1999). Their body condition is already markedly better even at the fledgling stage (E. Korpimäki, unpublished data) and their survival prospects are relatively high. Because vole cycles are fairly predictable, a selective advantage could be expected for those owl parents that invest the most in reproduction during the increase phase rather than the decrease phase, because their contribution to the future gene pool of the population is highest at that time.

10.4.   Mortality factors

The Finnish nationwide ringing data comprised a total of 1085 ring recoveries during 1973–2010 in which a boreal owl was found as deceased. In 546 (50.3%) cases, the cause of death remained unknown. This left a total of 539 ring recoveries with likely cause of death, but one should still bear in mind that these are often based on the subjective

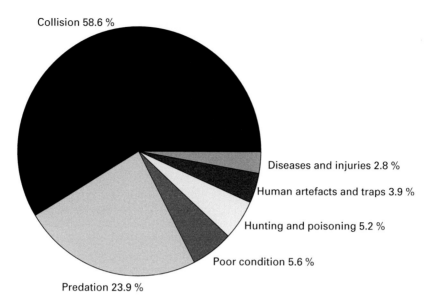

Collision 58.6 %

Diseases and injuries 2.8 %

Human artefacts and traps 3.9 %

Hunting and poisoning 5.2 %

Poor condition 5.6 %

Predation 23.9 %

Fig. 10.4. Distribution of most important causes of death of boreal owls according to a total of 539 ring recoveries from Finland during 1973–2010.

judgement of the member of the public finding the owl's corpse. Therefore, they should be interpreted with caution.

Collision was the most frequent cause of death (Fig. 10.4). Collision with a vehicle was the main cause (139 cases, 25.8%), followed by collision with a window (80, 14.8%), entering a man-made structure (54, 10.0%), drowning in a water container (20, 3.7%), collision with a building (bridge, wall, etc., 11, 2.0%), and collision with a thin cable or mast (10, 1.9%). Predators killed approximately one-quarter of boreal owls, among which larger diurnal raptors and owls were the worst enemies (78 cases, 14.5%), followed by domestic or feral mammalian predators (mainly cats, 24, 4.5%). Poor body condition due to starvation was a relatively infrequent cause of death, although one would expect it to be much more common during the frequent lean periods and severe winters experienced by owls in the north. It is probable that some of the owls killed by predators or vehicles or by entering buildings, etc. were already starving and thus came to hunt in risky places. Hunting, trapping and poisoning still play a small role in causing the deaths of boreal owls. It is also remarkable that one of the two oldest boreal owls ever recorded, ringed in Thüringen, Germany in 1981, was shot 13 years later on the west coast of France (Meyer W. 2010). Some owls also succumb due to contact with human artefacts that are either still in use or no longer used, including entanglement in barbed wire, fish-nets, netting on fish ponds, fruit-netting, and even mouse traps. Diseases and injuries have been recorded only infrequently as causes of mortality.

The total number of mortality cases based on ring recoveries varied widely between years during 1973–2010 (Fig. 10.5) and followed a similar long-term trend as the population density estimates of boreal owls in Finland (see Fig. 13.7, lower panel). Both the number of deaths and population density indices increased towards a peak in the late 1980s,

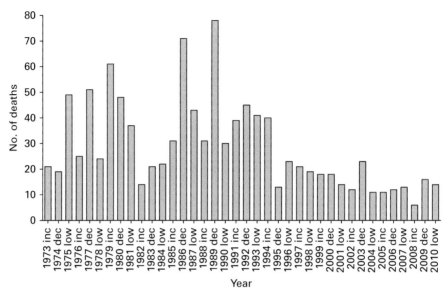

Fig. 10.5.   Yearly variation in number of deaths of boreal owls recorded in Finland during 1973–2010 (inc = the increase phase, dec = the decrease phase and low = the low phase of the vole cycle).

and thereafter they started to decline. A closer look at the annual numbers of mortality cases revealed that mortality was usually higher in the decrease phase than in the increase phase of the vole cycle (for example, 1976 < 1977, 1982 < 1983, 1985 < 1986, 1988 < 1989, 1991 < 1992) with a few exceptions (1973 > 1974, 1979 > 1980 and 1994 > 1995) (Fig. 10.5). These data tentatively indicate that the mortality of boreal owls was indeed greater in the decrease than in the increase phase of the vole cycle and thus suggest that it was the decreasing amount of main food supply (voles) that induced this high mortality.

Boreal owls that starved to death in boreal forest with little human occupation were more likely to remain undetected than individuals hit by vehicles on roads or those that entered buildings, etc. Therefore, their corpses in the forest would probably have been scavenged almost straight away and therefore ring recoveries by members of the public would have been less likely. These factors probably underestimate poor condition due to starvation as a cause of mortality, although boreal owls have a high fasting endurance. With the restrictions set by small body size and water economy (owls rely entirely on water either ingested with food or produced metabolically), the boreal owl has probably taken fasting endurance to the extreme (Hohtola et al. 1994).

In our study area, 2 of the 11 boreal owl males fixed with radio-transmitters in 2005 (Chapter 3.5) succumbed when delivering food for their young in the nest. One was hit by a vehicle on the main road nearby and the other was killed by a goshawk. These scanty data appear to be consistent with the importance of different mortality factors revealed by Finnish ring recoveries (Fig. 10.4). In addition, 6 of the 24 adult boreal owls monitored from late January to August in Idaho (USA) using radio-telemetry died, two due to starvation and three due to predation (Hayward et al. 1993). However, there is still an urgent need for further studies, for example by fitting large numbers of radio-transmitters and/or satellite transmitters to boreal owls to reveal their long-distance movements and to confirm their mortality causes by finding fresh corpses for autopsy. In this sense, the study of Taylor I. (1994) on the mortality factors of barn owls is really one of the best examples, because he and his co-workers found many fresh barn owl carcasses for autopsy to reveal the real causes of mortality. Starvation was the main cause of death of Scottish barn owls. For instance, of the 66 adults found freshly dead, 43 were severely famished, with no signs of injury or poisoning (Taylor I. 1994). Moreover, 36% of 131 tawny owl young fitted with radio-transmitters in Denmark died between the fledging and independence period due to avian and mammalian predators (mainly red foxes, goshawks and common buzzards; Sunde 2005). But young tawny owls leave their nest-holes when they are not able to fly, whereas boreal owls are able to fly when fledging. Therefore, one could suggest that juvenile boreal owls are less vulnerable to mammalian predators, but perhaps more vulnerable to avian predators due to their smallness, than are tawny owls.

11 Old forests increase survival and lifetime reproductive success

11.1. Forest age structure in the study area

Landscape modification and habitat fragmentation are of special concern to conservation biologists (Haila 2002, Fazey et al. 2005), because these two processes pose the most important threats to global biodiversity among several taxonomic groups (Sala et al. 2000, Foley et al. 2005). Thus there is an increasing need to investigate whether anthropogenic changes in habitat quality threaten the viability of natural populations (Hanski and Gilbin 1997). This especially concerns species living in old forests (reviewed in Bunnell 1999). In Finland, for example, area-sensitive old-growth-specialist birds have been decreasing while generalist species have been increasing (reviewed in Esseen et al. 1997). Several owl species, including boreal owls, have also been considered old-forest associates (Mikkola 1983, Hakkarainen et al. 1997b, Hayward 1997, Blakesley et al. 2005) and the recent decline of boreal owl populations in Finland has been attributed to the decrease in the amount of old forest (Korpimäki et al. 2008, 2009, Saurola 2009; see Chapter 13.4). Forest loss and changes in forest age structure owing to modern forestry practices probably have the potential to worsen the welfare of species relying on forest resources. For such species, clear-cut areas and young forests are probably poor habitats. Old forests instead offer, for example in many owl species, adequate shelter, food and natural cavities (Mikkola 1983, Hayward 1997; see also Chapter 14). One classic and thoroughly studied example of a rare species that is dependent upon old-growth forest is the northern spotted owl *Strix occidentalis caurina* (Franklin et al. 2000, Roberge et al. 2008). Its apparent survival and reproductive rate are positively associated with older forest types close to the nest or primary roost site (Dugger et al. 2005). Such forest-dwelling species are threatened owing to habitat destruction through excessive logging. In the boreal owl, however, in addition to our studies, breeding performance and survival have not been examined with respect to forest age.

Human impact on north European forests first became apparent about 5000 years ago (cattle grazing, agriculture and slash-and-burn agriculture), but the rapid expansion of the sawmill industry in the second half of the nineteenth century and subsequent repeated loggings resulted in local overexploitation of the available pine timber. During recent decades, the growth of the pulp industry has also had large-scale impacts on forest landscapes throughout Finland and elsewhere in northern Europe.

Modern forest management includes clear-cutting as the main renewal method, followed by the planting of saplings (mainly conifers) which during their young

Fig. 11.1.   Forest management includes clear-cutting as the main renewal method of old-growth spruce forests (upper panel). This is followed by the planting of pine saplings (lower panel). Photos: Pertti Malinen.

successional stages have to be thinned (Fig. 11.1). In southern and central Finland, taking commercially grown conifers into the mature and late-serial stages occurs in a cycle of 80–100 years or longer. As a result, the mean volume of timber in the mature forests is over 200 m^3/ha (Tomppo et al. 1999) when the forests are voluminous enough for harvesting and timber production.

Because of intensive logging and modern forestry, old-growth forests are scarce in our study area (< 1%). Therefore, in this book we cannot speak about pristine or virgin forests because those are almost exclusively to be found in protected areas in Finland, which are totally lacking in our study area. Instead, we speak about 'old forests', which refers to forests more than 80 years old (forests over 200 years old are extremely rare). However, our 'old forests' resemble old-growth forests in many ways. They contain large, old, living and dead trees and snags, at least to some extent, and the canopy of trees and relatively high density of tree trunks offer shelter to forest dwellers. Young forests, in contrast, are more open and vulnerable habitats for so-called forest owls, and probably have different compositions and abundances of prey species compared with old forests. Hence, the wide-ranging deforestation of old forests may have a negative impact on forest-dwelling owl species. To discover whether this holds true for the boreal owl, we examined whether the cover of old coniferous forests increases survival and lifetime reproductive success of site-resident male boreal owls (for further details, see Laaksonen et al. 2004, Hakkarainen et al. 2008).

11.2.   Survival effects

Although a number of theoretical studies suggest that loss of habitat reduces the survival of organisms (e.g. Fahrig 1999), sophisticated survival models employing individual-level empirical data are rare, especially for species dwelling in old forests. Therefore, we analysed the survival of male owls with respect to the proportion of forests within their home ranges, using the mark–recapture program MARK (for further details, see Hakkarainen et al. 2008).

Our sample included 209 males, each occupying a separate home range for 10 years (i.e. three vole cycles in 1987–96). We used this time period for recruiting males because our satellite data on land use and forest resources were obtained in the middle of the period, in 1991. In our study population, old forests (>80 years old; >152 m^3/ha of timber) cover approximately 12% of home range size (Fig. 11.2). The percentage

Fig. 11.2.   This old-growth (>80 years old) spruce forest includes >152 m^3 timber per hectare. Photo: Pertti Malinen.

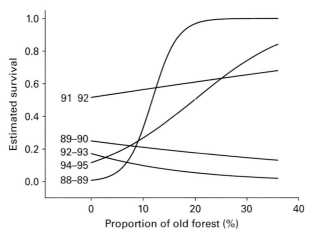

Fig. 11.3. Predicted survival of parent males over five winters in relation to the proportion of old-growth forest in the home range. Winters 1988–9, 1991–2 and 1994–5 represent survival estimates for the increasing phase of the vole cycle, whereas in winters 1989–90 and 1992–3 vole numbers declined steeply in the field (see Hakkarainen et al. 2008 for further details).

coverage of middle-aged (40–80 years; 102–152 m³/ha) and young (20–40 years; 13–101 m³/ha) forests in home ranges was 12% and 37%, respectively. We measured habitat structure at the home range scale (a radius of 1000 m from the nest, covering 314 ha). This spatial scale is considered the most relevant for breeding boreal owls (Hakkarainen et al. 2003).

When we modelled the proportions of these three different forest age classes (old-growth, middle-aged and young) separately, it turned out that the old-growth model was the best. The cover of old forest in the territory was positively associated with the survival of male boreal owls, and outweighed the roles of middle-aged and young forests (Fig. 11.3). Interestingly, these positive relationships between the proportion of old forests and survival occurred while vole numbers in the field were increasing (i.e. 1988–9, 1991–2, 1994–5).

We suggest two mechanisms through which the relationship is likely to arise. First, old forests may maintain high prey numbers, such as shrews, small passerine birds and bank voles. Second, old forests may provide refuges against large avian predators.

During harsh winter conditions when deep snow cover prevails for 5 months and temperatures may drop to as low as −30°C, prey abundance is pivotal. Old forests maintain high populations of small passerine birds (e.g. Edenius and Elmberg 1996), which are important prey during adverse food conditions in winter (Chapter 5.7). Furthermore, forest-dwelling bank voles are common in boreal coniferous forests of northern Europe, and their proportion in the diet of boreal owls is high, especially in low and increasing vole years (Chapter 5.6). The densities of bank voles are higher in old-growth spruce forests than in younger ones and pine-dominated forests (Chapter 4.1.3). Bank voles move on top of snow cover and may even climb trees in search of food, making them vulnerable to avian predation. In accordance with this, nearly 60% of the diet identified from regurgitated pellet samples in nest-boxes prior to breeding during midwinter and early spring consists of bank voles, passerines and shrews, while the role of *Microtus* voles increases after the snow melts (see Chapter 5.7).

The second mechanism through which the cover of old forest may increase survival is that old forests probably provide sheltered refuges against large avian predators such as Ural owls, which are probably less capable of hunting in thick old forests with less hunting space and low visibility compared with younger forest age classes. Nocturnal Ural owls may decrease the occupancy of nest-boxes and impair the breeding success of boreal owls up to 2 km from Ural owl nests (Hakkarainen and Korpimäki 1996; Chapter 14.4.1). Hooting male boreal owls may be particularly vulnerable to predation by larger owls in late winter and early spring when they advertise their nest-holes to attract females. Accordingly, one study has documented hooting locations of boreal owls in different-aged forests in the north-western USA by comparing hooting locations of males with random locations within the study area. All hooting males were detected in areas with old-forest characteristics, including high basal areas of trees, tall snags, many large downed logs and a tall overstorey canopy (Herren et al. 1996). Old forests may thus provide shelter (and nesting cavities) against avian predators during the mating season. Similarly, it has been suggested that old forests provide cover that allows northern spotted owls to avoid predation by larger avian predators, such as the barred owl *Strix varia* (Franklin et al. 2000, Roberge et al. 2008). Furthermore, there is evidence that relatively large diurnal predators such as goshawks are capable of hunting parent owls in our study population. Two fledglings, two females and one male boreal owl were found in prey remains in the vicinity of goshawk nests in our study area. It is likely that these individuals were roosting in open and unsheltered forest stands, making them vulnerable to hawk attacks. In the future there is a need to examine the impact of predation on forest-dwelling species, because modern forestry probably decreases the amount of sheltered forest habitats, and hence predation may have an ever-increasing impact on forest-dwelling owls (for further details see Chapter 4.2). Furthermore, radio-tracking studies would give information on habitat choice of owls in the different phases of the vole cycle: do owls prefer old forests for hunting and roosting when compared with younger forest age classes within a home range under varying food conditions?

11.3.   Lifetime reproductive success

Lifetime reproductive success (LRS) is a good measure of an individual's reproductive success, as it covers a long period. Reproductive success measured over a shorter period, such as one breeding season, does not necessarily reflect the variation between individuals over a full lifespan (see Chapter 8.7). Therefore we examined the probable association between the proportion of old forests and LRS of boreal owls, because LRS is the best measure of individual fitness and combines both survival and reproductive success into a single measure (see Chapter 8.7). Therefore, in addition to survival, we also examined how the proportion of old forests in the territory was associated with the

LRS of males in our study area, which typically consists of a mosaic of different-aged forests and agricultural areas. In our study, LRS was measured by summing the total number of fledglings produced during a lifetime. Our data included 209 males that entered our breeding population during 1987–92 (Laaksonen et al. 2004).

Our main finding was that the LRS of males increased with the proportion of old forests in the home range owing to a higher number of breeding attempts there. The proportion of old forests correlated more strongly with the number of breeding attempts (r_s = 0.24, p = 0.0005) than the proportion of middle-aged forests did (r_s = 0.12, p = 0.07). In other words, the number of breeding attempts increased more obviously on home ranges in which the proportion of old forests was high. Furthermore, LRS of males increased significantly with the proportion of old forests (χ^2 = 6.24, p = 0.012), but not with the proportion of middle-aged forest (χ^2 = 1.98, p = 0.16). For more detailed analyses of LRS and habitat characteristics, see Laaksonen et al. (2004). We performed principal component analysis (PCA) to decrease the number of variables in our statistical models and to create new uncorrelated variables (for further details, see Laaksonen et al. 2004). The best component for the interpretation of the results represented increasing proportions of old and middle-aged forest (i.e. proportion of forest in Fig. 11.4), and decreasing proportions of agricultural land, young forest and clear-cut areas within a radius of 1000 m around the nest. LRS was not associated with the proportion of middle-aged and old forests within the home range of males that bred for the first time at the age of 1 year (χ^2 = 6.31, p = 0.11; Fig. 11.4, upper panel). Among older males that first bred at the age of ≥2 years, on the other hand, LRS increased with the proportion of middle-aged and old forest within a territory (χ^2 = 6.31, p = 0.12; Fig. 11.4, lower panel). Hence, old forests seem to be important in determining the LRS of males when compared with younger forest age classes, at least if the male owls breed in the population for the first time at a relatively older age. The reason for this owl-age-specific association between forest age and LRS, however, remains largely unknown. We suspect that older males are better able to take advantage of their territory owing to their greater experience.

The LRS of male boreal owls decreased with the proportion of agricultural land (χ^2 = 8.38, p = 0.004), probably because fledgling production crashed in the field-dominated areas in the decreasing phase of the vole cycle (Hakkarainen et al. 2003). This is probably due to the fact that in the decrease vole years, vole populations crash earlier in agricultural areas owing to a high predation impact by the many predator species there (Korpimäki and Norrdahl 1991a; Chapter 5.9.2).

Studies on the effects of habitat loss and fragmentation have traditionally been conducted at the population and community levels. One of the few examples where these effects have been studied at the individual level is a study on food provisioning and reproductive success of saw-whet owls in the aspen parkland of Alberta, Canada (Hinam and St. Clair 2008). Male saw-whet owls nesting in areas with less forest cover and long inter-patch distances had smaller home ranges, spent more time perch-hunting, and supplied their families less frequently. These males also showed higher levels of physiological stress (estimated by heterophil/lymphocyte H/L ratios) and fledged fewer young. Home range size, feeding rates and number of fledglings tended to be higher in

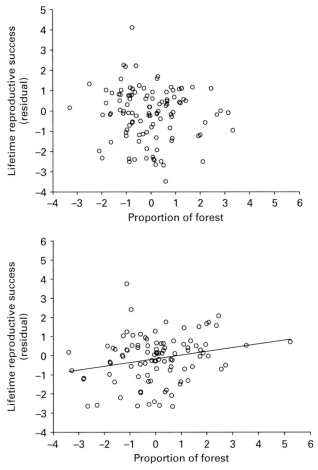

Fig. 11.4.   Lifetime reproductive success (LRS) in relation to the proportion of old-growth forest in the home range in males that bred for the first time at the age of 1 year (upper panel) and at the age of 2 or more years (lower panel). LRS is presented as the residuals from a model with vole cycle phase, age at first breeding, and proportion of agricultural land (see Laaksonen et al. 2004 for further details).

landscapes with an intermediate to large amount of forest cover (Hinam and St. Clair 2008). These results suggest that low levels of forest loss and fragmentation due to clearcutting may be beneficial to saw-whet owls, potentially by increasing food abundance, whereas higher levels of forest loss may in fact reduce the hunting efficiency of males, increase their stress levels and decrease their reproductive output (Hinam and St. Clair 2008).

To the best of our knowledge, our study yielded the first evidence to show that the LRS of forest-dwelling vertebrate species increases with the proportion of old forest in the territory. In the boreal owl, high LRS resulted mainly from a high number of breeding attempts in old-forest areas. We propose two mechanisms for the high LSR of males dwelling in old forests.

1. Old forests with dense forest structure probably offer better refuges than younger forests against avian predators such as the Ural owl (Hakkarainen and Korpimäki 1996) and the goshawk (Sunde et al. 2003, Sunde 2005).
2. Old forests may be better feeding areas for boreal owls, which would improve the probability of survival to the next breeding season.

Our explanations for high LRS and survival in old forests are thus parallel to those presented in the previous section because of the similar nature of those fitness measurements (i.e. high survival in old forests increases the number of breeding attempts, which in turn increases LRS).

In the future, there is a need to examine the effects of forest age, as well as other characteristics of the forest landscape, on LRS in so-called old-growth forest associates, because it is only during the last two decades that measures of LRS have become available. At the same time, methods for accurate and sophisticated habitat analyses at appropriate landscape levels have been intensively developed (e.g. geographic information systems, GIS). This will enable us to study in more detail whether the effects of anthropogenic changes in habitat quality threaten the viability of natural populations.

Our results provide an interesting insight into human influence at the landscape level caused by forestry. The essential measures of fitness – survival and LRS – were both significantly associated with the proportion of old forest within a home range. We suggest that, at the home range scale, the protection of about 15–20% of old forest containing over 150 m^3/ha of timber can be taken as an approximate threshold level, at least for survival (Fig. 11.3), and this is probably also enough to maintain the lifetime reproduction of male boreal owls at an adequate level (for further details on the conservation of the boreal owl, see Chapter 15).

12 Family planning under fluctuating food conditions

Animals are assumed to allocate resources between their own somatic maintenance, growth and survival, and reproductive investment to maximise reproductive output. This is often achieved by allocating as much to the current reproductive event as can be spared without a substantial decrease in future fecundity and survival (Williams 1966). Accordingly, an increase in reproductive output obviously increases the costs of reproduction. The studies on costs of reproduction can be classified into two categories depending on how trade-offs are considered: either intra-individual or intergenerational trade-offs (see Stearns 1989, 1992). The first examines the effects of current reproductive investment on the future parental survival and reproductive output of an individual (Williams 1966). The second links current reproductive effort with the future survival and breeding prospects of the offspring. To date, the costs of reproduction have been almost solely investigated from the intra-individual point of view, and mostly in stable environmental conditions (e.g. Smith C. and Fretwell 1974, Winkler and Wilkinson 1988). In contrast, studies on the survival prospects and breeding opportunities of independent offspring are scarce (Stearns 1989, 1992). This is a pity, because intergenerational trade-offs represent a large proportion of the different trade-off types (see Stearns 1989, 1992).

The ideas gained from research carried out in stable ecological environments have resulted in a narrow view of different life-history strategies, underlining the significance of optimisation with respect to different life-history traits, although the magnitude of variation in life-history traits has been considered to be an important factor in determining fitness (Giesel 1976). Individuals may respond averagely to a range of environmental variation, or they may evolve phenotypic plasticity. In the late 1970s and early 1980s, this kind of plasticity was also included in life-history theory, explaining reproductive strategies under varying environments (e.g. Hirshfield and Tinkle 1975, Giesel 1976, Stearns 1976, Goodman 1979, Carlisle 1982). In an environment that fluctuates in a predictable manner, the importance of investment in parallel with future changes in food abundance and offspring survival prospects has been stressed. This is relevant because, in many animal groups, variation (for example, in offspring size) has been considered to be an adaptation to variable environmental conditions (e.g. Capinera 1979, Kaplan 1980, Brody and Lawlor 1984). In this context, there is an absence of empirical studies on reproductive adjustment in heterogeneous environments, although theoretically this has been taken into account.

Reproductive effort is the proportion of the total energy budget of an animal that is devoted to reproduction at the expense of other functions (Hirshfield and Tinkle 1975).

It consists of mating effort and parental effort. We estimated parental effort of owls by egg size, clutch size, nest defence intensity, and feeding rate of the young. Nest defence can be seen as a good indication of short-term parental effort, as it is costly in terms of energy and involves a risk of injury or even death (Montgomerie and Weatherhead 1988). Feeding rate, on the other hand, can be seen as a measure of long-term investment. In this chapter we examine whether boreal owls are able to adjust their reproductive effort, i.e. to plan their family size and quality, in response to largely varying environmental conditions induced by wide fluctuations in main food abundance.

12.1.   Plastic responses in parental effort under fluctuating food conditions

We have been interested in the variation in important life-history traits and tactics of boreal owls, which face large variations in their main resource levels during their life cycle. Pronounced environmental fluctuations in vole numbers could create large temporal variation in the parental effort of boreal owls, which encounter quite predictable 3-year population cycles of voles in our study population (Fig. 1.1) and elsewhere. Under such conditions, there might be a role for phenotypic plasticity in reproductive traits. The question then is, are owls able to adjust their important life-history traits such as clutch size, egg size, risk taking in offspring defence and feeding effort in parallel with varying offspring survival prospects, or do boreal owls base their parental effort only on current benefits (i.e. on prevailing food conditions)? It has been suggested that animals may use some cue in predicting the future breeding probabilities (Murdoch 1966, Lindström E. 1988, Pilson and Rausher 1988), when adjustment in parallel with future resources might be adaptive.

This would be beneficial, especially if the survival prospects of offspring vary greatly over time. Due to wide annual variation in vole abundance, independent owl offspring face widely varying food conditions during their first year of life. Offspring born in the low and increase phases of the vole cycle will encounter increasing food densities during the independence period and during their first potential breeding season. In contrast, offspring born in the initially high but decreasing phase of the vole cycle will face a dramatic decline in food abundance soon after fledging, when most offspring will probably die. Some of them may survive until the next autumn migration, which also plays a part in decreasing the survival prospects of offspring. The idea that offspring survival prospects are highest in the increase phase of the vole cycle is supported by data on the proportion of fledged owlets that subsequently recruited to the whole Finnish population: this increased from the decrease (0.5%) through the low (0.7%) to the increase phase (1.6%) of the vole cycle (Korpimäki and Lagerström 1988). These values were taken as estimates of survival chances of offspring during the course of the vole cycle. Also, the recent data on the proportion of nesting attempts producing recruits in our study population suggested similar cycle-phase-related differences in offspring survival rate (Chapter 10.3).

12.2.   Variation in egg and clutch size

Clutch size is considered to be a central life-history trait (e.g. Stearns 1976, Bell 1980). Egg size may also be a fitness-related character, as it is known to affect the early survival and development of offspring (Parsons 1970). Accordingly, egg size was weakly but significantly correlated with the number of fledglings produced, which mainly determines the lifetime reproductive success of boreal owls (Korpimäki 1992b). During 1981–90, consisting of three 3-year vole cycles, boreal owls produced the largest clutches and eggs in the increase phase of the vole cycle (Table 8.2), despite the fact that voles were more abundant and egg-laying started earlier in the decrease phase than in the increase phase of the vole cycle (Korpimäki and Hakkarainen 1991, Hakkarainen and Korpimäki 1994b). Furthermore, there were only minor differences in clutch size between the low and decrease phases of the vole cycle during the same laying periods (Table 8.2), though the vole supply during egg-laying and in the preceding winter was poor in the low phase and good in the decrease phase. This suggests that boreal owls invest most in a clutch in the increase phase, when the reproductive value of eggs is largest because of the high survival of yearlings, and owls may even invest more in offspring in the low than in the decrease phase.

Females laid earlier and eggs were smaller in the decrease phase than in the increase phase (Korpimäki and Hakkarainen 1991, Hakkarainen and Korpimäki 1994b), which suggests that in the initially high but decreasing food conditions it may be beneficial to reproduce as early as possible, even at the expense of egg number and size. At that time, laying date obviously affects fitness more than egg size, because offspring survival is apparently more determined by length of time to the following vole crash than by egg size.

Female age had no obvious effect on egg size, but partners of older males produced large eggs in the increase phase and partners of younger males in the decrease phase of the vole cycle. The partners of adult males also decreased their egg volume from the increase to the decrease phase, while the partners of yearling males produced larger eggs in the decrease phase than in the increase phase (Hakkarainen and Korpimäki 1994b). These different age effects might be explained by the fact that in increasing vole years adult males invest a great deal in courtship feeding because of the high reproductive value of eggs at that time. In contrast, in decrease years, adult males may reduce the courtship feeding, due to assessment of the future crash of vole populations, which apparently diminishes the survival prospects of their independent offspring. In the early breeding season, female boreal owls are dependent on male food provision because females then have to put on weight to lay eggs. Therefore, differences in egg size are apparently a result of male courtship feeding, which is known to affect egg size in some birds (Nisbet 1973).

12.3.   Variation in nest defence intensity

Nest defence is an essential life-history trait, increasing the probability of successful breeding (e.g. Blancher and Robertson 1982). Nest defence can be considered a good example of reproductive investment, as it is costly in terms of energy and, above all, it includes the risk of injury or even death (e.g. Brown M. 1942, Montgomerie and Weatherhead 1988). We first tried to perform our nest defence experiments using a stuffed pine marten, but we then replaced it with a living American mink because it turned out that some owls reacted only mildly to a stuffed pine marten, although it was placed close to the entrance hole of the nest-box. Some males even delivered food to the nest, only a few centimetres away from the teeth of the stuffed pine marten! Hence, some owls probably perceived that a stuffed predator without active movements is not dangerous. This should be taken into account in future studies on nest defence intensity in birds of prey, because nest defence responses may vary greatly depending on the predator model used (i.e. stuffed or alive).

In our studies on the intensity of nest defence, a caged brown American mink was used as a proxy for a nest predation attempt during 1990–2. In the experiment ($n = 47$), a caged mink was placed on the roof of the nest-box and the defence behaviour of both sexes was recorded for 10 minutes. We used a defence index which varied from '1' = no reaction to the mink, to '5' = extremely vigorous defence, including direct strikes in which the owl touched the cage (for further details, see Hakkarainen and Korpimäki 1994c). Observations were made on light nights (from 22:00 to 02:00 hours).

Males defended their offspring more in the increase vole year than in the decrease vole year, whereas in females such a relationship was not found. The intensity of male and female nest defence was unrelated to brood size alone, but males defended high-quality broods with high survival prospects (brood size × mean recruitment probability of an offspring) more than low-quality broods with low survival prospects (Fig. 12.1). Females, however, had no response to brood quality (Fig. 12.1). In our brood quality estimates, mean recruitment probabilities of offspring were derived according to Korpimäki and Lagerström (1988). Therefore, males apparently invested a great deal in offspring defence in the increasing vole year, owing to the high recruitment rate of nestlings at that time. It seems that males are able to adjust their risks in nest defence according to the reproductive value of their offspring, while brood size alone seemed to be irrelevant to nest defence efficiency. To the best of our knowledge, this was the first study which has taken into account the reproductive value of offspring in the nest defence behaviour of parents. Generally, brood size alone has been considered. We suggest that in future studies on nest defence, especially under fluctuating environmental conditions, temporal variation in recruitment rate and survival of offspring should be taken more into consideration.

Surprisingly, a negative relationship was found between male age and the intensity of nest defence in a decrease vole year (Fig. 12.2). This suggests that before a vole crash, old males may have a higher residual reproductive value than young ones, or that old males may reduce their investment at a time when a large investment in offspring is not beneficial. Also, it is possible that parental effort is lowered in the decrease year due to the high costs of a previous breeding season. However, this is unlikely, as there is no clear evidence for the costs of reproduction in boreal owls (see Chapter 8.5.2), and the increasing amount of food in the increase phase of the vole cycle obviously compensating for the costs of parental effort. Anyway, this result is interesting and worthy of future study.

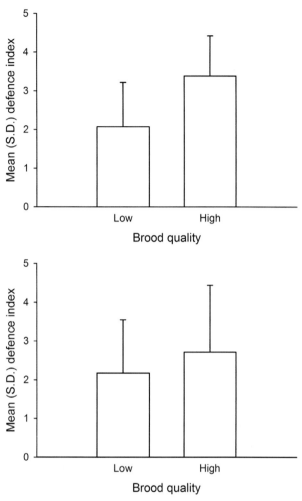

Fig. 12.1.   Mean (± s.d.) nest defence index of males (upper panel) and females (lower panel) caring for low-quality and high-quality broods (Hakkarainen and Korpimäki 1994c).

Why then did male owls adjust their nest defence according to the future survival prospects of their offspring, while such a relationship was not observed for females? This probably results from the different dispersal patterns of the two sexes. Natal dispersal of males is much shorter than that of females (Chapter 9.1). Furthermore, most males are site-tenacious after their first breeding attempt and occupy their territories continuously throughout their lives, whereas females disperse widely up to >500 km between successive breeding seasons (Chapter 9.3). Therefore, we suggest that females may not have adapted to prevailing fluctuations in vole abundance, because the amplitude and the length of the vole cycle largely varies in space (Hanski et al. 1991, Sundell et al. 2004). Males, instead, can probably assess the direction of changes in vole abundance within their home ranges and can foresee whether vole populations are increasing or decreasing

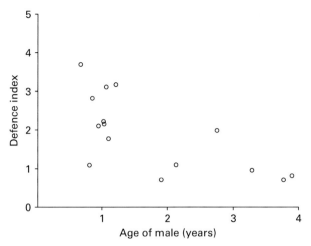

Fig. 12.2.   Nest defence index in relation to male age in a decrease vole year (Hakkarainen and Korpimäki 1994c).

in the future. Therefore, males can adjust their reproductive effort in relation to the reproductive value of their offspring (high in the increase phase, low in the decrease phase), as the parental investment theory predicts.

12.4.   Variation in feeding rate

Nest defence measures short-term parental investment, whereas feeding rate can be considered as a measure of long-term investment. Because males are the main food providers for the whole family during the breeding season, male hunting efficiency is a major determinant of breeding success in boreal owls. Hence, in the course of a vole cycle consisting of low, increase and decrease years of vole abundance, we measured the provisioning rate of males, using direct observations in the field and by counting prey remains accumulating at the bottom of the nest (for further details, see Hakkarainen and Korpimäki 1994a).

Prey deliveries of 56 males were recorded during a 4-hour period at night (22:00–02:00 hours) during 1990–2 (Fig. 12.3). Brood size in the observational nests increased from the low (mean ± s.d. 2.70 ± 0.95) through the increase (3.50 ± 1.65) to the decrease vole year (4.67 ± 1.62). Observations were made from a hide near the nest-hole when nestlings were 2 weeks old. At that time, the female broods the young in the nest-box and the male hunts for the whole family. To obtain estimates of long-term prey numbers delivered to the nest, we examined prey remains: a layer of pellets and other prey remains accumulates at the bottom of the nest-hole during the last 2 weeks of the nestling period (Chapter 3.6). These layers were collected after the breeding season to determine the number of different prey items. These two methods were comparable and probably gave reliable estimates of feeding efficiency of males,

Table 12.1. Vole-cycle-related differences in the mean (s.e.) number and biomass of food (g) delivered by males per offspring, and in the total number and mass of prey per offspring determined from the pellets and other prey remains accumulated at the bottom of the nest-hole. Dependent variables were log-transformed because of non-normal distribution (data from Hakkarainen and Korpimäki 1994a).

| | Phase of the vole cycle | | | | | | ANCOVA |
| | Low | n | Increase | n | Decrease | n | |
Dependent variable							p
No. of prey items brought per offspring	3.2 (0.7)	13	1.4 (0.2)	22	0.6 (0.1)	21	<0.001
Mass of prey items brought per offspring (g)	34.5 (7.8)	10	26.6 (2.4)	21	12.1 (1.3)	18	<0.001
No. of prey items per offspring at the bottom of the nest-box	29.3 (3.4)	27	24.7 (1.6)	87	18.8 (1.8)	117	<0.001
Prey mass per offspring at the bottom of the nest-box (g)	359.6 (53.5)	27	423.7 (26.3)	87	294.3 (28.1)	117	<0.001

Note. Clutch size minus the number of fledglings was used as a covariate to remove the effects of annual changes in food abundance in the course of the breeding season (see Hakkarainen and Korpimäki 1994a).

Fig. 12.3. Prey deliveries of males were recorded during a 4-hour period at night (22:00–02:00 hours). Photo: Pertti Malinen.

because the number of prey animals delivered by males to their nests during a 4-hour period was positively correlated with the total number of prey items found in prey remains accumulated at the bottom of the nest-hole ($r = 0.35$, df $= 50$, $p = 0.01$). This indicates that 4-hourly recording of food deliveries is sufficient to reveal the feeding frequency of individual males during the nestling period.

Our direct observations showed that males brought significantly more prey items per offspring in a low vole year than in the increase and decrease vole years (Table 12.1), although in the decrease vole year, food was more abundant in early spring. The data on

Table 12.2. The first food delivery of males after sunset (minutes; mean with s.e. in parentheses) and the last food delivery of males before sunrise in the different phases of the vole cycle (data from Hakkarainen and Korpimäki 1994a).

	Phase of the vole cycle						
	Low	n	Increase	n	Decrease	n	ANOVA
After sunset	16.1 (11.2)	13	36.8 (8.8)	21	76.5 (9.0)	20	$p < 0.001$
Before sunrise	28.9 (19.9)	13	120.6 (15.7)	21	153.7 (16.1)	20	$p < 0.001$

prey remains also showed that both prey number and prey mass per offspring were larger in the low and increase vole years than in the decrease year (Table 12.1), although in the decrease phase, food was more abundant than in the other phases. In the late evening (between 22:00 hours and midnight), males fed their family equally in the increase and decrease vole years. In the early morning hours (00:01–02:00 hours), males reduced their feeding rate compared with the late evening in the decrease year but not in other years (Hakkarainen and Korpimäki 1994a). Furthermore, in the increase vole year, males started to feed their family sooner after sunset than in the decrease year (Table 12.2). We suggest that males reduced their feeding effort in the decrease vole year, because they might have used more time for resting and their own feeding after midnight. Site-tenacious males may anticipate the future changes in vole abundance and predict the coming vole crash, when the reproductive value of their offspring will decrease to a large extent. When offspring survival is predicted to be poor, owing to deteriorating food abundance, it may be adaptive for an individual to invest in its own survival, rather than the survival in offspring with low life expectancies. Hence, a large investment in many offspring at that time could only decrease the residual reproductive value and future breeding prospects of the male.

12.5.   How to anticipate future changes in food abundance

Many studies have shown that birds can adjust their clutch size in parallel with the present food supply in the field (e.g. Perrins 1965, Arcese and Smith J. 1988, Järvinen A. 1989; and Chapter 8.1.1), whereas evidence for clutch size and egg size adjustment to future conditions is mostly lacking (Clutton-Brock 1991). Instead, some plastic responses have been detected in insects: for example, several herbivorous insects are known to lay smaller clutches of eggs in host plants which provide fewer resources for the developing young (e.g. Pilson and Rausher 1988). In poor conditions with low survival of offspring, carabid beetles *Aconum* sp. did not breed, apparently to increase their probability of reproducing in the next season, when conditions may have improved (Murdoch 1966). Adjustment in the hatching time of the Atlantic herring *Clupea harengus* might also be linked with the moment when plankton abundance peaks, ensuring good feeding conditions for the developing newly hatched larvae (Cushing 1975). The reproductive effort of red foxes *Vulpes vulpes* was found to be related to the future abundance of fluctuating vole prey rather than to current food abundance

(Lindström E. 1988). These examples seem to be in accordance with the results found in our study population. Furthermore, both American and Eurasian tree squirrels (*Tamiasciurus hudsonicus* and *Sciurus vulgaris*) anticipate mast years and increased reproduction and population growth rates before a masting event (Boutin et al. 2006). Accordingly, there seems to be increasing evidence that animals living in environments with conditions that vary in a predictable manner may adjust their reproductive effort to future changes in the environment or to survival prospects of their independent offspring.

How do owls anticipate future changes in food abundance? For boreal owls we propose the following explanation. Males, which do most of the hunting, may use some cue in determining the direction in the density changes of local vole populations. Winter reproduction of voles of the genus *Microtus* usually happens in the increase phase of the population cycle (e.g. Hansson 1984, Korpimäki et al. 1991, Norrdahl and Korpimäki 2002b). Therefore, resident male owls may record that winter vole densities are increasing and culminate their parental effort in the next breeding season when the expected survival prospects of offspring are high. Also, during the egg-production period, owls may anticipate future density changes of voles on the basis of the voles' body size and the proportion of breeding females versus juvenile voles in the field. Lindström E. (1988) suggested that, in mammalian predators, there may be a hormonal linkage between prey and predator, so that a physiological by-product of the gonadotropic hormones provided by the ingested voles may reveal the future changes in food supply. This seems unlikely in avian predators because of differences in their hormonal characteristics compared with mammalian prey. Anyway, this would need further study dealing with the physiological responses of individuals under varying food conditions.

The fact that male boreal owls reduced their parental effort in initially high but decreasing years of vole abundance could be due to the changing age composition of the owl population during the vole cycle, as the proportion of breeding young males is highest in decrease vole years (Korpimäki 1988c). We discount this possibility because, in the decrease phase of the vole cycle, young males invested in courtship feeding and nest defence more than adult males, counterbalancing the effect of age on phase-related differences in parental effort. Because density of owl populations is high in the initially high but decreasing years of the vole cycle, density-dependent effects, such as competition, might explain low reproductive effort in that phase. This may be valid for studies on clutch size, egg size and feeding effort, but not for the nest defence experiment, because the intensity of nest defence was highest in the increase vole year and not in the decrease year, when intraspecific competition within the owl population is probably high. Cycle-phase-related differences in the locomotion activity of voles might partly explain the higher reproductive effort of males in the increase phase than in the decrease phase of the vole cycle. However, Halle and Lehmann (1992) showed that field voles, which are one of the main foods of boreal owls (see Chapter 5.4), did not shift their activity pattern between the increase and decrease phases of the vole cycle, while the low phase differed from the other phases of the vole cycle. The results for egg and clutch size, as well as for provisioning rate, might be affected by possible annual differences in the vulnerability of voles. For example, the age composition of the vole population probably changes over the course of the vole cycle. However, the intensity of nest defence is apparently independent of such effects.

12.6. High reproductive effort with improving survival prospects of offspring?

Up until now, reproductive effort and the costs of reproduction have been investigated almost solely from the point of view of the reproducing individual, and generally under relatively stable environmental conditions. In contrast, studies on the survival prospects and breeding chances of independent offspring are scarce. In this context the boreal owl is a good model species, as it lives under highly varying food conditions.

Natural selection has probably promoted phenotypic plasticity in the reproductive traits of boreal owls. Accordingly, low repeatability and heritability of egg size and clutch size (Chapter 8.2), as well as flexibility in the intensity of nest defence (Chapter 12.3) and feeding effort (Chapter 12.4) present an opportunity to maximise fitness at any stage in a fluctuating environment. The first-year survival of owlets born in the increase phase of the vole cycle was twice as high as in the low and decrease phases of the cycle (Chapter 10.3). Thus, it may be advantageous for parents to invest more in offspring in the increase phase than in the decrease phase of the cycle. Accordingly, clutch size, egg size, the intensity of male nest defence and feeding effort culminated in the increase phase of the vole cycle, rather than in the decrease phase, when food was most abundant in the previous autumn. In the decrease phase, when survival prospects of offspring were low, older males reduced the intensity of nest defence compared with yearling males. This suggests that, in decrease vole years, older males had better experience in prevailing fluctuations of food abundance or higher residual reproductive value than yearling males. Similarly, the intensity of courtship feeding by older males may vary annually, as their partners decreased egg size from the increase to the decrease phase, while the partners of yearling males produced the largest eggs in the decrease phase of the vole cycle.

In conclusion, most of our results on the parental effort of boreal owls suggest that old males may be able to adjust their reproductive effort not only to prevailing food conditions but to the survival prospects of their offspring. These findings are consistent with the life-history theory predicting that in varying but predictable environments, reproductive effort should be high when survival chances of offspring are good in order to maximise lifetime reproductive success (Hirshfield and Tinkle 1975, Goodman 1979). Although male boreal owls have not heard anything about this grandiose theory, they appeared to be able to adjust their reproductive effort as the theory predicts. So far there is a paucity of studies on the adjustment of reproductive effort parallel with changes in the survival prospects of off-spring, although models derived from Ural owls suggest that this may also be the case in this species living under rather similarly varying environmental conditions in southern Finland (Brommer et al. 2000). The cyclic food, residual reproductive value of parents, and reproductive value of offspring, however, may vary in a complex way in the course of the vole cycle, and hence more intensive studies examining the fate of radio-marked parents and offspring should be conducted over appropriate temporal and spatial scales.

13 Population dynamics

13.1.   Vole cycle induces wide variation in population dynamics

Knowledge of the factors that regulate population dynamics and density may help us to disentangle how local populations would respond to natural or human-induced environmental changes, and why many populations are declining, and some also increasing, in temporally and spatially varying environments.

During 1966–2010, we found a total of 1605 nests of boreal owls in our study area (Table 1). In the core area of our boreal owl nest-box network, covering 400 km^2 (20 × 20 km), the mean breeding density estimate varied from 0 to 16.5 nests (mean 4.8 ± 4.4 nests) per 100 km^2 during 1977–2009 (Fig. 13.1). We have usually used the number of nests found per 100 nest-boxes in our whole study area, covering approx. 200–1300 km^2 during 1973–2009, as an estimate when studying the between-year variation of boreal owl breeding density and its relationship to abundance of main foods (voles) (e.g. Korpimäki et al. 2008, 2009). The question then is whether this breeding density estimate really reflects the between-year variation in the breeding density of boreal owls in our study population. The answer to this question appears to be yes, because these two breeding density estimates were closely positively correlated (Fig. 13.1; $r_s = 0.95, n = 33, p < 0.0001$). Therefore, it appears to be appropriate to use the number of nests found per 100 nest-boxes to compare owl breeding densities between years.

There was a large amount of variation in breeding density estimates during the study period: in years of low vole abundance only 2–4 owl nests were found in 100 nest-boxes, whereas in years of high vole abundance 20–33 nests were found, with a mean of 9.7 nests per 100 boxes during 1973–2009 (Fig. 13.2, upper panel). In the poorest years, such as in 1998, 2001 and 2010, only 2–4 nests were found in our 490–500 nest-boxes, whereas in the best years the total number of nests was from 99 (1988) to 163 (1992). It is remarkable that the highest nest number was recorded in 1992, when the eminent authority on the ecology of North American owls, the late Professor Carl D. Marti (1944–2010; Weber State University, Ogden, Utah, USA; see Kochert et al. 2010), visited our study area and spent some never-to-be-forgotten days trapping and ringing boreal owl parents and owlets with us.

In the Seinäjoki region, from 83 to 507 (mean 305) nest-boxes suitable for boreal owls were annually inspected and from 0 (2001 and 2004) to 84 (1992) nests were found in these boxes during 1973–2009. The number of nests per 100 boxes varied widely, from 0

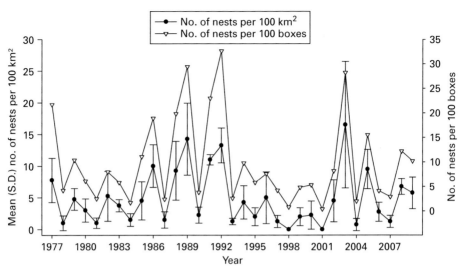

Fig. 13.1.   Mean (± s.d.) breeding density (no. of nests per 100 km²) in four 10 × 10 km squares (Kauhava, Huhmarkoski, Orava and Ojutjärvi landscape maps) in the middle of the 1300 km² study area of boreal owls, and the number of boreal owl nests per 100 nest-boxes in the whole study area covering 1300 km² during 1977–2009.

to 28 (1992), with a mean of 7.1 nests per 100 boxes during 1973–2009 (Fig. 13.2, lower panel). A total of 183–449 (mean 310) boreal owl nest-boxes were annually inspected and from 0 (2001) to 58 (2003) nests were detected during 1973–2009 in the Ähtäri region. There also was a large between-year variation in the occupancy rate of the boxes, from 0 (2001) to 17, with a mean of 5.3 nests per 100 boxes during 1973–2009 (Fig. 13.3, upper panel). In the Keuruu region, from 60 to 283 (mean 176) similar nest-boxes were annually checked and 0 (1987) to 31 (1986) nests were found in these boxes during 1973–2009. Also in this area, the density estimate varied greatly between years, from 0 to 22 nests per 100 boxes during 1973–2009 (mean 4.9 nests; Fig. 13.3, lower panel).

In all four study areas, the increase in owl breeding densities from low to peak numbers usually took place over 2 years, whereas the decline happened in 1 year (Figs. 13.2 and 13.3). At the Kauhava, Seinäjoki and Ähtäri study sites, the number of owl nests per 100 boxes was closely positively correlated with the abundance indices of *Microtus* and bank voles in the current spring, but less so with the abundance indices of these voles in the preceding autumn (Fig. 13.2 and upper panel of Fig. 13.3). At the Keuruu study site, this breeding density estimate of owls was only associated with the abundance indices of bank voles in the preceding autumn (Fig. 13.3, lower panel). Moreover, the number of hooting males per 10 boxes (data from Fig. 3.16) was related to the abundance index of bank voles in the current spring during 1979–92 ($r_s = 0.73, n = 14, p = 0.003$) and with the abundance index of *Microtus* voles in the preceding autumn ($r_s = 0.71, n = 14, p = 0.004$). Therefore, we conclude that it is the abundance of *Microtus* and bank voles in late winter that determines both the number of hooting males and the subsequent breeding density and size of local boreal owl populations.

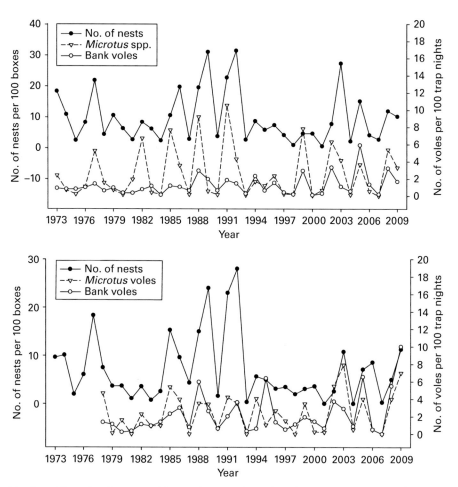

Fig. 13.2.   Number of boreal owl nests per 100 nest-boxes and number of *Microtus* and bank voles snap-trapped per 100 trap-nights in May during 1973–2009 in the Kauhava region (upper panel) and the Seinäjoki region (lower panel), western Finland. (Kauhava region: Spearman rank correlation $r_s = 0.68$, $n = 37$, $p < 0.001$ and $r_s = 0.71$, $n = 37$, $p < 0.001$ for abundance indices of *Microtus* and bank voles in the current spring, respectively, and $r_s = 0.60$, $n = 36$, $p < 0.01$ and $r_s = 0.61$, $n = 36$, $p < 0.01$ for abundance indices of these voles in the preceding autumn, respectively). (Seinäjoki region: $r_s = 0.65$, $n = 31$, $p < 0.001$ and $r_s = 0.57$, $n = 31$, p 0.001 for abundance indices of *Microtus* and bank voles in the current spring, and $r_s = 0.08$, $n = 32$, $p = 0.65$ and $r_s = 0.17$, $n = 32$, $p = 0.34$ for abundance indices of these voles in the preceding autumn, respectively). Correlations between number of nests per 100 boxes and study year: $r_s = -0.11$, $n = 37$, $p = 0.53$ for the Kauhava region, and $r_s = -0.18$, $n = 37$, $p = 0.29$ for the Seinäjoki region.

13.2.   Inter-areal temporal variability in population fluctuations

One of the most general patterns of animal species diversity is the increase in the number of terrestrial species from the earth's poles towards the equator (e.g. Begon et al. 1996). This pattern is complicated by the variety of landforms (e.g. mountains), proximity to the

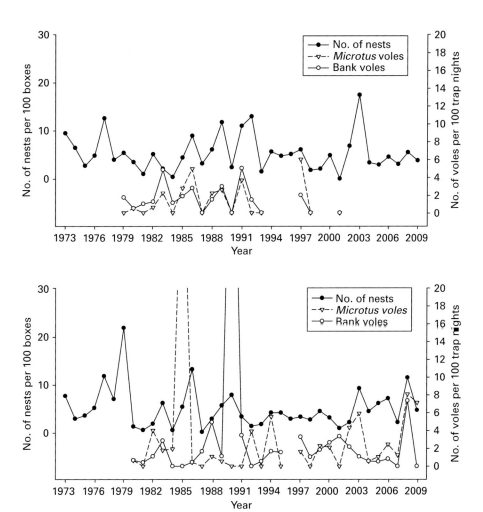

Fig. 13.3.   Number of boreal owl nests per 100 nest-boxes and number of *Microtus* and bank voles snap-trapped per 100 trap-nights in May during 1973–2009 in the Ähtäri region (upper panel) and the Keuruu region (lower panel), western Finland. (Ähtäri region: Spearman rank correlation $r_s = 0.48$, $n = 18$, $p = 0.04$ and $r_s = 0.54$, $n = 18$, $p = 0.02$ for abundance indices of *Microtus* and bank voles in the current spring, and $r_s = 0.44$, $n = 17$, $p = 0.08$ and $r_s = -0.21$, $n = 18$, $p = 0.41$ for abundance indices of these voles in the preceding autumn, respectively. Keuruu region: $r_s = 0.28$, $n = 28$, $p = 0.15$ and $r_s = 0.23$, $n = 28$, $p = 0.25$ for abundance indices of *Microtus* and bank voles in the current spring, and $r_s = -0.19$, $n = 32$, $p = 0.30$ and $r_s = 0.42$, $n = 32$, $p = 0.02$ for abundance indices of these voles in the preceding autumn, respectively). Correlations between number of nests per 100 boxes and study year: $r_s = -0.10$, $n = 37$, $p = 0.55$ for the Ähtäri region, and $r_s = -0.03$, $n = 37$, $p = 0.87$ for the Keuruu region.

coasts, season of the year, and the scale used when looking at these patterns. The species diversity of European and North American birds generally follows this pattern (e.g. Järvinen O. 1979, Rabenhold 1993). The stability (constancy in time) of animal populations also varies with latitude, so that southern populations are more stable in time than northern ones. For example, southern and central European land-bird assemblages are relatively stable between years compared with northern European assemblages (Järvinen O. 1979). Moreover, populations of small mammals (voles, mice and shrews) and small game (hare and grouse) are usually relatively stable between years in western and central Europe, just showing seasonal changes, with low densities in spring and high densities in autumn. In northern Europe, however, small mammal populations fluctuate in a cyclic manner, with peaks usually separated by 3–4 years (Hansson and Henttonen 1988, see also Chapter 1). The question then is whether the cyclicity or amplitude of the population fluctuations of boreal owls also shows any regional trends within Europe associated with the differences in the diversity patterns and stability of the main prey populations in Europe.

There are 20 local ornithological societies in Finland covering the whole country (Fig. 3.1). Bird-watchers and ringers have monitored the breeding densities of birds of prey in all these areas from 1986 onwards (Saurola 2008, 2009, Honkala et al. 2010) and, in particular, the breeding densities of hole-nesting owls have been studied by annually inspecting thousands of nest-boxes from 1977 onwards (Fig. 13.4 and Table 3.4). We used these long-term, large-scale data, which are based on hours of unstinting effort put into fieldwork by voluntary ringers and bird-watchers, to analyse the temporal variability of fluctuations in breeding density estimates of boreal owls during 1977–2009. For the area covered by the Ornithological Society of Suomenselkä, four time series (the Kauhava, Seinäjoki, Ähtäri and Keuruu regions) during 1973–2009 were used (Figs. 13.2 and 13.3). In addition, we compiled all the published records of fluctuations in breeding populations of boreal owls elsewhere in Europe (a total of 19 time series). This gave a total of 42 time series extending from northern Italy and Switzerland to northern Finland and northern Norway (46° to 69°N; Table 3.4). Only series covering 4 years or more were accepted, as vole cycles in northern parts of Fennoscandia commonly extend over 3 years (Sundell et al. 2004). A shorter number of years could have resulted in a possible error in cyclicity indices for owls in these northern areas. Only data sets from fluctuations in the number of nests were used, because the monitoring of hooting boreal owl males does not give reliable information about density variations in breeding populations (the majority of hooters being bachelor males; Chapter 7.3.2). In addition, we attempted to use only those data sets in which the number of suitable nest-boxes or holes for the owl in different breeding seasons has been relatively constant. The amplitude of density variations in owls was estimated by computing coefficients of variation (CV = s.d. × 100/mean, hereafter called the cyclicity index) for different local populations.

The temporal constancy of breeding owl populations decreased towards both the north and the east within Europe, but the amplitude of population fluctuations was more closely correlated with longitude (Fig. 13.5). Therefore, the regional trend for cyclicity indices was similar to regional trends for diet width indices of boreal owls, which also tended to decrease with longitude (Fig. 5.13). In the earlier analyses including cyclicity indices for 30 local populations of boreal owls up to 1984 in Europe (figure 1 in Korpimäki 1986b), cyclicity indices within northern Europe were more closely correlated with mean annual period with snow and mean maximum snow depth than with longitude and latitude. However, this was not true in the present larger data set ($r_s = 0.06$, $n = 29$, $p = 0.75$ and

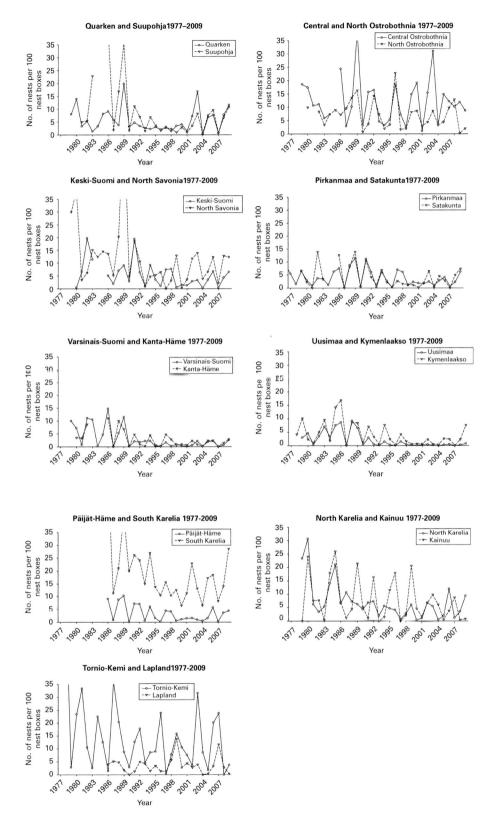

Fig. 13.4. Yearly number of boreal owl nests per 100 boxes in the districts of 18 local ornithological societies in Finland (see their locations in map shown in Fig 3.1) during 1977–2009.

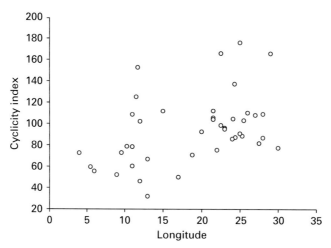

Fig. 13.5.   Inter-areal cyclicity indices of the number of nests for 42 local populations of boreal owls (data from Table 3.4) plotted against the longitude (°E) within Europe ($r_s = 0.44$, $n = 42$, $p = 0.004$; $r_s = 0.36$, $n = 42$, $p = 0.02$ for latitude).

$r_s = -0.05$, $n = 29$, $p = 0.98$, respectively), perhaps because snow cover and period have been decreasing in northern Europe from the early 1990s onward.

We found that temporal constancy of boreal owl populations decreased towards the east and north within Europe, although the data used in the analyses were derived from several studies conducted at different altitudes, from different study periods, and from populations of various sizes and numbers of nest-boxes inspected annually. Unfortunately, northernmost records from the margins of boreal owl distribution in northern Europe are still scarce, as they are from southern Sweden, where vole populations do not show any marked multi-annual cyclic fluctuations, probably due to heavy impacts by generalist predators (Erlinge et al. 1983, 1984). Environmental heterogeneity appears to be another source of bias, because, for example, the proportion of agricultural land is clearly higher in southern and central Finland than at the same latitudes of Sweden. The diversity of habitats may stabilise population fluctuations of both voles (Hansson 1979) and owls (Korpimäki 1986b). Because the amplitude of population cycles of voles becomes larger northwards and eastwards in Europe (Hansson and Henttonen 1985) and also within Finland (Sundell et al. 2004), we conclude that the geographical trends observed in owl fluctuations were mostly due to differences in the variations in vole densities. Interspecific synchronism between population fluctuations of small mammals decreases from north to south (Hansson and Henttonen 1985) and the densities of breeding birds increase in the same direction (Järvinen O. and Väisänen 1980). Furthermore, in central Europe two species of wood mice *Apodemus* spp. are important alternative prey for boreal owls (Chapter 5.4). Climate also becomes more continental towards the east and north in Europe, and a deep snow layer and long cold periods there can increase the mortality of boreal owls through starvation. These factors probably decrease the stability in north-eastern European owl populations in comparison with south-western European populations.

13.3.   Inter-areal spatial synchrony in population fluctuations

Large-scale spatial synchrony of population fluctuations covering hundreds of square kilometres at northern latitudes is a fascinating phenomenon. It has most often been documented for species showing cyclic population dynamics, including lepidopterans (e.g. Klemola et al. 2002), game birds (Ranta et al. 1995) and small to medium-sized mammals (Sundell et al. 2004, Huitu et al. 2003b). Small rodents and their predators at northern latitudes, for example, show synchronous fluctuations across distances of 80–100 to 500–600 km (Sundell et al. 2004, Huitu et al. 2008). Spatial synchrony of population fluctuations of small rodents has been suggested to be attributable to three mutually non-exclusive mechanisms:

- spatially correlated climatic environmental perturbations (i.e. the Moran effect; see Ranta et al. 1995)
- predation by mobile predators, including nomadic owls and diurnal raptors (e.g. Ydenberg 1987, Korpimäki and Norrdahl 1989, Norrdahl and Korpimäki 1996)
- dispersal movements of the focal animals themselves (e.g. Paradis et al. 2000).

We computed inter-areal cross-correlation coefficients for all the 23 time series of Finnish local owl populations to explore spatial synchrony (Figs. 13.2, 13.3, 13.4 and Table 13.1). These correlation coefficients showed that significant spatial synchrony of owl population fluctuations extended up to 250 km (Fig. 13.6). Looking at the data in greater detail revealed that the spatial synchrony of owl cycles in western, south-western and southern Finland extended over even larger distances (Figs. 13.2, 13.3, 13.4 and Table 13.1). For example, owl populations in our study area fluctuated in close synchrony with all the surrounding areas at distances of 50–150 km (Figs. 13.2, 13.3 and 13.4). In fact, the spatial synchrony of owl population fluctuations in our study area extended to the south-western and southern coast, including Varsinais-Suomi in the vicinity of Turku, Uusimaa in the vicinity of Helsinki (380 km away), Kymenlaakso in the vicinity of Kotka, and South Karelia in the vicinity of Lappeenranta (200–400 km away). Towards the east, the spatial synchrony extended only to North Savonia near Kuopio and northwards only as far as Central Ostrobothnia near Kokkola and Ylivieska (150–200 km away). A similar spatial pattern was also found for most other western, south-western and southern boreal owl populations. In the easternmost province of Finland (North Karelia) and in Kainuu, North Ostrobothnia, Kemi-Tornio and Lapland, owl populations appeared to fluctuate in their own synchronous cycles, in which owl population densities peaked when they were in the low phase in western and southern Finland. They were thus in the low phase when owls in the south peaked. Nationwide bird-ringing data from vole-eating boreal, Ural and long-eared owls, as well as from rough-legged buzzards *Buteo lagopus*, during 1973–2000 also showed that fluctuations in breeding densities were synchronous over large areas and extended as far as 500–600 km (Sundell et al. 2004). The spatial synchrony of population fluctuations of owls is therefore evident over much larger areas than have been reported for north European vole cycles (80–120 km; Huitu et al. 2003b, 2008), but proper large-scale,

Table 13.1. Spearman rank correlation matrix between the number of boreal owl nests per 100 boxes in different localities and provinces of Finland (see Fig. 3.1 for their locations and Figs. 13.2, 13.3, 13.4 and Table 3.4 for data) during 1973–2009. The localities or provinces have been arranged in order of increasing distance from our study area (the Kauhava region in western-central Finland). Spearman rank correlation (r_s, n, two-tailed p value) between the number of owl nests per 100 boxes and year (1973–2009) in parentheses in the left column.

Locality or province	2	3	4	5	6	7	8	9	10	11	12	13	14	15	16	17	18	19	20	21	22	23
1 Kauhava (−0.11,37,0.53)	0.84	0.84**	0.46**	0.63**	0.76**	0.70**	0.40*	0.58**	0.82**	0.64**	0.49**	0.60**	0.42*	0.60**	0.66**	0.27	0.65**	0.34	0.11	0.27	−0.06	−0.16
2 Seinäjoki (−0.18,37,0.29)		0.74**	0.38*	0.70**	0.77**	0.53**	0.47*	0.62**	0.75**	0.52**	0.52**	0.76**	0.42*	0.59**	0.69**	0.30	0.58**	0.53**	0.15	0.34	0.20	−0.08
3 Ähtäri (−0.10,37,0.55)			0.37	0.48**	0.54**	0.62**	0.35	0.54**	0.67**	0.42*	0.44*	0.59**	0.24	0.43*	0.52**	0.06	0.39*	0.20	0.08	0.24	0.02	−0.18
4 Keuruu (−0.03,37,0.87)				0.36*	0.46*	0.33	0.20	0.38*	0.49**	0.40*	0.16	0.17	0.13	0.44	0.34	0.44**	0.41*	0.02	−0.20	−0.28	−0.31	−0.43*
5 Quarken (−0.11,31,0.55)					0.62**	0.67**	0.16	0.40*	0.58**	0.54**	0.44*	0.45*	0.36*	0.42*	0.57**	0.54**	0.65**	0.59**	0.17	0.27	0.01	−0.14
6 Suupohja (−0.27,27,0.17)						0.46*	0.73**	0.76**	0.83**	0.79**	0.77**	0.84**	0.77**	0.84**	0.77**	0.35	0.72**	0.57**	−0.09	0.12	−0.12	−0.17
7 Central Ostrobothnia (−0.08,30,0.69)							−0.00	0.30	0.44	0.48**	0.28	0.27	0.20	0.25	0.29	0.24	0.55**	0.28	0.37**	0.58**	0.13	−0.07
8 Keski-Suomi (−0.32,28,0.09)								0.75**	0.69**	0.54**	0.77**	0.86**	0.73**	0.78**	0.67**	0.06	0.25	0.24	−0.28	−0.20	−0.19	−0.18
9 Pirkanmaa (−0.24,37,0.15)									0.84**	0.70**	0.86**	0.92**	0.77**	0.85**	0.64**	−0.05	0.48**	0.31	−0.13	−0.03	−0.29	−0.17
10 Satakunta (−0.18,29,0.36)										0.81**	0.77**	0.84**	0.72**	0.84**	0.80**	0.24	0.64**	0.36	−0.09	0.01	−0.30	−0.21
11 Varsinais-Suomi (−0.41,31,0.02)											0.78**	0.72**	0.75**	0.78**	0.75**	0.17	0.61	0.39*	−0.04	0.19	−0.22	−0.12
12 Kanta-Häme (−0.37,27,0.06)												0.90**	0.94**	0.89**	0.70**	0.08	0.38*	0.29	0.08	−0.07	−0.22	−0.12
13 Päijät-Häme (−0.31,24,0.15)													0.92**	0.91**	0.72**	0.07	0.49**	0.48**	−0.11	0.02	−0.16	0.00
14 Uusimaa (−0.61,31,<0.001)														0.89**	0.81**	0.11	0.59**	0.53**	0.06	0.04	−0.19	−0.12
15 Kymenlaakso (−0.41,32,0.02)															0.84**	0.18	0.57**	0.45*	−0.13	−0.08	−0.37*	−0.25
16 South Karelia (−0.36,24,0.08)																0.32	0.58**	0.64**	−0.07	−0.04	−0.30	−0.27
17 South Savonia (0.18,24,0.39)																	0.49**	0.53**	0.03	−0.03	−0.16	−0.07
18 North Savonia (−0.35,31,0.06)																		0.71**	0.25	0.33	0.05	0.00
19 North Karelia (−0.48,31,0.007)																			0.26	0.37*	0.32	0.17
20 Kainuu (−0.20,31,0.29)																				0.60**	0.58**	0.25
21 North Ostrobothnia (−0.27,29,0.16)																					0.75**	0.39
22 Kemi-Tornio (−0.32,33,0.07)																						0.45*
23 Lapland (−0.16,24,0.27)																						

*two-tailed $p < 0.05$, **$p < 0.01$, $n = 24$–37.

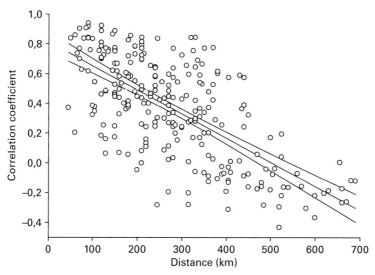

Fig. 13.6.   Linear regression and 95% confidence intervals of inter-areal cross-correlation coefficients (Spearman rank correlation) from Table 13.1 against the distance between areas (regression equation $y = 0.815 - 0.0016x$, $F_{1,251} = 221.8$, $r^2 = 0.47$).

long-term data from abundance estimates of voles are lacking in Finland and elsewhere in northern Europe.

The crucial question is what might be the most likely explanation for the spatially correlated population fluctuations of owls in Finland. Our results of synchrony investigations over large areas up to at least 250 km apart suggest that some explanations are more probable than others. We believe that the most likely explanation is vole-supply-dependent dispersal movements of owls. This suggestion is also supported by the finding that numbers of boreal owl nests found in the study areas of Harz Mountain in Germany, and eastern Belgium (approx. 380 km apart), fluctuated in close synchrony during 1977–2006 ($r_s = 0.42$, $n = 30$, $p = 0.02$) (see Table 3.4). The scale of spatial synchrony of population fluctuations of voles is considered to be much less in temperate Europe than in northern Europe (Chapter 1). Both juvenile females and juvenile males as well as adult female boreal owls have been reported to move between their birth site and first breeding site and between successive breeding attempts at the scale that was observed in the synchrony of owl population fluctuations. Both natal and breeding dispersal are also more extensive in the decline than in the increase phase of the vole cycle (Chapter 9.2 and 9.3). Even polyandrous females were reported to move up to 200 km, and probably even more, between successive breeding attempts within a season (Chapter 7.3.7). Therefore, we conclude that vole-supply-dependent dispersal from declining abundance areas to increasing abundance areas is the most likely explanation for the large-scale synchrony of population fluctuations of owls found in Finland. Nomadic owls and diurnal raptors can then, in turn, operate as synchronising agents at the scale of spatial synchrony of population fluctuations of small mammals (Chapter 5.9.2). Spatially correlated climatic environmental perturbations, such as long and heavy snowfall or sleet periods, can also act as complementary synchronising factors over large areas by protecting voles against boreal owl hunting and thus decreasing the survival of owls via starvation.

13.4.   Long-term trends in breeding populations

There were no obvious long-term trends in the breeding density estimates for boreal owls in our study area (Fig. 13.2), in the Seinäjoki region (Fig. 13.2), in the Ähtäri region (Fig. 13.3) or in the Keuruu region (Fig. 13.3) during 1973–2009. However, for all these areas the breeding density estimates tended to decline slightly with time (Table 13.1). The declining trend for breeding density estimates over time also emerged for data from all the other local ornithological societies of Finland, apart from South Savonia, where the annual numbers of nest-boxes inspected remained low (Fig. 13.4 and Tables 3.4, 13.1). The declining trends with time were significant or almost significant for the areas in the vicinity of the southern (Uusimaa, Kymenlaakso and South Karelia) and south-western coasts (Varsinais-Suomi and Kanta-Häme; Fig. 13.1), as well as for the central-eastern region (Keski-Suomi, North Savonia and North Karelia) and for Kemi-Tornio in the north. In the remaining area in western and northern Finland, the declining trend appeared not to be close to significant.

When pooling all the 23 time series of breeding density estimates in Finland, a highly significant decrease in boreal owl population was evident (Fig. 13.7, upper panel). The annual decrease rate in owl population was 2.1% during 1973–2009. This corresponds well with the annual decline rate (2.3%) for population indices estimated on the basis of numbers of boreal owl territories per 100 km^2 in 'raptor and owl grid' surveys during 1982–2009 (Fig. 13.7, lower panel). In each year of this period, 60–120 'raptor and owl grids' were surveyed nationwide, and ringers and bird-watchers have expended much effort in finding all the territories and nests of birds of prey (Honkala et al. 2010; see also Saurola 1986). We also analysed long-term trends (>20 years) in the five other local populations of Europe: Umeå (Sweden), Harz Mountain (Germany), eastern Belgium, Jura Mountains (Switzerland) and West Jura Mountains (Switzerland; Table 3.4). Correlation coefficients for five of these populations were negative but none of them appeared to be close to significant. For the eastern Belgium population, a temporal increase was found ($r_s = 0.61$, $n = 33$, $p < 0.001$).

What then are the probable reasons for the drastic declines in the Finnish boreal owl population over the last four decades (1970s to 2000s)? Our preliminary results indicate that even a moderate degree of clear-cutting within owl territories (on average 49% of total forest area clear-cut or sapling areas, range 35–70%) may increase rather than decrease the breeding success of boreal owls (Hakkarainen et al. 1996b), probably by creating new grassy habitats for their main prey, *Microtus* voles. Our recent results using modern GIS analyses, however, showed that over-winter survival and lifetime reproductive success of male owls decreased with reduction in the cover of old-growth forests (Chapter 11). In addition, owls preferred to hunt in spruce and pine forests and avoided hunting in open clear-cut and agricultural areas (Chapter 4.1.1). Therefore, our favourite hypothesis is that the decreasing area of old-growth forests is the most likely reason for

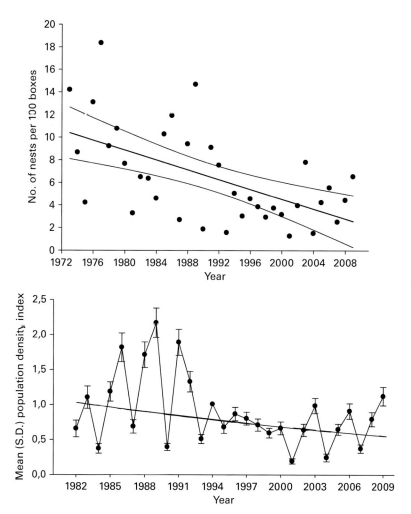

Fig. 13.7.   Linear regression and 95% confidence intervals for yearly breeding density estimates (number of nests per 100 boxes) of boreal owls against the year (1973–2009) in Finland (regression equation $y = 438.43 − 0.22x$, $F_{1,35} = 16.32$, $p < 0.001$). The upper panel shows data from Figs. 13.2, 13.3 and 13.4). Yearly population indices for boreal owls calculated using the program TRIM and the numbers of occupied territories found in the 100-km^2 'raptor and owl grids' during 1982–2009 are shown in the lower panel. The base year is 1994 with an index value of 1.0. Vertical bars show standard errors and the line connects annual indices to show the year-to-year trajectory. The thick line is the log-linear regression line (redrawn from Honkala et al. 2010).

declining boreal owl populations. Reduction of old-growth forests lowers the densities of bank voles and small birds, including willow and crested tits, which are important foods of boreal owls in winter (Chapter 5.6). In addition, old-growth forests offer refuges against Ural and tawny owls and goshawks, which are the primary avian enemies of boreal owls in coniferous forests of Europe. According to the nationwide 'raptor and owl grid' monitoring study, breeding densities of Ural and tawny owls have substantially increased in Finland during the period 1982–2009 (Björklund et al. 2009, Saurola 2009,

Honkala et al. 2010), and the increase has been most pronounced in southern and south-western parts of the country, where local populations of boreal owls have decreased most rapidly. Reduction in the important food resources and increased avian predation risk probably markedly decrease the juvenile and adult survival of boreal owls, thus inducing relatively rapid decreases in their population abundance. Consistent with our results from boreal owls and our favourite hypothesis, male saw-whet owls breeding in areas with little forest cover and large inter-patch distances spent more time perching, sustained smaller home ranges, provisioned their nests less frequently, and produced fewer fledglings (Hinam and St. Clair 2008). Home range size and provisioning rates levelled off in landscapes with moderate to high forest cover. Male owls nesting in areas with low forest cover also showed high levels of chronic stress (Hinam and St. Clair 2008). These results suggest that low levels of forest loss and fragmentation may be beneficial to saw-whet owls by potentially increasing the abundance of main prey (small rodents), while high levels of forest loss and fragmentation appeared to decrease hunting success and thus fledgling production (Hinam and St. Clair 2008).

14 Population regulation

14.1. Site-dependent population regulation?

Because of the overwhelming influence of habitat quality on fitness (Newton 1998), individuals benefit from recognising probable spatial variation in habitat quality and settling accordingly (Orians and Wittenberger 1991). One of the most useful models of individual settlement in territorial species is probably the 'ideal despotic distribution' model (Fretwell and Lucas 1969), which predicts that the best-quality individuals (older, dominant, or the first to settle in) occupy the best-quality sites. Territorial behaviour prevents access to these high-quality sites by lower-quality (young, subordinate, or later arriving and settling) individuals, which are then consigned to poorer territories. At the population level, habitat heterogeneity, combined with an ideal despotic process of settlement, makes three predictions.

1. The percentage of low-quality territories that is occupied increases with population density.
2. The percentage of low-quality individuals entering the breeding population increases with population density.
3. This continuing increase in the use of poor-quality territories and settlement of lower-quality individuals leads to a decrease in the mean per capita productivity of the population, resulting in density-dependent reproduction which may act to regulate the population (i.e. site-dependent population regulation; Rodenhouse et al. 1997, Sergio and Newton 2003).

Here we test these predictions and examine the most important factors: food supply, natural enemies, nest-sites and habitat quality, which may limit or regulate boreal owl populations.

Previously it has been shown that the occupancy of boreal owl nest-sites varied in a non-random way. In a 10-year period, more nest-sites were occupied in only one year or in at least five years than expected by chance (Korpimäki 1988d). Although this method of grading nest-sites is not independent of occupant quality, we also found marked differences in habitat composition, and abundances of main and alternative prey were higher in high-quality than low-quality nest-sites (Korpimäki 1988c, Hakkarainen et al. 1997a; see also Chapter 3.2 and Chapter 8.1.5). Therefore, the grading of nest-sites on the basis of occupancy appears to be justified.

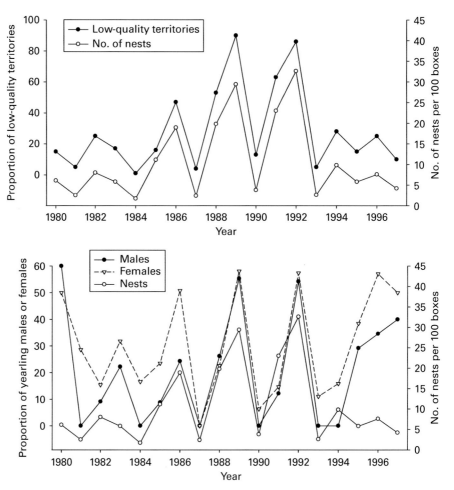

Fig. 14.1. Upper panel: Annual proportion of low-quality home ranges (grades 0–2; see Chapter 8.1.5) occupied by breeding boreal owls and the breeding density estimate (number of nests per 100 boxes) during 1980–97 ($r_s = 0.47$, $n = 18$, $p = 0.049$). Lower panel: The same but for the annual proportion of yearling males of all male parents and yearling females of all female parents of known age in the breeding population ($r_s = 0.57$, $n = 18$, $p = 0.02$ for males; $r_s = 0.47$, $n = 18$, $p = 0.06$ females).

We analysed our data from 1980–97 on the occupancy of different-quality nest-sites, age of female and male parents, breeding density and reproductive success to find out whether our boreal owl population behaved as the theory on site-dependent population regulation would predict. We found evidential support for the first two predictions, because both the occupancy of poor-quality nest-sites and the proportion of yearling male parents increased with breeding population density (Fig. 14.1). The proportion of yearling female parents in the population also tended to increase along with the breeding density. We did not find any obvious support for the third prediction, however, and – if anything – our results appear to be in some disagreement with it. Annual mean laying date of the owl population tended to advance with an increasing proportion of low-quality nest-sites occupied during 1980–97 ($r_s = -0.38$, $n = 18$, $p = 0.12$). Yearly mean clutch sizes and numbers of fledglings produced also appeared not to decrease with a rising proportion of poor nest-sites occupied ($r_s = 0.21$, $n = 18$, $p = 0.41$ and $r_s = 0.14$, $n = 18$, $p = 0.57$, respectively). These results suggest that the overwhelming temporal variation in vole abundance largely overrides the spatial habitat heterogeneity by increasing main food resources to the level at which even yearling males are able to feed their partner sufficiently for her to produce eggs even in lower-quality nest-sites, particularly in the increase and decrease phases of the 3-year vole cycle. Yearling male owls have generally poorer performance (Chapter 8.4) than older males but they can still successfully attract a mate in good vole years. This interpretation was further supported by the fact that there were close positive correlations between the yearly proportions of yearling male and female parents in the breeding population and abundance indices of *Microtus* voles in the previous autumn ($r_s = 0.68$, $n = 18$, $p = 0.002$ for males and $r_s = 0.69$, $n = 18$, $p = 0.002$ for females). In addition, the annual proportion of low-quality nest-sites occupied by owls increased with increasing densities of bank voles in the spring of settlement during 1980–97 ($r_s = 0.48$, $n = 18$, $p = 0.046$).

In a review of 22 studies of territory occupancy in 17 bird species including the eagle owl and the flammulated owl and 10 diurnal raptor species, territory occupancy always deviated from a random pattern and was each time associated with reproductive success and/or some other estimate of territory quality (Sergio and Newton 2003). However, all these studies were conducted with species occupying temperate or even more southern areas with stable, or having only minor temporal variation, environmental quality, at least when compared with northern areas with high-amplitude multi-annual fluctuations of herbivore populations such as small rodents. Consistent with our results, the yearly productivity of a black kite *Milvus migrans* population in northern Italy was not obviously correlated with the proportion of poor-quality territories occupied, although a higher proportion of poor territories was occupied in years of high than low density (Sergio and Newton 2003). These studies thus lend only partial support to the theory of site-dependent population regulation. One important reason contributing to the failure of support for site-dependent population regulation in boreal owls is probably their weak territoriality. As in other nomadic, vole-eating northern owls, males only defend a restricted area near their nest-holes (Korpimäki 1992c), although a strong territoriality is one of the underlying assumptions of the ideal despotic distribution model (Fretwell and Lucas 1969).

14.2. The phenomenal importance of food limitation

In this book we have presented compelling evidence that food supply and its multi-annual, seasonal and regional variation is by far the most important factor regulating boreal owl populations. The vital role of food supply in regulating boreal owl populations was evidenced by the close relationships between breeding densities of owls and density estimates of voles in our study area and elsewhere in Finland (Chapter 13.1) and northern

Sweden (Hörnfeldt et al. 2005). There were also vast interregional differences in the amplitude of breeding densities and in the degree of site-tenacity of breeding populations within Europe. In south-western and central Europe, where vole populations remain relatively stable between years and alternative prey types (for example, wood mice and small birds) are more readily available, particularly in winter, most adult male and female owls appeared to be year-round residents (Schwerdtfeger 1990, 1997), while juveniles may disperse widely (Franz et al. 1984, Schwerdtfeger 2008). In these conditions, cyclicity indices for yearly breeding density estimates of owl populations usually ranged from 40% to 120% (Chapter 13.2, Fig. 13.5). In northern and north-eastern parts of Europe with high-amplitude (100- to 300-fold) 3–4-year population cycles of voles, and with low availability of alternative prey types, particularly in winter, cyclicity indices ranged from 80% to 180% (Chapter 13.2, Fig. 13.5). In these places, a substantial proportion of adult females, and perhaps even a small proportion of males, may disperse widely between successive breeding attempts (i.e. they are nomadic; Chapter 9.3). An earlier study also suggested that snow conditions in winter are an important factor governing annual movements (i.e. dispersal) of boreal owls, thus limiting breeding densities of owl populations (Korpimäki 1986b). This is reasonable because a deep snow layer effectively protects small mammals against aerial predators, and thus decreases the hunting success of boreal owls (see also Chapter 5.3, Fig. 5.5). Therefore, food limitation by boreal owl populations is probably even more crucial than in many other bird species living in areas with less permanent snow cover (reviewed in Newton 1998).

A strong limiting effect of food availability on the reproduction of boreal owls was further supported by our following main findings. The yearly number and weight of prey items in the food stores that accumulated at the bottom of nest-boxes was greater with the increasing abundance of voles in the field (Chapter 5.3), and this in turn was reflected in a higher number of eggs per clutch (Chapter 5.3). The annual mean laying dates in our study population were postponed by approx. 1 month in poor vole years (Chapter 8.1.1, Fig. 8.1), and clutch sizes were 2–3 eggs less, and brood sizes at fledging 2–3 owlets less, than in good vole years (Chapter 8.1.1, Figs. 8.2, 8.5). Moreover, experimental food supplementation prior to and during the egg-laying period substantially advanced laying date and increased clutch size (Chapters 8.1.2, 8.1.3). Food supplementation from hatching onwards decreased nestling mortality of boreal owls in one of the two years with lower natural main food abundance, but sexing of hatchlings using molecular techniques revealed that nestling mortality was decreased only among the smaller male owlets and remained unaffected for larger female owlets (Hipkiss et al. 2002a). Supplementary feeding at 11 nests from hatching to fledging stage in the decline phase of the vole cycle (in 2009) resulted in a substantial decrease in prey delivery rate by male parents and increased body condition and haematocrit levels of females and chicks at the fledging stage, but did not have obvious positive effects on nestling survival (Santangeli et al. 2012). In an another surplus feeding experiment, with a considerably smaller food amount provided, mass loss of female parents during the brooding stage was substantially lower for fed than for non-fed control nests. Food supplementation appeared not to increase nestling survival, and had no obvious effects on body mass and wing length of nestlings (Eldegard and Sonerud 2010). All these observational and experimental results summarised in this and the two previous paragraphs point to strong food limitation effects on the timing of breeding, body condition and health status of female parents, as well as clutch and brood sizes and reproductive success of boreal owls. This conclusion is further supported by the fact that lifetime reproductive success was improved in male owls that initiated their breeding lifespan in the increase phase of the vole cycle rather than in the decrease phase (Chapter 8.7).

Periodic fluctuations in main food abundance induced wide variation in mating systems, dispersal, survival, recruitment rate and autumn movements of boreal owls. A considerable proportion (5–20%) of male owls were polygynous in good vole years, whereas no cases of polygyny were recorded in the low phase of the vole cycle (Chapter 7.3.3, Figs. 7.3, 7.4). Successive polyandry of female owls was mainly recorded in the increase phase of the vole cycle, with improving food conditions in the course of the summer, while it was only seldom recorded in the decrease phase and remained undetected in the low phase of the vole cycle (Chapter 7.3.7, Fig. 7.7). Natal dispersal of female owls was twice as extensive in the decrease than in the increase phase of the vole cycle (Chapter 9.2). Female parents also dispersed more between successive breeding attempts in the decrease than in the increase and low phases of the vole cycle (Chapter 9.3). Similar vole cycle phase-related differences in natal and breeding dispersal distances were not documented for male owls. In addition, the long-distance autumn movements of juveniles and adult females could also be considered as responses to the lack of food in the previous breeding areas (Chapter 9.5). Over-winter survival of males varied widely (from 25–40% to 60–75%) and improved with increasing abundance indices of voles in the preceding autumn (Chapter 10.2). The proportion of nests that produced recruits to the nationwide owl population in the increase phase of the vole cycle was three times as high as in the decrease phase, and twice as high as in the low phase of the vole cycle (Chapter 10.3). These results showed a substantially better survival of owlets hatched in improving rather than in deteriorating or poor food conditions. All these findings in fact show that most of the variables of population growth rate contributing to the demography of boreal owls, including fecundity, mortality (survival), emigration and immigration, are largely governed by cyclic fluctuations in main food abundance.

14.3.   Food competition

The mechanisms of interspecific competition include consumptive, territorial and encounter competition (Schoener 1983). Consumptive or exploitative competition occurs if resource use by one individual depresses the abundance of common resources or deprives others of those resources. The latter may arise through the hunting behaviour of a predator species. If predation by one species changes the behaviour or habitat choice of a prey, it may reduce the prey's vulnerability to other predator species (equivalent to resource depression as discussed in Charnov et al. 1976). Territorial competition occurs when a species defends a particular space containing vital resources against other species.

Encounter competition results from direct interactions between mobile animals, including food piracy, fighting, or even death by predation (Schoener 1983). An extreme form of encounter competition is intra-guild predation, which we deal with in Chapter 14.4.1.

A comparison of the diet of boreal owls with other owl species and diurnal vole-eating raptor species in our study area and surrounding areas is presented in Table 14.1. All the food samples were collected in or near the nest during or after the breeding season and the sampling period covers at least 4 years in order to gather data from diet composition in both good and poor vole years. The main foods of boreal owls, voles of the genera *Microtus* and *Myodes*, were also the main foods of all the other seven owl species (pygmy, hawk, long-eared, short-eared, tawny, Ural and eagle owls) breeding in the long-term study areas of the Ornithological Society of Suomenselkä (Table 14.1). Even the largest owl species, the eagle owl, which is capable of killing very large prey including, for instance, hares, forest grouse and goshawks, took a considerable proportion of *Microtus* voles in good vole years. Among the diurnal raptors, Eurasian kestrels and common buzzards also mainly feed on voles of these two genera. Another relatively frequently breeding diurnal raptor that is known to mainly subsist on *Microtus* voles is the hen harrier (review in Korpimäki 1985d), but unfortunately long-term data on their diet are not available for our study area.

The diets of boreal and pygmy owls in the breeding season include many more bank voles than those of the other co-existing owls and raptors (Table 14.1). Boreal owls diverged from other avian predators by often preying upon shrews, whereas the most important alternative prey of pygmy owls was small birds. The dietary diversity (diet width) of boreal owls was considerably higher than that of the open-country-hunting hawk, long-eared and short-eared owls that bred in high densities only in the good vole years and took high numbers of *Microtus* voles. The diet width index indicates that the dietary diversity of boreal owls is similar to the larger owl species, including tawny, Ural and even eagle owls, and also similar to the dietary diversity of open-country-hunting kestrels. Tawny, Ural and eagle owls diverged from boreal owls by feeding more on larger rodents, including water voles and brown rats (only eagle owls). Among ten co-existing bird of prey species, common buzzards have the most diverse diet that frequently consists of *Microtus* and bank voles, larger rodents (water voles), shrews, various-sized birds, snakes and frogs.

Niche overlap has been widely used as an indicator of competition (MacArthur and Levins 1967), although its suitability as such has met with criticism (May 1975, Lawlor 1980, Abrams 1983). Food overlap measures indicated that the closest food competitors of boreal owls during the breeding season were the hawk, long-eared and short-eared owls (Table 14.1) but, as open-country hunters, these species show habitat separation that obviously relaxes food competition. Habitat separation and different hunting times may also decrease food competition between boreal owls and Eurasian kestrels. In addition, most long-eared owls, short-eared owls and kestrels are migratory in Finland, which probably further decreases food competition with boreal owls in winter, when food resources are more limiting. Among forest-dwelling species, tawny and Ural owls emerged as the keenest food competitors of breeding boreal owls, and they may compete for food even more in winter, when the availability of voles is decreased by snow cover

Table 14.1. The diet composition (proportion of prey number) of boreal (Bor), pygmy (Pyg), hawk (Haw), long-eared (Lon), short-eared (Sho), tawny (Taw), Ural (Ura) and eagle (Eag) owls and vole-eating diurnal raptors (Eurasian kestrels, Kes, and common buzzards, Buz) during the breeding season in the Kauhava, Seinäjoki, Ähtäri and Keuruu regions (see map in Fig. 3.1), western Finland, in the late 1960s to early 1990s. Diet width (dietary diversity) was calculated according to Levins' (1968) formula. Food niche overlap of the diet of boreal owls with other co-existing owls and raptors was calculated using the MacArthur and Levins' (1967) measure.

Prey species	Bor	Pyg	Haw	Lon	Sho	Taw	Ura	Eag	Kes	Buz
European hedgehog *Erinaceus europaeus*	–	–	–	–	–	–	0.0	1.8	–	–
Common shrew *Sorex araneus*	27.4	1.3	2.6	10.7	15.9	2.4	12.3	0.3	11.3	5.5
Masked shrew *S. caecutiens*	0.0	–	0.2	–	–	–	0.6	–	–	0.2
Pygmy shrew *S. minutes*	0.1	0.2	0.1	0.5	–	–	0.8	–	0.3	0.3
Taiga shrew *S. isodon*	0.8	0.8	–	–	–	–	0.3	–	–	–
Least shrew *S. minutissimus*	–	0.0	–	–	–	–	–	–	–	–
Shrew spp. *Sorex* spp.	0.0	1.1	0.2	0.3	1.1	1.3	–	–	0.1	1.6
Water shrew *Neomys fodiens*	0.3	–	0.1	0.1	0.1	–	0.7	–	0.0	–
Shrews total	28.7	3.5	3.2	11.5	17.1	3.8	14.7	2.1	11.8	7.6
Northern bat *Eptesicus nilssoni*	–	–	–	–	–	0.2	–	0.0	–	–
Bat spp. Vespertilionidae	–	0.1	–	–	–	0.2	–	–	–	–
Least weasel *Mustela nivalis*	–	–	0.3	–	0.1	–	0.6	0.4	0.0	0.8
Stoat *Mustela erminea*	–	–	–	–	–	–	0.1	0.4	–	0.4
American mink *Mustela vison*	–	–	–	–	–	–	–	0.2	–	0.1
Red squirrel *Sciurus vulgaris*	0.0	–	0.2	–	–	–	0.4	1.2	–	1.1

Table 14.1. (cont.)

Prey species	Bor	Pyg	Haw	Lon	Sho	Taw	Ura	Eag	Kes	Buz
Flying squirrel	–	–	–	–	–	–	0.2	0.7	–	0.4
Pteromys volans										
Wood lemming	–	–	0.2	–	–	–	–	–	–	–
Myopus schisticolor										
Bank vole	26.3	28.2	10.3	10.2	7.8	6.7	11.3	1.7	8.3	3.6
Myodes glareolus										
Water vole	0.2	–	2.3	1.0	3.5	11.3	17.4	23.1	1.4	16.8
Arvicola terrestris										
Muskrat	–	–	–	–	–	–	–	1.4	–	–
Ondatra zibethicus										
Field vole	9.3	19.7	15.9	11.6	15.5	42.0	17.2	–	3.7	5.1
Microtus agrestis										
Sibling vole	14.5	2.1	12.8	51.1	40.0	–	1.9	–	12.1	0.5
M. rossiaemeridionalis										
Microtus spp.	7.8	–	41.9	8.0	6.2	–	13.4	20.3	20.4	16.2
Microtus spp. total	31.6	21.8	70.6	70.8	61.7	42.0	32.6	20.3	36.2	21.8
Vole spp.	–	–	8.5	–	–	–	2.9	0.7	–	5.2
Myodes/Microtus spp.										
Voles total	58.1	50.0	91.8	82.0	73.0	60.0	64.1	47.3	46.0	47.4
Brown rat	0.1	–	0.2	0.3	0.1	3.3	2.1	33.1	0.0	0.5
Rattus norvegicus										
Harvest mouse	1.7	0.2	1.7	2.6	4.8	2.4	0.2	0.6	1.9	0.1
Micromys minutes										
House mouse	1.4	0.2	–	1.5	0.6	2.7	0.3	0.3	4.7	–
Mus musculus										
Mouse spp.	–	–	–	0.0	0.0	–	–	–	0.1	–
Micromys/Mus spp.										
Murids total	3.1	0.4	1.9	4.4	5.5	8.4	2.6	34.0	6.7	0.6
Hare spp.	–	–	–	0.1	<0.1	–	3.4	3.2	–	2.7
Lepus europaeus/L. timidus										
Other mammal spp.	–	–	–	–	–	–	–	–	–	0.5
Mammals total	89.3	54.0	97.6	98.1	95.8	72.7	86.2	89.5	64.4	61.5

Birds										
Anseriformes	–	–	–	–	–	0.2	0.2	1.7	–	0.8
Galliformes	0.0	–	0.2	–	0.0	0.2	1.3	2.9	0.3	8.0
Accipitriformes	–	–	–	–	–	–	–	0.2	–	0.1
Falconiformes	–	–	–	–	–	–	0.0	0.1	0.4	–
Gruiformes	–	–	–	–	–	–	–	–	–	0.1
Charadriiformes	–	–	–	–	–	–	0.2	0.9	0.9	0.7
Columbiformes	–	–	–	–	–	–	0.2	0.0	–	–
Cuculiformes	–	–	–	–	–	–	0.2	0.0	–	–
Strigiformes	0.5	–	0.3	0.7	0.3	–	0.3	0.4	0.0	0.7
Caprimulgiformes	–	–	–	–	–	–	0.0	–	–	–
Apodiformes	0.1	0.1	–	–	–	–	–	–	–	–
Piciformes	0.0	0.3	–	–	–	–	0.3	0.1	0.0	0.4
Passeriformes total	9.4	44.1	1.5	0.9	1.2	18.7	6.5	2.0	8.0	20.7
Goldcrest size[a]	0.0	0.9	0.3	0.0	0.0	0.0	0.0	0.0	0.0	0.0
Coal tit size	0.7	5.3	0.0	0.0	0.0	0.0	0.0	0.0	0.0	0.0
Willow warbler size	0.1	0.1	0.4	0.1	0.2	0.7	1.3	0.0	0.8	1.1
Willow tit size	0.3	2.8	0.0	0.0	0.0	1.1	0.0	0.0	0.1	0.0
Common redpoll size	0.4	7.1	0.0	0.1	0.2	0.0	0.6	0.0	0.1	0.3
Whinchat size	0.1	5.8	0.0	0.0	0.1	0.0	0.0	0.0	0.3	0.2
Dunnock size	0.2	0.3	–	0.0	0.0	–	–	0.0	0.2	0.4
Great tit size	0.3	1.1	0.2	0.2	–	0.9	–	–	0.6	0.5
Chaffinch size	5.2	17.9	0.4	0.5	0.6	6.4	2.5	0.2	2.3	6.3
Greenfinch size	0.0	0.0	–	–	–	–	–	–	0.2	–
Yellowhammer size	0.4	1.8	–	0.1	–	–	–	–	0.3	0.9
Red crossbill size	0.0	0.7	–	–	0.0	0.2	–	–	0.0	0.2
Parrot crossbill size	0.0	–	–	–	–	–	–	–	–	–
Small thrush size	1.1	0.2	0.1	0.0	0.0	1.6	1.8	–	1.7	4.8
Starling size	0.1	–	–	–	–	–	–	0.3	0.1	0.5
Blackbird size	0.1	–	–	0.0	0.0	0.2	–	0.3	0.2	1.3
Fieldfare size	0.3	–	–	–	–	7.1	–	–	0.9	1.4
Mistle thrush size	0.0	–	–	–	–	–	–	–	0.0	0.1
Eurasian jay size	–	0.1	0.1	–	–	0.4	–	0.1	–	1.7
Magpie size	–	–	–	–	–	–	–	0.3	–	0.1
Hooded crow size	–	–	–	–	–	–	0.4	0.8	–	0.9
Birds total	9.9	44.5	2.0	1.6	1.5	19.1	9.0	8.1	9.7	31.5

Table 14.1. (cont.)

Prey species	Bor	Pyg	Haw	Lon	Sho	Taw	Ura	Eag	Kes	Buz
Reptiles total	–	1.3	0.2	–	0.0	–	–	–	0.5	5.8
Frogs total	–	–	0.2	–	–	7.8	4.7	2.2	0.3	1.1
Fish total	–	–	0.1	–	–	0.4	–	0.1	–	0.1
Insects total	0.2	0.3	–	0.3	2.7	–	–	0.1	25.1	–
No. of prey items	16 811	2240	1209	3760	4695	450	2531	3871	2631	1001
Diet width	4.04	5.81	1.96	1.91	2.40	4.70	5.90	4.80	4.60	10.18
Diet overlap	–	0.28	0.90	0.91	0.79	0.54	0.42	0.26	0.47	0.28
Source	1	2	3	4	5	6	7	8	9	10

[a] Body masses of goldcrest and the other birds mentioned here are found in Tables 5.3 and 5.5.

Source: 1. Korpimäki (1988b), 2. Kellomäki (1977), 3. Korpimäki (1972) and unpublished data, 4. Korpimäki (1992a), 5. Korpimäki and Norrdahl (1991b), 6. Mikkola (1977), 7. Korpimäki and Sulkava S. (1987), 8. Korpimäki et al. (1990), 9. Korpimäki (1985b), 10. Reif et al. (2001).

and only a few over-wintering birds are present at northern latitudes. In addition, food competition between forest-dwelling boreal and pygmy owls may be high in winter, although during the breeding season they showed clear dietary separation.

Food competition needs particular circumstances in order to exist (Martin T. 1986, Wiens 1989): important resources have to be limited, and potential competitors should depress the abundance of common resources. We summarised our strong evidence that food is indeed limiting to reproductive success, survival, dispersal and thus breeding density of boreal owl populations in Chapter 14.2. We provided observational and experimental evidence that synergetic impacts of mammalian and avian predators including boreal owls regulate the densities of *Microtus* and bank voles in Chapter 5.9.2. This clearly indicates that vole-eating predators depress common food resources, in this case vole abundance. Further evidence for food competition among owls is given by our results that when breeding in the neighbourhood of other vole-eating birds of prey, nests of boreal owls tended to contain smaller food stores than 'non-neighbouring' boreal owl nests (Chapter 5.3, Fig. 5.6). This result appeared to be true even in good vole years, when the densest vole patches were probably those near where boreal owls bred close to other vole-eating birds of prey, and poorer vole patches were those near where only non-neighbouring boreal owls bred. Additional indication of food competition among vole-eating birds of prey was also found in our study area, where Eurasian kestrels and long-eared owls often bred close to each other. In both species, neighbouring pairs fed less on main prey (*Microtus* voles) and more on alternative prey than did non-neighbouring pairs. Consistent with the competition theory, diet similarity was lower in poor vole years than in good ones and neighbouring pairs showed less diet overlap than non-neighbours (Korpimäki 1987c). In addition, food overlaps of neighbouring long-eared and tawny owl pairs were lower than those of non-neighbouring pairs (Nilsson I. 1984). Probably because of food competition, the neighbouring pairs of both kestrels and long-eared owls produced fewer offspring than the non-neighbours (Korpimäki 1987c). A similar decrease in offspring production was also imposed by the presence of larger tawny owls on smaller long-eared owls but not vice versa (Nilsson I. 1984). Only further studies will reveal whether food competition imposed by, for example, co-existing pygmy owls, kestrels, long-eared owls and short-eared owls decreases the reproductive success of boreal owls. We suggest, however, that exploitative (food) competition between these only somewhat larger birds of prey and boreal owls is an essential factor reducing fitness, whereas interference competition imposed by Ural and tawny owls is more important in the interactions between boreal owls and these larger owls (Chapter 14.4.1).

Finally, we would like to take issue with the old dogma that has explained the diet composition and dietary shifts of rodent-eating birds of prey only in the context of simple opportunistic foraging (Jaksic and Braker 1983, Marks and Marti 1984, Steenhof and Kochert 1985). The main reason for this explanation has been that it has been believed that there is no apparent competition for 'superabundant' (*sensu* Lack 1946) small mammal foods. Our results show that food competition may be an important factor even in good vole years at northern latitudes, where the seasonal low phase of voles during the breeding season of birds of prey and number of alternative prey are lower than in more southern areas. Because most co-existing avian and mammalian predators concentrate their predation impacts on small rodents, they can markedly depress this food resource, which probably increases exploitative competition and thus induces dietary shifts among birds of prey.

14.4.   Natural enemies

14.4.1.   Intra-guild predators

Intra-guild predation (alternatively defined as food web omnivory; see Aunapuu et al. 2010) is the killing of species that use similar resources (Polis et al. 1989, Polis and Holt 1992) and puts emphasis on the risk of death exerted by one of the competitors on the other, or by both competitors on each other. A recent review on intra-guild predation among birds of prey found 39 empirical and experimental studies on 63 populations belonging to 11 killer species and 15 victim species. These studies suggested that intra-guild predation was a widespread phenomenon and that the killer species is on average three times as large as the victim species (Sergio and Hiraldo 2008). As one of the smallest members of the avian predator guild, boreal owls have often been reported as being a common victim species. Mikkola (1983) compiled a total of 91 cases in Europe in which boreal owls were killed by larger birds of prey. The most common killers were eagle owls (36 cases), followed by goshawks (26), Ural owls (12), tawny owls (6) and hawk owls (3). This order, is of course, largely determined by how many dietary studies have been conducted on various bird of prey species and how many prey items have been identified in their diets. In Yukon, Canada, the most frequent killers of boreal owls were red-tailed hawks *Buteo jamaicensis* and great horned owls *Bubo virginianus* (Doyle and Smith J. 2001). Intra-guild predation was most frequent in the decrease phase of the 10-year population cycle of snow-shoe hares, and boreal owls seldom gave hoots during the peak period of intra-guild predation risk. On one occasion when recorded boreal owl hoots were broadcasted, an adult great horned owl flew silently over the speaker (Doyle and Smith J. 2001). On the other hand, the only species that boreal owls have been documented to kill are the smaller pygmy owls: two cases in Mikkola (1983), one mentioned in Chapter 5.4 and the other shown in Fig. 14.2 (both these events happened in winter).

Among birds of prey, eagle and Ural owls and goshawks are probably the worst enemies of boreal owls in our study area, because it is not unusual to find boreal owls among the prey items of these larger birds of prey (Korpimäki and Sulkava S. 1987, Korpimäki et al. 1990, Tornberg et al. 2005). Goshawks are day-active, however, and boreal owls can decrease goshawk risk by temporal segregation. In large areas of Eurasian coniferous forest belt, boreal owls co-exist with Ural and eagle owls, and all of them are night-active and use small mammals as their main food. We examined the effects of predatory and competitive interactions between these three owl species on reproductive success and population composition of boreal owls, both experimentally and observationally. We asked whether predation risk and interspecific competition due to eagle and Ural owls reduced the breeding density and fitness of boreal owls, and whether these interactions increased intraspecific competition for safe nesting sites

Fig. 14.2. A boreal owl with a freshly killed pygmy owl on 17 April 2009 in Tornio, southern Lapland of Finland. Photo: Matti Suopajärvi. (For colour version, see colour plate.)

among boreal owls. We manipulated breeding densities of potentially competing owls by erecting nest-boxes; the controls being boxes in areas where breeding attempts by competing owl species were absent (Hakkarainen and Korpimäki 1996).

During the 5-year experimental period, control nest-boxes in sites with no eagle and Ural owl territories, and nest-boxes within eagle owl territories (<2 km distant), were used by breeding boreal owls more than nest-boxes within Ural owl territories (<2 km distant; Fig. 14.3). The mean (s.d.) breeding percentage of boreal owl nest-boxes in the presence of Ural owls was only 3% (7%), whereas in the presence of eagle owls it was 15% (17%) and in control sites without larger owls 12% (12%). Most breeding attempts of boreal owls near Ural owls failed during the courtship period, while the majority of their nesting attempts were successful near eagle owls and at the control sites.

The observational data from 17 years revealed that breeding frequency of boreal owls was lower within 2 km of Ural owl nests than in control nests far from Ural owl nests (9 ± 17% vs. 25 ± 25%). The start of egg-laying was also delayed by an average of 11 days in the presence of Ural owl nests in comparison with the absence of these owls. In addition, male boreal owls at these nests were younger and paired more often with short-winged (i.e. young and generally subdominant) females than when further away from Ural owl nests. During the nestling period, boreal owls were found as prey in 13% (6 of 47) of Ural owl nests in our study area (Korpimäki and Sulkava S. 1987 and unpublished data). Boreal owls represented only 0.2% (6 of 2647) of the total number of prey items of Ural owls. These values are probably underestimates of the real impact of Ural owl predation on boreal owls, because these diet data were only available during the nestling period of Ural owls, when nearly all offspring of boreal owls are still in sheltered nest-boxes. Ural owl predation on boreal owls will probably be increased when boreal owl offspring leave their safe nest-cavities and are begging for food when hungry.

Our field experiment revealed that the occupancy of boreal owl nest-boxes was reduced by the presence of Ural owls, whereas the presence of eagle owls did not have

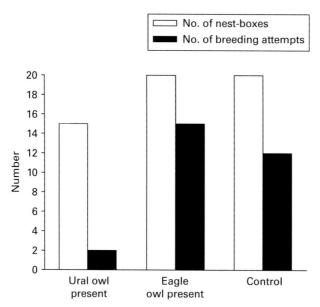

Fig. 14.3.   Number of nest-boxes (white columns) and number of breeding attempts (black columns) of boreal owls near Ural and eagle owl territories and in the absence of larger owl species, both potential predators and food competitors of boreal owls. Pooled data from 1989–94 in central Finland (Hakkarainen and Korpimäki 1996).

such a detrimental effect. During the courtship period, more male boreal owls seemed to hoot near eagle owls than near Ural owls, and the mating success was also higher near to eagle owls. Breeding frequency and breeding success in the presence of eagle owls did not differ from the same characters in the absence of larger owls, although egg and nestling mortality tended to be higher in pairs neighboured by eagle owls than in pairs non-neighboured by these owls. This suggests that the presence of eagle owls may slightly reduce the food provision rate of breeding boreal owls.

The substantial 11-day delay in the start of boreal owl egg-laying in the presence of Ural owls suggests that the time of pair formation is retarded near to Ural owls, because in vole-eating raptors laying date is positively correlated with the order in which females choose males (Village 1985, Palokangas et al. 1992). Male boreal owls near Ural owl nest-sites may not hoot as much as males in Ural-owl-free sites, which may delay their mating. In addition, males under Ural owl predation risk may not be able to invest as much in courtship feeding, which further delays the start of egg-laying because females produce eggs on the basis of the food provided by males (Korpimäki 1981). Clutch size and the number of fledglings were unaffected by the presence of Ural owls, which may be because our data are mostly from good vole years when food abundance may compensate for the adverse effects of competition during the incubation and nestling periods. An additional fitness loss in the presence of Ural owls is evident during the fledging and post-fledging periods. Begging and clumsy fledglings are easy for Ural owls to locate and kill.

We found some evidence for intraspecific competition for safe nesting habitats between boreal owls. Short-winged females bred more frequently near Ural owls than did long-winged females. This may be because long-winged females are superior in inter-female interactions and thus can choose the best mating options, whereas short-winged females are driven to mate with males in suboptimal habitats. Breeding males were also younger near Ural owls than far from these owls. Intraspecific competition probably drives young males to occupy poor sites, where predation risk is high.

Predation risk and food competition induced by Ural owls reduced the breeding frequency and delayed the start of egg-laying in boreal owls, whereas the larger eagle owls did not have such obvious adverse effects. Although eagle owls may sometimes kill boreal owl offspring, they may in fact be beneficial for boreal owls in areas with increasing densities of Ural owls, because eagle owls probably offer a protective umbrella for boreal owls. Eagle owls often remove Ural owls from their territories, thus creating Ural-owl-free refuges for boreal owls to survive and breed. In addition, Ural owls tend to spatially avoid eagle owls.

Among owls, intra-guild predation is probably size-related, as eagle owls are probably too large to harm full-grown boreal owls, which can quite easily find refuges in dense forests. The smaller Ural owls are more likely to encounter and even kill boreal owls, because they are adapted to hunt in even quite dense spruce forests. This is also supported by the fact that in the diets of eagle and Ural owls the size of the most common owl and raptor prey is about 10% of their own body mass (Mikkola 1983). It should also be noted that even close interspecific interactions between owls are not always immediately lethal for the smaller species, because they can avoid intra-guild predators by moving in different microhabitats and at different times. It is also known that a boreal owlet has fledged from the nest of the hawk owl (Törmälehto and Korpimäki 1978). In addition, at least one boreal owl male and one nestling survived for 3 weeks when a Ural owl pair usurped the nest-box of the boreal owl and began to breed there (Lagerström 1978). In all these cases, the main foods of these owls (voles) were at their peak density, when the need for alternative prey was minimal.

We conclude that the extreme mechanism of interspecific competition, the encounter competition which can even include death by predation (equivalent to intra-guild pre-dation) probably had a major role in the interactions between boreal and Ural owls (Fig. 14.4). Despite limited food resources among vole-eating owls (Chapter 14.3), exploitative competition for food was apparently less important than interference com-petition in this case, as the number of prey items in the food stores of boreal owls remained unaltered by the presence of the larger Ural owls (Hakkarainen and Korpimäki 1996). However, food was probably more limited in low vole years and in winter when the voles are protected by snow cover. We suggest that exploitative competition for food may also play an important role in the interactions between boreal and saw-whet owls, in particular in winter, while encounter competition – even including death by predation – might be more frequent in the interactions between boreal owls and barred owls in the boreal forests of North America.

We are well aware that our experimental and observational results do not allow us to fully evaluate the relative importance of intra-guild predation and exploitative

Fig. 14.4.   Ural owls proved to be the worst avian enemies of boreal owls in north European coniferous forests. Photo: Aku Kankaanpää. (For colour version, see colour plate.)

competition. However, direct interspecific interactions inducing reduced vocal activity and hunting effort, repeated escapes to refuges by boreal owls after Ural owl detection, and death on the talons of Ural owls probably have more adverse fitness consequences than exploitative competition for food. The availability of nest-holes is an essential feature for boreal owls, and the scarcity of suitable nest-holes has probably promoted selection against home range shift in males (Chapter 9.1). Our results suggest that inexperienced male boreal owls are forced to establish their territories near Ural owls, where they often paired with subdominant females. The areas near to Ural owl nests (at least within a radius of 2 km) are thus suboptimal habitats for boreal owls, while those near to eagle owls may even offer protection against the worst enemies of boreal owls. Therefore, predatory and encounter competition interactions from Ural owls decrease the breeding population size of boreal owls by reducing the number of suitable habitats (Hakkarainen and Korpimäki 1996). This was the first experimental evidence that such interactions may decrease the fitness of birds of prey and other vertebrate predators. More recently, the detrimental effects of goshawks on the reproductive success of common buzzards have also been demonstrated both experimentally and observationally (Krüger 2002, Hakkarainen et al. 2004).

14.4.2.   Nest predators

Among the natural mammalian predators of boreal owls in Eurasia and North America, martens (*Martes* spp.) are probably the most important predators of nest contents (eggs and owlets) and females. Pine squirrels *Tamiasciurus hudsonicus* have also been listed as probable egg predators in Idaho (Hayward and Hayward 1993). All three marten species co-existing with boreal owls, the pine marten, the stone marten *Martes foina* and the American marten *M. americana*, use natural cavities and nest-boxes for daytime roosting, food caching and sometimes even for nesting (Fig. 14.5). Boreal owls probably have

Fig. 14.5.   Pine martens use natural cavities and nest-boxes for daytime roosting, food caching and sometimes even for nesting. Photo: Pierre Henrioux.

a long evolutionary history of co-existence with martens, which can easily enter their nest-holes. Incubating female owls carry pellets far from the nest-box each morning (Korpimäki 1981), and after hatching, eggshells are also transported at a distance from the nest (Mikkola 1983). These behaviours have been considered to be protective ploys against martens and other mammalian predators (Sonerud 1985b). Moreover, incubating and brooding female boreal owls are also well adapted to escape the attacks of martens, because when someone scratches the trunk of a nest-tree, female owls suddenly appear at the entrance hole, watch very closely, and escape when the intruder is 0.5–3 m away. This behaviour has been interpreted as self-defence against martens (März 1968), although 17% of 178 females sat very tightly on their eggs or chicks and did not appear at the entrance hole of the box when the next-tree was scraped (Korpimäki 1981). Despite this, we have only two records during 1973–2009 in which pine martens have been able to kill a female in her nest-box. On the other hand, 21% of females were already looking out of the entrance hole of the box when we were not in the immediate vicinity of the nest-tree. Perhaps these females had earlier experience of marten attacks or were inherently very nervous.

Many consider predation as one of the main factors reducing the nesting success of birds (Martin T. 1995). In small birds such as passerines, predation has usually been assessed to be less important among hole-nesting than open-nesting species (Lack 1954, von Haartman 1968). In larger birds, this difference may be less apparent, mainly because more predators can gain entrance to larger nest-boxes and holes. It has been suggested that pine martens memorise the location of nest-holes in their home ranges and incorporate them into their foraging trips, which increases predation risk in holes with previous nesting failure due to marten attacks (Sonerud 1985a).

In most north European studies, predation rates of boreal owl nests by pine martens were low (5–15%; Korpimäki 1987e, 1994a with references), but in south-eastern Norway it was 47% (Sonerud 1985a). In addition, young fledged in only 38% of black woodpecker nests in old cavities compared with 71% in new ones (Nilsson S. et al. 1991). This difference was mainly due to pine marten predation. In south-eastern Norway, the pine marten predation risk of boreal owl nests was reduced in relocated boxes compared with the control boxes, which lent some obvious support to the predation risk hypothesis (Sonerud 1989, 1993). However, even in this study area with a high predation risk, the breeding frequency of relocated boxes was no higher than that of control boxes (6 nests in 14 relocated boxes vs. 5 nests in 14 control boxes), as predicted by the predation risk hypothesis. In our study area, the frequency of nest predation by pine martens was independent of nest-hole type, nest-box age, breeding density of owls, and success of previous clutches in the box. Therefore, these results cannot be considered to give consistent support for the predation risk hypothesis either. This suggests that pine martens did not regularly revisit holes and boxes found previously when hunting, since their main foods (small rodents) and their most important alternative prey (squirrels; see Pulliainen and Ollinmäki 1996, Zhou et al. 2011) are probably much more abundant than eggs and chicks in roosting cavities (Korpimäki 1987e). Pine martens instead memorise the locations of their roosting cavities and consume any nest contents that happen to be there.

We conclude that the level of marten predation risk may be difficult to assess by female boreal owls beforehand, because in most cases female owls have just immigrated to the area and breed there only once. However, the assessment of marten predation risk might be easier for site-tenacious male owls. Earlier we provided some experimental evidence for increased dispersal distances of males as a response to the treatment where a caged mink, imitating the visit of a marten to the nest-box, was put on the roof of the nest-box (Chapter 9.4). In addition, mammalian predators mostly eat nest contents during the egg-laying and incubation period, but very rarely kill incubating and brooding female owls (see above), because their escaping behaviour is well developed to reduce the mammalian predation risk (Korpimäki 1981, Sonerud 1985b). Therefore, there is probably enough time for re-nesting later in the season. We suggest that setting up new nest-boxes facilitates the reduction of predation risk of boreal owl nests in areas with a high nest-robbing risk by martens.

14.4.3. Parasites

Parasites may impact bird populations in different ways, including the limitation of population numbers and even periodic vast reductions of population size (e.g. Newton 1998). The importance of avian parasites as morbidity and mortality factors affecting bird populations may thus be relevant to conservation issues, particularly when dealing with decreasing and threatened bird species such as boreal owls.

We found blood-parasite infections of boreal owls particularly in years of food scarcity, whereas in years of food abundance the prevalence of blood-parasite infections was considerably decreased in female owl parents but not in male parents. Blood parasitism appeared to be costly, as clutch sizes of females infected with leucocytozoids were reduced in comparison with non-infected females, and trypanosome-infected male owls defended their offspring less vigorously than non-infected ones (Chapter 8.6 and

Fig. 8.17). This suggested that leucocytozoids could cause anaemia in female owls and compete for nutritional resources that are used for egg production. Therefore, increased parental effort may make owl hosts susceptible to blood-parasite infections, and the level of natural food supply can considerably modify vulnerability to blood-parasite infections. Our results further showed that the prevalence of trypanosomes in females can be reduced by experimental food supplements, suggesting that females invested a proportion of extra food resources in parasite clearance (Chapter 8.6), because immunosuppression is probably costly. All in all, blood parasites probably have direct detrimental impacts on reproductive success in boreal owls, but their possible effects on future survival, fitness and population size – and thus on population regulation of owls – remains an open question.

In the future, it will be a challenge to discover the possible detrimental impacts of blood-sucking black flies and other ectoparasites on the survival of boreal owls and other birds of prey living in boreal forests. For example, juvenile survival of great horned owls in the boreal forests of Yukon, Canada was high during the peak of the snow-shoe hare cycle but decreased in the first year of decline of hare prey (Rohner and Hunter 1996). Anaemia induced by leucocytozoids and blood-sucking black flies was identified as a major cause of mortality in young great horned owls during declining food conditions (Hunter et al. 1997). Exposure to a large number of sucking black flies leads to external lacerations, and the blood loss may contribute to reduced haematocrit levels and other symptoms of anaemia, and finally lead to reduced survival (Rohner and Hunter 1996). According to our observations, young boreal owls hatched in late nests can carry high numbers of black flies and mosquitos full of blood prior to and during the fledging phase. We suggest that these blood-sucking ectoparasites can have particularly adverse effects on owlets in years of food shortage with late hatching dates, and thus substantially reduce their first-year survival. Therefore, their possible detrimental effects on the survival of young and adult owls deserve to be studied in future, because it appears to be an important conservation issue.

14.5. Nest-sites

A shortage of nest-sites can keep the number of some bird species, including owls and diurnal raptors, in specific areas below the level that food abundance would allow (Newton 1998). There are two kinds of evidence for this contention. Breeders may be absent from areas that suffer from a shortage of nest-sites but which are suitable in other aspects. An addition of artificial nest-sites can lead to an increment in breeding density, while reduction in the number of suitable nest-sites can lead to a decline in breeding density.

As boreal owls are obligate cavity-nesters, the number of high-quality nest-sites may be a strong limiting factor for their breeding densities, at least in Eurasian coniferous forest belt. Only one large woodpecker species, the black woodpecker, makes holes that are large enough for breeding boreal owls. The holes excavated by green woodpeckers in central Europe are too small for large clutches and many stored prey items. Commercial forestry in most parts of Eurasian boreal forests substantially decreases the availability of suitable natural cavities for boreal owls (Hakkarainen et al. 1997b). The breeding densities of black woodpeckers, even in pristine coniferous forests in southern Finland and Sweden, are relatively low (13 pairs/100 km^2, Virkkala et al. 1994; 15–16 pairs/ 100 km^2, Tjernberg et al. 1993), and there are many species that usually nest in these holes (Chapter 4.3.2). In addition, for various reasons, a considerable proportion of natural cavities are of low quality for breeding owls (Chapter 4.3.2). On the other hand, there are many large woodpecker species in North America that excavate suitable nest-cavities for boreal owls. Although many species in this region use large natural cavities for breeding and roosting, it appeared that nest-site limitation is probably less important there: only two of the five nest-box supplementation experiments appeared to be successful (both in Alaska; see Chapter 4.3.3).

Limitation of boreal owl breeding densities by a shortage of high-quality nest-holes is likely in large areas of Eurasia where no network of suitable nest-boxes is maintained. This conclusion is supported by all the nest-box supplementations conducted in Europe (Chapter 4.3.3 and Table 3.4). In addition, non-breeding bachelor males in our study population possessed suitable nest-holes for breeding (Chapter 7.3.2), which showed that a shortage of nest-sites can be solved by the erection and maintenance of a nest-box network, the density of which corresponds to the abundance of black woodpecker cavities in pristine coniferous forests. Further experimental evidence for the importance of nest-sites determining breeding density was obtained in the avian predator reduction experiment. Nest-boxes and natural cavities suitable for boreal owls in forests surrounding five farmland areas (2.5–3.0 km^2 each) were closed for 5 years, while five comparable control areas with a network of suitable nest-sites served as controls. Breeding densities of boreal owls were significantly reduced in the predator reduction areas in comparison with control areas (Norrdahl and Korpimäki 1995a). As far as we know, this is the only experiment where avian ecologists have attempted to reduce the breeding densities of birds of prey, but this experiment was done to find out the limiting impacts of avian predators on 3-year population cycles of voles. In our study area, boreal owls preferred nest-boxes made of boards to those hollowed out of logs, but tree-holes were used as frequently as board boxes (Chapter 4.3.3). The owls chose large nest-boxes, especially in good vole years, when large clutches needed plenty of space. Breeding success was also higher in large boxes than in smaller ones, and in boxes than in natural cavities (Chapter 4.3.4), The avoidance of nest predation by pine marten did not appear to affect the choice of nest-site in our study area, with low marten predation risk (Chapter 4.3.4). Boreal owls preferred new or relocated boxes (Sonerud 1985a, 1989), or chose to breed in boxes protected against marten predation in areas with high marten predation risk (Ravussin 1991, Ravussin et al. 2001b).

15 Conservation of boreal owl populations

15.1. Rapid decline of boreal owl populations

Apex vertebrate predators have traditionally fascinated humans. Conservationists and managers have often utilised their charisma as flagship, indicator or umbrella species in order to increase environmental awareness, obtain financial support, and plan networks of protected areas (reviews in Simberloff 1998, Sergio et al. 2008a). Public concern can be directed towards a charismatic flagship species as a symbol of nature (Sergio et al. 2005, 2008b), while an indicator species reflects the quality of a certain habitat, or changes in populations of other species (Roberge and Angelstam 2006). An umbrella species is relatively demanding on the size of the area and probably also on certain habitat types (Simberloff 1998, Branton and Richardson 2010), and as such has the comprehensive spatial aspect for management of sustainable habitats and for planning of conservation areas.

Although the boreal owl was probably the most common bird of prey in the coniferous forests of Finland and Sweden in the 1950s and 1960s (Chapter 2.4), the recent status of its populations in Finland, and probably elsewhere in Europe, has proved to be dramatically different. We showed that there was an evident negative growth rate (−2.1% to −2.3% per year) of the local boreal owl population in Finland during the 1980s to 2000s (Chapter 13.4). We predicted that if this annual decrease rate continues for 20–30 years, the Finnish – and probably also the majority of the north European – local populations will be dangerously close to extinction, unless there is a major influx of boreal owls from Russia. However, such an influx is very unlikely because wide-scale clear-cutting of boreal forests is also occurring in Russia. The recent Red List of Finnish species also recognised the decreasing trend of boreal owls (Rassi et al. 2010). The boreal owl was assessed as 'near threatened' in Finland and was also close to fulfilling the criteria for 'vulnerable' species (i.e. decrease of population by 50% during the last 10 years or three generations).

We have presented compelling evidence that the decreasing area of old-growth forests was the main reason for the decline of the boreal owl populations. Owls prefer to hunt in spruce and pine forests and avoid hunting in open clear-cut and agricultural areas (Chapter 4.1.1). The over-winter survival and lifetime reproductive success of male owls decreased with reduction in the cover of old-growth forests (Chapter 11). The loss of old-growth forests reduces the densities of important foods of boreal owls (bank voles, shrews and small birds; Chapter 5.6 and Fig. 15.1). In particular, adult male owls

Fig. 15.1.   Forest-dwelling bank voles and common shrews are more readily available for over-wintering boreal owls in dense spruce forests, because there is less snow and because snowmelt is earlier underneath large spruce trees. Photo: Pertti Malinen.

that try to stay resident on their territories after their first breeding attempt, and are thus of great importance for the persistence of local populations, may be detrimentally affected by a decreased food supply in winter. Old-growth forests offer sheltered habitats against intra-guild predators (Ural and tawny owls and goshawks; Chapter 14.4.1). In particular, juvenile owls are probably vulnerable to avian predation after leaving safe nest-holes that those larger birds of prey cannot enter (Hakkarainen and Korpimäki 1996; Chapter 10.4). Therefore, interactive or additive effects of reduced food resources and increased avian predation risk due to loss and fragmentation of old-growth forests probably decreases both the juvenile and adult survival of boreal owls and thus induces drastic decreases in their population abundances. Increasing forest loss and fragmentation may also ultimately decrease the population size of saw-whet owls in Alberta, Canada by lowering reproductive success (Hinam and St. Clair 2008).

A fading out of vole cycles and decreasing vole densities in spring due to mild and wet winters brought about by climate change (Hörnfeldt et al. 2005; see also Solonen 2004) could in future also be an additional threat to boreal owls in the north. At present, this seems less important in Finland, however, because the high-amplitude vole cycle is still going strong in western, central and eastern Finland and even in southern Finland (Fig. 1.1; see also Lehikoinen et al. 2009, 2011, Brommer et al. 2010), where boreal owls are decreasing most rapidly.

15.2.   Conservation of old coniferous forests and biodiversity

It is evident that the destruction of boreal forests also decreases the number of suitable natural cavities for boreal owls to breed in. But at least one study indicates that the present degree of fragmentation in Sweden does not have any obvious detrimental effects on the breeding densities and reproductive success of black woodpeckers (Tjernberg et al. 1993), the main natural provider of nest-sites for owls in Eurasia. In addition, members of local ornithological societies in Finland have erected and maintained >10 000 nest-boxes for boreal owls, and these boxes are annually inspected and most nestlings ringed. Even in the highest peak years of vole populations, only about 10% of nest-boxes in Finland have been occupied by breeding boreal owls (Björklund et al. 2009, Honkala et al. 2010). Therefore the marked declines in breeding densities of boreal owls cannot be attributed to a decrease in the number of nest-sites only. Other possible causes of declines in boreal owl populations include the destruction of coniferous forests due to acid rain in central Europe, which is likely to reduce the habitat suitable for owls. Direct detrimental effects of heavy metal contamination of small mammals on owls also seems unlikely, because in a study performed in a pronounced heavy metal pollution gradient in northern Sweden, no direct metal toxicity effects on the breeding success of owls were found. Instead, a decreased food supply probably caused poor reproductive success of owls breeding in the vicinity of a heavy metal and sulphur dioxide pollution source (Hörnfeldt and Nyholm 1996).

What management efforts are then needed to prevent further declines in local boreal owl populations and to sustain vital populations in northern Europe and elsewhere? We think that the first – and by far the most important – management effort should be to conserve sufficient areas of old-growth forests, particularly spruce forests with an adequate number of natural cavities excavated by black woodpeckers. We have shown that about 15–20% of old forest containing over 150 m^3/ha of timber can be taken to be an approximate threshold that sustains the higher survival and lifetime reproductive success of boreal owls (Chapter 11). Maybe a considerable proportion of this forest cover could also be middle-aged forest with >100 m^3/ha of timber (Laaksonen et al. 2004). The same threshold proportion of old forest would also be sufficient for other old-forest associates including, for example, Eurasian treecreepers (Suorsa et al. 2005). Treecreepers have also been shown to suffer from the loss of old spruce forests because their physiological stress levels and nest predation rates are elevated when that happens (Suorsa et al. 2003, Huhta et al. 2004).

At least in Finland, forest managers usually plant pine saplings in areas where old-growth spruce forest has been cut down (Fig. 11.1), but we suggest that more spruce saplings should be planted instead, because then the young forest will be thicker and would offer more refuges for boreal owls and other forest-dwelling species. A review on occupancy of bird territories showed that it is important to prioritise conservation efforts towards high-quality sites (Sergio and Newton 2003). Occupancy may be a reliable method of quality assessment, especially for populations in temperate areas in which not all territories are always occupied, or for species in which checking occupancy is easier than finding nests. However, this does not hold true for boreal owls living under temporally fluctuating environmental conditions, where low-quality nest-sites become productive in the increase phase of the vole cycle experienced by northern owls in general (Chapter 14.1). Therefore, it seems unwise to concentrate conservation efforts only on good territories of boreal owls.

Vulnerable or threatened vertebrate species have usually been used as indicator or umbrella species for assessing the conservation value of coniferous forest stands in northern Europe. The Siberian flying squirrel (Selonen et al. 2001, Hurme et al. 2008), the three-toed woodpecker *Picoides tridactylus* (Roberge and Angelstam 2006) and the Siberian jay *Perisoreus infaustus* (e.g. Eggers et al. 2005, Lillandt 2009), which are year-round resident old-forest species, have most often been used as indicators of biodiversity value. However, we suggest that the boreal owl could be an appropriate, or even better, umbrella species for the planning and management of sufficient networks of old-growth boreal forests, because it appears to fulfil all the criteria required. Boreal owls dwell in coniferous forests all year round and are thus sensitive to the loss and fragmentation of the forest environment. They need a large home range (150–230 ha) in the breeding season (Chapter 4.1.3), and probably range over even larger areas during the winter. They are also relatively demanding with respect to the forest habitat, because the boreal owl is clearly an old-forest associate (Chapter 11). Although the boreal owl is clearly not the top predator in boreal forests, its sympathetic appearance has fascinated humans for centuries, and this may help in raising awareness and in raising funds for the conservation of old forests. More importantly, because thriving boreal owl populations rapidly respond to increases in vole densities, they can effectively dampen the peak densities of voles (Chapter 5.9.2). During peak densities in winter, most of the enormous damage caused by voles becomes apparent (Fig. 15.2), and boreal owls can thus have an essential role in preventing increased vole-induced damage to forestry and agriculture. For example, considering the costs of replanting tree saplings alone, the financial impact of vole

Fig. 15.2.   During vole peaks in winter, most of the enormous amount of damage caused by voles to tree saplings and agricultural products becomes apparent, but thriving boreal owl populations can act as biocontrol agents with regard to vole damage. Aspen saplings debarked by field voles in April 2003 in our study area. Photo: Erkki Korpimäki.

damage to forest management during winter 2005/6 was €2.2–4.0 million in Finland alone (Huitu et al. 2009). During the peak density of voles in winter 2008/9, the costs of vole damage would probably have been many times bigger.

The comparison of biodiversity values recorded at the breeding sites of goshawks and five owl species, including boreal, pygmy, scops, long-eared and tawny owls, with those of control sites in the Italian Alps revealed that biodiversity levels were consistently higher at sites occupied by these birds of prey (Sergio et al. 2006). They estimated biodiversity as the richness of bird, butterfly and tree species, and also found that sites occupied by avian predators sustained higher densities of individual birds and butterflies than control sites (Sergio et al. 2006). Although these results have also attracted some criticism (see, e.g., Sergio et al. 2008b), we suggest that the underlying idea of using the occurrence of breeding sites of birds of prey as surrogates for the biodiversity value of the ecosystem in order to plan landscape-level conservation should be thoroughly tested in northern coniferous forests.

We surmise that the boreal owl could be a better indicator species of biodiversity value than many other birds of prey because it has been suggested that the species richness of breeding birds could be higher near boreal owl nests than farther away from these nests. This suggestion was supported by results from field experiments on breeding-site selection of migrant pied flycatchers *Ficedula hypoleuca* in the multi-predator environment of our study area. Pied flycatchers avoided breeding near the nests of diurnal pygmy owls, while such avoidance behaviour was not shown in the presence of nocturnal boreal owl nests (Morosinotto et al. 2010). Pied flycatchers also responded to pygmy owl risk with 4-day delay in the initiation of egg-laying, because their nest-building period was substantially prolonged. Flycatchers also laid smaller clutches under pygmy owl risk, even if laying date was controlled for, while no breeding delay or reduction in clutch size was detected under the risk imposed by boreal owls (Morosinotto et al. 2010). Therefore, day-active pied flycatchers obviously perceived pygmy owls as being a greater threat than boreal owls. We suggest that other day-active forest-dwelling small birds including, for instance, tits *Parus* spp., chaffinches and willow warblers, would also try to avoid breeding near pygmy owl nests. In addition, resident willow and crested tits and other passerine birds also attract migrating forest birds to establish their territories nearby because they are indicators of high-quality and safe breeding sites in coniferous forests (Mönkkönen et al. 1990, 1996). This heterospecific attraction mechanism seems also to work in North American boreal forests (Mönkkönen et al. 1997). Pygmy owls seemed to avoid breeding close to boreal owls (<0.8 km, E. Korpimäki, unpublished data), where they were only able to amass smaller larders due to food competition and the predation risk imposed by boreal owls, and preferred to live far away from boreal owls in order to survive over the winter (Suhonen et al. 2007). This indicates that boreal owls may offer a year-round protective umbrella for resident forest-dwelling small birds and also for migrant small passerine birds in coniferous forests. This effect may even be reinforced by the heterospecific attraction of migrant forest-dwelling birds to those sites with high numbers of resident forest birds. This might increase the biodiversity value of boreal owl nest-sites in comparison at least with pygmy owl sites. This idea is worthy of further testing against field data on the distribution of forest-dwelling birds in relation to the distance from boreal and pygmy owl nests in spring and summer and from their home ranges in winter.

15.3.   Construction and setting up of nest-boxes

Constructing and setting up nest-boxes for boreal owls is an effective way of reducing competition for high-quality nest-sites, because there are many potential users of larger natural cavities in boreal forests (Chapter 4.3.2). Modern forestry also largely removes the large trees hollowed by woodpeckers or leaves them isolated in the middle of clear-cut areas, where they are mostly useless. According to our long-term experience, wooden nest-boxes for the boreal owl should be made from board at least 2 cm thick, in order to give good insulation against cold spells. The boreal owl box should have the following dimensions: inner width of the square-shaped bottom 19–21 cm, total height of the box 55–60 cm, diameter of the entrance hole 7.0–8.5 cm placed about 40–50 cm above the bottom, and a waterproof roof at least 3–5 cm bigger than the box so that the water cannot flow into the box (Fig. 4.6). There should be no cracks in the walls, not even tiny ones, in order to increase the effectiveness of insulation, because boreal owls may start to lay eggs very early under the harsh conditions of late winter (Chapter 8.1). The area of the bottom of the box should be at least 400 cm^2 so that fat females, large clutches and big larders have enough space even in good vole years (Chapter 4.3.4). One should put a 5–10 cm layer of sawdust or some other wooden material, moss or hay in the bottom of the box because boreal owls do not carry any nest material into the box. The diameter of the entrance hole is important if the worst competitors for nest-sites are to be prevented from breeding in the boxes. For example, common goldeneyes probably cannot enter the box if the diameter of the entrance hole is <7.5 cm (Chapter 4.3.2). The corresponding minimum entrance diameters are <9.0 cm for hawk owls and <10.0 cm for tawny owls.

Nest-boxes for boreal owls can be set up 1–2 km apart on coniferous trees at 5–6 m above the ground (Fig. 3.4). The boxes can easily be fixed with plastic ropes (diameter 3–4 mm) that do not become rotten and do not penetrate inside the growing tree trunk. The dense twigs of spruce trees protect nest-boxes against rain and other weather extremes better than other tree species in areas with only low marten predation risk. Where the predation risk is high, boreal owl boxes should be fixed on branchless trees 6–8 m above ground. It is worth noting that setting up nest-boxes for boreal owls and other cavity-nesting species is a long-term task because nest-boxes should preferably be inspected and cleaned annually and boxes found to have cracks should be regularly renewed. Although carefully made wooden board nest-boxes with a waterproof roof do not crack and become decayed during 10 or even 25 years in the field, their occupancy rate by boreal owls will undeniably decrease because boreal owls prefer new boxes (Chapter 4.3.5), particularly in areas with co-existing marten foes (Chapter 14.4.2). Therefore, the annual setting of some new nest-boxes in new sites is of overall importance as a conservation tool for local boreal owl populations.

In central Europe, particularly Germany, Switzerland and Belgium, there is a long tradition of protecting boreal owl nest-boxes and natural cavities against martens in areas where losses of eggs and owlets are frequent. Because martens can jump no more than 4 metres, the entrance holes of nest-boxes for boreal owls should be set up at least 4 m away from the nearest trees in high marten risk areas. Bird-watchers, managers and conservationists have used great ingenuity in developing designs of these nest-boxes to protect against predation by pine martens and other mammalian predators (Fig. 15.3). The simplest one which includes a half-metre wide metal-sheet ring around the nest-box tree some two metres below the nest-box, was described more than 40 years ago (März 1968). If the nest-box or natural cavity is on a tree that is <4 m away from neighbouring trees, it will also need some sort of sheet-metal ring above the box. However, probably an easier solution is to protect the nest-box itself so that martens cannot climb up to the entrance hole. This can be done with a metal sheet covering the front wall so that the sheet extends >10 cm from the sides and bottom of the front wall and leaves only the entrance hole open. In addition, a metal sheet should cover the roof and extend >10 cm from the sides of the roof and preferably even more over the front wall (see Fig. 15.3).

Perhaps the most complicated and effective design has been that constructed by Pierre-Alain Ravussin and co-workers (2001b) from PVC tubes in Switzerland. This nest-box was made from a hard PVC pipe tube of 25 cm outer diameter (upper panel of Fig. 15.4). The wall thickness was 4.9 mm. The bottom of the nest-box was made of a 2.7-cm-thick wooden plate and was 23.5 cm in diameter, treated and pierced with 6 mm holes to reduce humidity. The bottom was screwed on using four wood-screws with conical heads, allowing no hold on the exterior of the box. A wooden octagonal tube placed inside the pipe doubles the nest-box wall (lower panel of Fig. 15.4). This octagon was made of wood panelling glued and stapled together. This tube was placed inside the bottom of the nest-box and its height reached to the bottom of the nest-hole, ensuring good insulation. The top of the box was closed by the PVC bung normally provided for this type of pipe, onto which an aluminium cone was fixed. This cone was made of aluminium sheet of 0.5–1 mm thickness. The roof was attached to the nest-box with a cable from which it stayed suspended during the inspections. Two wooden studs were fixed vertically on the back of the nest-box to allow for better stability. The whole structure was fixed on the tree with a plastic sheathed cable, enabling quick fixation or removal for an easy change of position if needed (Ravussin et al. 2001b).

When building and setting up nest-boxes for boreal owls and other birds for scientific purposes, it is generally important to report the details of nest-box design including size, shape and material. Despite requests more than 15 years ago to increase the reporting of such information (e.g. Møller 1994), more than 60% of recent publications on owls and diurnal raptors did not provide any details on nest-box design (Lambrechts et al. 2012). Similar results have also been reported for hole-nesting passerines (Lambrechts et al. 2010). Boreal owls in our study population produced larger clutches in medium-sized and large nest-boxes than in small nest-boxes in good vole years, apparently because there was more space for prey stored in the larger nest-box types (Korpimäki 1985a). This relationship was not observed in years with food constraints. The amount of food stored in nest-boxes may therefore influence egg formation directly when females consume these stores before or during the period of egg development (Korpimäki

Fig. 15.3.   Conservationists have used great ingenuity in developing designs of boreal owl nest-boxes that protect against predation by pine martens and other mammalian predators. These designs have been used in Switzerland (upper panel) and Belgium (lower panel). Photos: Pierre-Alain Ravussin (upper), Serge Sorbi (lower).

Fig. 15.4.   The PVC-tube nest-box constructed in Switzerland proved to be an effective way of protecting boreal owl nests against marten predation. Upper panel, exterior view; lower panel, interior view. Photos: Pierre-Alain Ravussin.

1987d). They can also be used as a cue for anticipating the abundance of food available at the time of rearing nestlings (Korpimäki 1987a, Hörnfeldt et al. 1990).

Our final take-home message is that the management and conservation of bird communities in boreal coniferous forests needs a global approach, given the possibly lethal interspecific interactions between birds of prey. Boreal forests should be managed so that smaller birds of prey such as boreal and saw-whet owls can also sustain their viable local populations. Care should be taken when erecting nest-boxes for larger owls (e.g. owls of the genus *Strix* in both Europe and North America) in areas where they have sustainable and possibly increasing populations, and when making artificial stick-nests for larger owls (e.g. great horned owls) and diurnal raptors (i.e. goshawks, common buzzards and red-tailed hawks). If their breeding densities are not limited by nest-sites, these large avian predators can substantially increase as a response to increases in the availability of nest-sites and thus reduce the amount of suitable breeding habitat for small birds of prey for many kilometres around their nest-sites (Chapter 14.4.1) We would like to emphasise that nest-boxes for Ural and tawny owls, and probably also artificial twig nests for diurnal raptors, should be at least 2 km away from boreal owl nest-boxes. For example, Ural owl predation risk results in reduced habitat quality for boreal owls for a radius of 2 kilometres (Hakkarainen and Korpimäki 1996), which means a total area of 12.5 km^2, around their nest-sites. When there are two Ural owl territories per 100 km^2, we estimate that 25% of the landscape is of low quality for boreal owls, and four Ural owl territories per 100 km^2 means that a half of the forest landscape is somewhat unsuitable for occupancy by boreal owls. All conservationists, bird-watchers, forest managers, nurserymen, farmers and gardeners interested in setting up nest-boxes for owls and in protecting their tree saplings and crops against damage caused by voles need to be well aware of this fact. They can unintentionally increase the lethal and non-lethal predation risks imposed by larger birds of prey on smaller ones if they establish too dense nest-box networks for larger owls. When setting up nest-boxes for boreal owls, it is also important to be aware that they may have detrimental effects on the smallest owl species, the pygmy owl in Europe, as well as the saw-whet owl, the northern pygmy owl *Glaucidium gnoma* and the flammulated owl in North America.

References

Abrams, P. 1983. The theory of limiting similarity. *Annu. Rev. Ecol. Syst.* **14**: 359–376.

Aebischer, N. J., Robertson, P. A. and Kenward, R. E. 1993. Compositional analysis of habitat use from animal radio-tracking data. *Ecology* **74**: 1313–1325.

Ahlbom, B. 1976. Slaguggla, pärluggla och sparvuggla, något om deras föda i Gästrikland och Hälsingland. *Fåglar Sörmland* **9**: 17–24.

Ahola, K. and Terhivuo, J. 1982. Breeding pine martens recorded in nest-boxes set out in southern Finland. *Mem. Soc. Fauna Flora Fenn.* **58**: 137.

Ahti, T., Hämet-Ahti, L. and Jalas, J. 1968. Vegetation zones and their sections in northwestern Europe. *Ann. Bot. Fenn.* **5**: 169–211.

Alatalo, R. V., and Lundberg, A. 1990. Polyterritorial polygyny in the pied flycatcher. *Adv. Study Behav.* **19**: 1–27.

Alatalo, R. V., Carlson, A., Lundberg, A. and Ulfstrand, S. 1981. The conflict between male polygamy and female monogamy: the case of the Pied Flycatcher. *Am. Nat.* **117**: 738–753.

Alatalo, R. V., Lundberg, A. and Glynn, C. 1986. Female pied flycatchers choose territory quality and not male characteristics. *Nature* **323**: 152–153.

Alatalo, R. V., Gustafsson, L. and Lundberg, A. 1989. Extra-pair paternity and heritability estimates of tarsus length in pied and collared flycatchers. *Oikos* **56**: 54–58.

Altenburg, W., Daan, S., Starkenburg, J. and Zijlstra, M. 1982. Polygamy in the marsh harrier, *Circus aeruginosus*: individual variation in hunting performance and number of mates. *Behaviour* **79**: 272–312.

Altmüller, R. and Kondrazki, B. 1976. Eine neue Methode zum Fang von Rauhfusskauz-Männchen (*Aegolius funereus*). *Vogelwarte* **28**: 306–308.

Altwegg, R., Roulin, A., Kestenholz, M. and Jenni, L. 2006. Demographic effects of extreme winter weather in the barn owl. *Oecologia* **149**: 44–51.

Andersson, M. 1980. Nomadism and site tenacity as alternative reproductive tactics in birds. *J. Anim. Ecol.* **49**: 175–184.

Andersson, M. 2005. Evolution of classical polyandry: three steps to female emancipation. *Ethology* **111**: 1–23.

Andersson, M. and Erlinge, S. 1977. Influence of predation on rodent populations. *Oikos* **29**: 591–597.

Andersson, M. and Norberg, R. Å. 1981. Evolution of reversed sexual size dimorphism and role partitioning among predatory birds, with size scaling of flight performance. *Biol. J. Linn. Soc.* **15**: 105–130.

Andrén, H. 1995. Effects of landscape composition on predation rates at habitat edges. In: Hansson, L., Fahrig, L. and Merriam, G. (eds.), *Mosaic Landscapes and Ecological Processes*. Chapman & Hall, pp. 225–255.

Appleby, B. M., Anwar, M. A. and Petty, S. J. 1999. Short-term and long-term effects of food supply on parasite burdens in Tawny Owls, *Strix aluco. Funct. Ecol.* **13**: 315–321.

Arcese, P. and Smith, J. N. M. 1988. Effects of population density and supplemental food on reproduction in song sparrows. *J. Anim. Ecol.* **57**: 119–136.

Arsenault, D. P., Stacey, P. B. and Hoelzer, G. A. 2002. No extra-pair fertilization in flammulated owls despite aggregated nesting. *Condor* **102**: 197–201.

Atkinson, C. T. and Van Riper, C., III. 1991. Pathogenicity and epizootiology of avian haematozoa: *Plasmodium, Leucocytozoon* and *Haemoproteus*. In: Loye, J. E. and Zuk, M. (eds.), *Bird–Parasite Interactions*. Oxford University Press, Oxford, pp. 19–48.

Aunapuu, M. and Oksanen, T. 2003. Habitat selection of coexisting competitors: a study of small mustelids in northern Norway. *Evol. Ecol.* **17**: 371–392.

Aunapuu, M., Oksanen, L., Oksanen, T. and Korpimäki, E. 2010. Intraguild predation and interspecific co-existence between predatory endotherms. *Evol. Ecol. Res.* **12**: 151–168.

Begon, M., Harper, J. L. and Townsend, C. R. 1996. *Ecology. Individuals, Populations and Communities*. Blackwell Scientific Publications, Oxford.

Beissinger, S. R. and Snyder, N. F. R. 1987. Mate desertion in the snail kite. *Anim. Behav.* **35**: 477–487.

Bell, G. 1980. The costs of reproduction and their consequences. *Am. Nat.* **116**: 45–76.

Bennet, G. F. 1970. Simple techniques for making avian blood smears. *Can. J. Zool.* **48**: 585–586.

Bennett, P. M. and Owens, I. P. F. 2002. *Evolutionary Ecology of Birds: Life Histories, Mating Systems and Extinction*. Oxford University Press, Oxford.

Bensch, S. and Hasselquist, D. 1991. Nest predation lowers the polygyny threshold: a new compensation model. *Am. Nat.* **138**: 1297–1306.

Bernard, E. and Fenton, M. B. 2003. Bat mobility and roosts in a fragmented landscape in central Amazonia, Brazil. *Biotropica* **35**: 262–277.

Birkhead, T. R. 1987. Sperm competition in birds. *Trends Ecol. Evol.* **2**: 268–272.

Birkhead, T. R. and Møller, A. P. 1992. *Sperm Competition in Birds: Evolutionary Causes and Consequences*. Academic Press, London.

Björklund, H. and Saurola, P. 2004. Petolintuvuosi 2003 – paikoittain huippuvuosi (Summary: Breeding and population trends of raptors and owls in Finland in 2003, a good vole year in western Finland). *Linnut Yearbook* **2003**: 58–72.

Björklund, H., Saurola, P. and Haapala, J. 2003. Petolintuvuosi 2002 – karvapalleroita popsittaviksi (Summary: Breeding and population trends of common raptors and owls in Finland in 2002, many new records saw the daylight). *Linnut Yearbook* **2002**: 28–40.

Björklund, H., Honkala, J. and Saurola, P. 2009. Petolintuvuosi 2008: eteläiset myyräkannat kasvaneet (Summary: Breeding and population trends of common raptors and owls in Finland 2008). *Linnut Yearbook* **2008**: 52–67.

Black, J. M. 2001. Fitness consequences of long-term pair bonds in barnacle geese: monogamy in the extreme. *Behav. Ecol.* **12**: 640–645.

Blancher, P. J. and Robertson, R. J. 1982. Kingbird aggression: does it deter predation? *Anim. Behav.* **30**: 929–930.

Blanckenhorn, W. U. 2000. The evolution of body size: what keeps organisms small? *Q. Rev. Biol.* **75**: 385–407.

Blakesley, J. A., Noon, B. R. and Anderson, D. R. 2005. Site occupancy, apparent survival, and reproduction of California spotted owls in relation to forest stand characteristics. *J. Wildl. Manage.* **69**: 1554–1564.

Boag, P. T. 1987. Effects of nestling diet on growth and adult size in zebra finches (*Poephila guttata*). *Auk* **104**: 155–166.

Boag, P. T. and van Noordwijk, A. J. 1987. Quantitative genetics. In: Cooke, F. A. and Buckley, P. A. (eds.), *Avian Genetics*. Academic Press, London, pp. 45–78.

Boerma, E., Groen, L. G., Voous, K. H. and Wight, H. J. 1987. The first Tengmalm's owls *Aegolius funereus* in the Netherlands. *Limosa* **60**: 1–8.

Boiko, N. S. and Shutova, E. V. 2007. Diets of the pygmy owl *Glaucidium passerinum* and Tengmalm's owl *Aegolius funereus* in the Gulf of Kandalaksha area, White Sea. In: Koskimies, P. and Lapshin, N. V. (eds.), *Status of Raptor Populations in Eastern Fennoscandia: Proceedings of the Workshop, Kostamuskha, Karelia, Russia, Pedrozavodsk, November 8–10, 2005*, pp. 23–29.

Bondrup-Nielsen, S. 1977. Thawing of frozen prey by boreal and saw-whet owls. *Can. J. Zool.* **55**: 595–601.

Bondrup-Nielsen, S. 1984. Vocalizations of the boreal owl, *Aegolius funereus richardsoni*, in North America. *Can. Field Nat.* **98**: 191–197.

Boonstra, R. and Krebs, C. J. 1977. A fencing experiment on a high-density population of *Microtus townsendii*. *Can. J. Zool.* **55**: 1166–1175.

Boutin, S., Krebs, C. J., Boonstra, R., Dale, M. R. T., Hannon, S. J., Martin, K., Sinclair, A. R. E., Smith, J. N. M., Turkington, R., Blower, M., Byrom, A., Doyle, F. I., Doyle, C., Hik, D., Hofer, L., Hubbs, A., Karels, T., Murray, D. L., Nams, V., O'Donoghue, M., Rohner, C. and Schweiger, S. 1995. Population changes of the vertebrate community during a snowshoe hare cycle in Canada's boreal forest. *Oikos* **74**: 69–80.

Boutin, S., Wauters, L. A., McAdam, A. G., Humphries, M. M., Tosi, G. and Dhondt, A. A. 2006. Anticipatory reproduction and population growth in seed predators. *Science* **314**: 1928–1930.

Bowler, D. E., and Benton, T. G. 2005. Causes and consequences of animal dispersal strategies: relating individual behaviour to spatial dynamics. *Biol. Rev.* **80**: 205–255.

Bowmaker, J. K., and Martin, G. R. 1978. Visual pigments and colour vision in a nocturnal bird, *Strix aluco* (tawny owl). *Vision Res.* **18**: 1125–1130.

Branton, M. and Richardson, J. S. 2010. Assessing the value of the umbrella-species concept for conservation planning with meta-analysis. *Conserv. Biol.* **25**: 9–20.

Breiehagen, T. and Slagsvold, T. 1988. Male polyterritoriality and female–female aggression in pied flycatchers *Ficedula hypoleuca*. *Anim. Behav.* **36**: 604–605.

Brittain, R. A., Meretsky, V. J., Gwinn, J. A., Hammond, J. G. and Riegel, J. K. 2009. Northern saw-whet owl (*Aegolius funereus*) autumn migration magnitude and demographics in southern-central Indiana. *J. Raptor Res.* **43**: 199–209.

Brody, M. S., and Lawlor, L. R. 1984. Adaptive variation in offspring size in the terrestrial isopod *Armadillium vulgare*. *Oecologia* **61**: 55–59.

Brommer, J. E., Pietiäinen, H. and Kolunen, H. 1998. The effect of age at first breeding on Ural owl lifetime reproductive success and fitness under cyclic food conditions. *J. Anim. Ecol.* **67**: 359–369.

Brommer, J. E., Kokko, H. and Pietiäinen, H. 2000. Reproductive effort and reproductive values in periodic environments. *Am. Nat.* **155**: 454–472.

Brommer, J. E., Pietiäinen, H. and Kokko, H. 2002a. Cyclic variation in seasonal recruitment and the evolution of the seasonal decline in Ural owl clutch size. *Proc. R. Soc. Lond. B* **269**: 647–654.

Brommer, J. E., Pietiäinen, H. and Kolunen, H. 2002b. Reproduction and survival in a variable environment: Ural owls (*Strix uralensis*) and the three-year vole cycle. *Auk* **119**: 544–550.

Brommer, J. E., Karell, P. and Pietiäinen, H. 2004a. Supplementary fed Ural owls increase their reproductive output with one year time lag. *Oecologia* **139**: 354–358.

Brommer, J. E., Gustafsson, L., Pietiäinen, H. and Merilä, J. 2004b. Single-generation estimates of individual fitness as proxies for long-term genetic contribution. *Am. Nat.* **163**: 505–517.

Brommer, J. E., Pietiäinen, H., Ahola, K., Karell, P., Karstinen, T. and Kolunen, H. 2010. The return of the vole cycle in southern Finland refutes the generality of the loss of cycles through 'climatic forcing'. *Global Change Biol.* **16**: 577–586.

Brown, J. S. and Kotler, B. P. 2007. Foraging and the ecology of fear. In: Stephens, D. W., Brown, J. S. and Ydenberg, R. C. (eds.), *Foraging Behavior and Ecology.* The University of Chicago Press, Chicago and London, pp. 436–480.

Brown, M. 1942. Golden eagle captures red-shouldered hawk. *Auk* **64**: 317–318.

Bülow, B. von and Franz, A. 1982. Rauhfusskauz-Brüten und -Gevöllen aus dem Siegerland mit Anmerkungen zur Auftrennung von Apodemus-Unterkiefern. *Natur u. Heimat* **42**: 119–130.

Bunnell, F. L. 1999. What habitat is an island? In: Rochelle J. A., Lehman, L. A., Wisniewski, J. (eds.). *Forest Wildlife and Fragmentation: Management Implications.* Brill, Leiden, pp. 1–31.

Bye, F. N., Jacobsen, B. V. and Sonerud, G. A. 1992. Auditory prey location in a pause-travel predator: search height, search time, and attack range of Tengmalm's owls (*Aegolius funereus*). *Behav. Ecol.* **3**: 266–276.

Cannings, R. J. 1993. Northern saw-whet owl (*Aegolius acadicus*). In: Poole, A. and Gill, F. (eds.), *The Birds of North America.* Academy of Natural Sciences and American Ornithologists' Union, Philadelphia and Washington, DC, pp. 1–17.

Capinera, J. L. 1979. Qualitative variation in plants and insects: effect of propagule size on ecological plasticity. *Am. Nat.* **114**: 350–361.

Carlisle, T. R. 1982. Brood success in variable environments: implications for parental care allocation. *Anim. Behav.* **30**: 824–836.

Carlsson, B.-G. 1991. Recruitment of mates and deceptive behavior by male Tengmalm's owls. *Behav. Ecol. Sociobiol.* **28**: 321–328.

Carlsson, B.-G., and Hörnfeldt, B. 1989. Trigyny in Tengmalm's owl *Aegolius funereus* induced by supplementary feeding. *Ornis Scand.* **20**: 155–156.

Carlsson, B.-G., and Hörnfeldt, B. 1994. Determination of nestling age and laying date in Tengmalm's owl: use of wing length and body mass. *Condor* **96**: 555–559.

Carlsson, B.-G., Hörnfeldt, B. and Löfgren, O. 1987. Bigyny in Tengmalm's owl *Aegolius funereus*: effect of mating strategy on breeding success. *Ornis Scand.* **18**: 237–243.

Carlsson, U. T. 1983. WOF:s unggleinventering 1981 och 1982. *Värmlandsornitologen* **11**: 28–32.

Caro, T. 2005. *Antipredator Defenses in Birds and Mammals.* University of Chicago Press, Chicago and London.

Castro, A., Munoz, A.-R. and Real, R. 2008. Modelling the spatial distribution of the Tengmalm's owl *Aegolius funereus* in its southwestern palearctic limit (NE Spain). *Ardeola* **55**: 71–85.

Castro, S. A. and Jaksic, F. A. 1995. Great horned and barn owls prey differentially according to the age/size of a rodent in northcentral Chile. *J. Raptor Res.* **29**: 245–249.

Caswell, H. 1983. Phenotypic plasticity in life history traits: demographic effects and evolutionary consequences. *Am. Zool.* **23**: 35–46.

Catlin, D. H. and Rosenberg, D. K. 2008. Breeding dispersal and nesting behavior of burrowing owls following experimental nest predation. *Am. Midl. Nat.* **159**: 1–7.

Catling, P. M. 1972. A study of the boreal owl in southern Ontario with particular reference to the irruption of 1968–69. *Can. Field Nat.* **86**: 223–232.

Catry, P. and Furness, R. W. 1999. The influence of adult age on territorial attendance by breeding great skuas (*Catharacta skua*): an experimental study. *J. Avian Biol.* **30**: 399–406.

Catry, P., Phillips, R. A. and Furness, R. W. 1999. Evolution of reversed sexual size dimorphism in skuas and jaegers. *Auk* **116**: 158–168.

Cavé, A. J. 1968. The breeding of the kestrel, *Falco tinnunculus* L., in the reclaimed area Oostelijk Flevoland. *Neth. J. Zool.* **18**: 313–407.

Chabloz, V., Patthey, P. and Kunzle, I. 2001. Trois nichées simultanées de Chouettes de Tengmalm *Aegolius funereus* dans le même arbre. *Nos Oiseaux* **48**: 227–228.

Charnov, E. L., Orians, G. H. and Hyatt, K. 1976. Ecological implications of resource depression. *Am. Nat.* **110**: 247–259.

Cheveau, M., Drapeau, P., Imbeau, L. and Bergeron, Y. 2004. Owl winter irruptions as an indicator of small mammal population cycles in the boreal forest of eastern North America. *Oikos* **107**: 190–198.

Christe, P., Keller, L. and Roulin, A. 2006. The predation cost of being a male: implications for sex-specific rates of ageing. *Oikos* **114**: 381–384.

Clutton-Brock, T. H. 1988. *Reproductive Success*. The University of Chicago Press, Chicago and London.

Clutton-Brock, T. H. 1991. *The Evolution of Parental Care*. Princeton University Press, Princeton, NJ.

Clutton-Brock, T. H. and Iason, G. R. 1986. Sex ratio variation in mammals. *Q. Rev. Biol.* **61**: 339–374.

Collett, R. 1878. On *Myodes lemmus* in Norway. *Linn. Soc. J. Zool.* **13**: 329–331.

Collin, O. 1886. *Suomessa tavattavien pöllöjen pesimissuhteista*. Hämeen Sanomat, Hämeenlinna.

Cote, M., Ibarzabal, J., St-Laurent, M.-H., Ferron, J. and Gagnon, R. 2007. Age-dependent response of migrant and resident *Aegolius* owl species to small rodent population fluctuations in the eastern Canadian boreal forest. *J. Raptor Res.* **41**: 16–25.

Craighead, J. J. and Craighead, F. C. 1956. *Hawks, Owls and Wildlife*. Wildlife Management Institute, University of Michigan.

Currie, D., Valkama, J., Berg, Å., Boschert, M., Norrdahl, K., Hänninen, M., Korpimäki, E., Pöyri, V. and Hemminki, O. 2001. Sex roles, parental effort and offspring desertion in the monogamous Eurasian curlew *Numenius arquata*. *Ibis* **143**: 642–650.

Cushing, D. H. 1975. *Marine ecology and fisheries*. Cambridge University Press, London.

Daan, S. and Dijkstra, C. 1988. Date of birth and reproductive value of kestrel eggs: on the significance of early breeding. In: Dijkstra, C. (eds.), *Reproductive tactics in the kestrel* Falco tinnunculus. PhD thesis, University of Groningen, Netherlands, pp. 85–114.

Darwin, C. R. 1871. *The Descent of Man, and Selection in Relation to Sex*. John Murray, London.

Davies, N. B. 1989. Sexual conflict and the polygamy threshold. *Anim. Behav.* **38**: 226–234.

Davies, N. B. 1997. Mating systems. In: Krebs, J. R. and Davies, N. B. (eds.), *Behavioural Ecology: An Evolutionary Approach*. Blackwell Science, Oxford, pp. 263–294.

Deerenberg, C., Apanius, V., Daan, S. and Bos, N. 1997. Reproductive effort decreases antibody responsiveness. *Proc. R. Soc. Lond. B* **264**: 1021–1029.

Degn, H. J. 1978. A new method of analysing pellets from owls etc. *Dansk Orn. Fören. Tidsskr.* **72**: 143.

Dejaifve, P.-A., Novoa, C. and Prodon, R. 1990. Habitat et densité de la Chouette de Tengmalm *Aegolius funereus* à l'extrémité orientale des Pyrenees. *Alauda* **58**: 267–273.

Dell'Arte, G. L., Laaksonen, T., Norrdahl, K. and Korpimäki, E. 2007. Variation in the diet composition of a generalist predator, the red fox, in relation to season and density of main prey. *Acta Oecol.* **31**: 276–281.

De Neve, L., Soler, J. J., Ruiz-Rodríguez, M., Martín-Gálvez, D., Pérez-Contreras, T. and Soler, M. 2007. Habitat specific effects of a food supplementation experiment on immunocompetence in magpie *Pica pica* nestlings. *Ibis* **149**: 763–773.

Desjardins, D., Maruniak, J. A. and Bronson, F. H. 1973. Social rank in house mice: differentiation revealed by ultraviolet visualizations of urinary marking patterns. *Science* **182**: 939–941.

Desser, S. S. and Bennett, G. F. 1993. The genera *Leucocytozoon*, *Haemoproteus* and *Hepatocystis*. In: Kreier, J. P. (eds.), *Parasitic Protozoa*. Academic Press, San Diego, CA, pp. 273–305.

Dhondt, A. A. 2002. Changing mates. *Trends Ecol. Evol.* **17**: 55–56.

Dickman, C. R. 1992. Predation and habitat shift in the house mouse, *Mus domesticus*. *Ecology* **73**: 313–322.

Dickman, C. R., Predavec, M. and Lynam, A. J. 1991. Differential predation of size and sex classes of mice by the barn owl, *Tyto alba*. *Oikos* **62**: 67–76.

Dieckmann, U., O'Hara, B. and Weisser, W. 1999. The evolutionary ecology of dispersal. *Trends Ecol. Evol.* **14**: 88–90.

Dijkstra, C., Daan, S., Meijer, T., Cavé, A. J. and Foppen, R. P. B. 1988. Daily and seasonal variations in body mass of the kestrel in relation to food availability and reproduction. *Ardea* **76**: 127–140.

Dow, H. and Fredga, S. 1983. Breeding and natal dispersal of the goldeneye, *Bucephala clangula*. *J. Anim. Ecol.* **52**: 681–695.

Doyle, F. I. and Smith, J. N. M. 2001. Raptors and scavengers. In: Krebs, C. J., Boutin, S. and Boonstra, R. (eds.), *Ecosystem Dynamics of the Boreal Forest: The Kluane Project*. Oxford University Press, Oxford, pp. 377–404.

Drent, R. H. and Daan, S. 1980. The prudent parent: energetic adjustments in avian breeding. *Ardea* **68**: 225–252.

Dugger, K. M., Wagner, F. and Anthony, R. G. 2005. The relationship between habitat characteristics and demographic performance of Northern spotted owls in southern Oregon. *Condor* **107**: 863–878.

Duke, G. E., Jegers, A. A., Loff, G. and Evanson, O. A. 1975. Gastric digestion in some raptors. *Comp. Biochem. Physiol.* **50A**: 649–656.

Duncan, J. R. 2003. *Owls of the World: Their Lives, Behavior and Survival*. Firefly Books, New York.

Duncan, J. R., Swengel, S. R. and Swengel, A. B. 2009. Correlations of Northern saw-whet owl *Aegolius acadicus* calling indices from surveys in southern Wisconsin, USA, with owl and small mammal surveys in Manitoba, Canada, 1986–2006. *Ardea* **97**: 489–496.

Edenius, L. and Elmberg, J. 1996. Landscape level effects of modern forestry on bird communities in north Swedish boreal forests. *Landscape Ecol.* **11**: 325–338.

Eggers, S., Griesser, M., Andersson, T., and Ekman, J. 2005. Nest predation and habitat change interact to influence Siberian jay numbers. *Oikos* **111**: 150–158.

Ekerholm, P., Oksanen, L., Oksanen, T. and Schneider, M. 2004. The impact of short term predator removal on vole dynamics in a subarctic-alpine habitat complex. *Oikos* **106**: 457–468.

Eldegard, K. and Sonerud, G. A. 2009. Female offspring desertion and male-only care increase with natural and experimental increase in food abundance. *Proc. R. Soc. B* **276**: 1713–1721.

Eldegard, K. and Sonerud, G. A. 2010. Experimental increase in food supply influences the outcome of within-family conflicts in Tengmalm's owl. *Behav. Ecol. Sociobiol.* **64**: 815–826.

Elton, C. S. 1942. *Voles, Mice and Lemmings: Problems in Population Dynamics*. Oxford University Press, Oxford.

Elton, C. S. and Nicholson, M. 1942. The ten-year cycle in numbers of the lynx in Canada. *J. Anim. Ecol.* **11**: 215–244.

Emlen, S. T. and Oring, L. W. 1977. Ecology, sexual selection and the evolution of mating systems. *Science* **197**: 215–223.

Endler, J. A. 1986. *Natural Selection in the Wild*. Princeton University Press, Princeton, NJ.

Eriksson, M. O. G. 1979. Aspects of breeding biology of the goldeneye *Bucephala clangula*. *Holarct. Ecol.* **2**: 186–194.

Erkinaro, E. 1975. Zeitpunkt und Dauer der Mauser des Rauhfusskauzes, *Aegolius funereus*, und der Sumpfohreule, *Asio flammeus*. *Beitr. Vogelkd.* **21**: 288–290.

Erlinge, S. 1981. Food preference, optimal diet and reproductive output in stoats *Mustela erminea* in Sweden. *Oikos* **36**: 303–315.

Erlinge, S., Göransson, G., Hansson, L., Högstedt, G., Liberg, O., Nilsson, I. N., Nilsson, T., von Schantz, T. and Sylvén, M. 1983. Predation as a regulating factor in small rodent populations in southern Sweden. *Oikos* **40**: 36–52.

Erlinge, S., Göransson, G., Högstedt, G., Jansson, G., Liberg, O., Loman, J., Nilsson, I. N., von Schantz, T. and Sylvén, M. 1984. Can vertebrate predators regulate their prey? *Am. Nat.* **123**: 125–133.

Errington, P. L. 1930. The pellet analysis method of raptor food habits study. *Condor* **32**: 292–296.

Errington, P. L. 1932. Technique of raptor food habits study. *Condor* **34**: 75–86.

Errington, P. L. 1946. Predation and vertebrate populations. *Q. Rev. Biol.* **21**: 144–177.

Errington, P. L. 1956. Factors limiting higher vertebrate populations. *Science* **124**: 304–307.

Espie, R. H. M., Oliphant, L. W., James, P. C., Warkentin, I. G. and Lieske, D. J. 2000. Age-dependent breeding performance in merlins (*Falco columbarius*). *Ecology* **81**: 3404–3415.

Esseen, P.-A., Ehnström, B., Ericson, L. and Sjöberg, K. 1997. Boreal forests. *Ecol. Bull.* **46**: 16–47.

Faaborg, J. 1986. Reproductive success and survivorship of the Galapagos hawk *Buteo galapagoensis*: potential costs and benefits of cooperative polyandry. *Ibis* **128**: 337–347.

Faaborg, J. and Patterson, C. B. 1981. The characteristics and occurrence of cooperative polyandry. *Ibis* **123**: 477–484.

Faaborg, J., Parker, P. G., DeLay, L., de Vries, T., Bednarz, J. C., Maria Paz, S., Naranjo, J. and White, T. A. 1995. Confirmation of cooperative polyandry in the Galapagos hawk (*Buteo galapagoensis*). *Behav. Ecol. Sociobiol.* **36**: 83–90.

Fahrig, L. 1999. Forest loss and fragmentation: which is the greater effect on persistence of forest dwelling animals? In: Rochelle, L., Lehman, L. and Wisniewski, J. (eds.), *Forest Fragmentation: Wildlife and Management Implications*. Brill, Leiden, pp. 87–95.

Fahrig, L. 2003. Effects of habitat fragmentation on biodiversity. *Annu. Rev. Ecol. Syst.* **34**: 487–515.

Falconer, D. S. 1981. *Introduction to Quantitative Genetics*, 2nd edn. Longman, London and New York.

Fang, Y., Tang, S.-H., Gu, Y. and Sun, Y.-H. 2009. Conservation of Tengmalm's owl and Sichuan wood owl in Lianhuashan Mountain, Gansu, China. *Ardea* **97**: 649.

Fargallo, J. A., Martinez-Padilla, J., Vinuela, J., Blanco, G., Torre, I., Vergara, P. and De Neve, L. 2009. Kestrel–prey dynamic in a Mediterranean region: the effect of generalist predation and climatic factors. *PLoS One* **4**(2): e4311, doi:10.1371/journal.pone.0004311.

Fazey, I., Fisher, J. and Lindmayer, D. B. 2005. What do conservation biologists publish? *Biol. Conserv.* **124**: 63–73.

Finnish Meteorological Institute 1994. *Monthly Climate Observations in Finland*. Finnish Meteorological Institute, Helsinki.

Foley, J. A., DeFries, R., Asner, G. P., Barford, C., Bonan, G., Carpenter, S. R., Chapin, F. S., Coe, M. T., Daily, G. C., Gibbs, H. K., Helkowski, J. H., Holloway, T., Howard, E. A., Kucharik, C. J., Monfreda, C., Patz, J. A., Prentice, I. C., Ramankutty, N. and Snyder, P. K. 2005. Global consequences of land use. *Science* **309**: 570–574.

Forslund, P. and Pärt, T. 1995. Age and reproduction in birds: hypotheses and tests. *Trends Ecol. Evol.* **10**: 374–378.

Forsman, D., Jokinen, M., Kaikusalo, A. and Korpimäki, E. 1980. Pöllöjen pesintä Suomessa 1979 (Summary: Breeding of owls in Finland in 1979). *Lintumies* **15**: 2–9.

Franklin, A. B., Anderson, D. R., Gutiérrez, R. J. and Burnham, K. P. 2000. Climate, habitat quality, and fitness in Northern spotted owl populations in northwestern California. *Ecol. Monogr.* **70**: 539–590.

Franz, A., Mebs, T. and Seibt, E. 1984. Zur Populationsbiologie des Rauhfusskauzes (*Aegolius funereus*) im südlichen Westfalen und in angrenzenden Gebieten anhand von Beringungsergebnissen. *Vogelwarte* **32**: 260–269.

Fredga, K. 1964. En undersökning av pärlugglans (*Aegolius funereus*) bytesval i Mellansverige. *Vår Fågelvärld* **23**: 103–118.

Fretwell, S. D. and Lucas, J. L. J. 1969. On territorial behaviour and other factors influencing habitat distribution in birds. I. Theoretical development. *Acta Biotheor.* **19**: 16–36.

Fridolfsson, A.-K., and Ellegren, H. 1999. A simple and universal method for molecular sexing of non-ratite birds. *J. Avian Biol.* **30**: 116–121.

Galeotti, P. 1998. Correlates of hoot rate and structure in male tawny owls *Strix aluco*: implications for male rivalry and female mate choice. *J. Avian Biol.* **29**: 25–32.

Galeotti, P. and Sacchi, R. 2001. Turnover of territorial scops owls *Otus scops* as estimated by spectrographic analyses of male hoots. *J. Avian Biol.* **32**: 256–262.

Galeotti, P. and Sacchi, R. 2003. Differential parasitaemia in the tawny owl (*Strix aluco*): effects of colour morph and habitat. *J. Zool. Lond.* **261**: 91–99.

Galeotti, P., Paladin, P. and Pavan, G. 1993. Individually distinct hooting in male pygmy owls *Glaucidium passerinum*: a multivariate approach. *Ornis Scand.* **24**: 15–20.

Galeotti, P., Sacchi, R. and Vicario, V. 2005. Fluctuating asymmetry in body traits increases predation risks: tawny owl selection against asymmetric woodmice. *Evol. Ecol.* **19**: 405–418.

Gasow, H. 1968. Über Gevölle, Beutetiere und Schutz des Rauhfusskauzes. *Beitr. Angew. Vogelkd.* **5**: 37–60.

Gebhardt-Henrich, S. G., and van Noordwijk, A. J. 1991. Nestling growth in the great tit. I. Heritability estimates under different environmental conditions. *J. Evol. Biol.* **3**: 341–362.

Gehlbach, F. R. 1989. Screech owl. In: Newton, I. (ed.), *Lifetime Reproduction in Birds*. Academic Press, London, pp. 315–326.

Gehlbach, F. R. 1994. *The Eastern Screech Owl: Life History, Ecology, and Behavior in the Suburbs and Countryside*. Texas A&M University Press, College Station, TX.

Giesel, J. T. 1976. Reproductive strategies as adaptation to life in temporally heterogeneous environments. *Annu. Rev. Ecol. Syst.* **7**: 57–79.

Glutz von Blotzheim, U. N. and Bauer, K. M. 1980. *Handbuch der Vögel Mitteleuropas*. Akademische Verlagsgesellschaft, Wiesbaden, Germany.

Goodman, D. 1979. Regulating reproductive effort in a changing environment. *Am. Nat.* **113**: 735–748.

Goszczynski, J. 1977. Connections between predatory birds and mammals and their prey. *Acta Theriol.* **22**: 399–430.

Goszczynski, J. 1983. Ecology of the bank vole: predators. *Acta Theriol.* **28**: 49–54.

Grant, B. R. and Grant, P. R. 1989. *Evolutionary Dynamics of a Natural Population: The Large Cactus Finch of the Galapagos.* The University of Chicago Press, Chicago.

Grant, P. R. 1972. Interspecific competition among rodents. *Annu. Rev. Ecol. Syst.* **3**: 79–106.

Graul, W. D. 1977. The evolution of avian polyandry. *Am. Nat.* **111**: 812–816.

Greene, H. W. and Jaksic, F. M. 1983. Food niche relationships among sympatric predators: effects of level of prey identification. *Oikos* **40**: 151–154.

Greenwood, P. J. 1980. Mating systems, philopatry and dispersal in birds and mammals. *Anim. Behav.* **28**: 1140–1162.

Greenwood, P. J. and Harvey, P. H. 1982. The natal and breeding dispersal of birds. *Annu. Rev. Ecol. Syst.* **13**: 1–21.

Haapala, J. and Saurola, P. 1986a. Petolintujen pesintä Suomessa 1986 (Summary: Breeding of raptors and owls in Finland in 1986). *Lintumies* **21**(6): 258–267.

Haapala, J. and Saurola, P. 1986b. Petolintukantojen seurantaprojekti 1982–1985 (Summary: Monitoring of common raptors and owls in Finland in 1982–1985). *Lintumies* **21**(3): 140–145.

Haapala, J. and Saurola, P. 1987. Petolintujen pesintä Suomessa 1987 (Summary: Breeding of raptors and owls in Finland in 1987). *Lintumies* **22**(6): 244–251.

Haapala, J. and Saurola, P. 1989. Petolintujen pesintä ja petolintukantojen seuranta Suomessa 1988 (Summary: Breeding and population trends of common raptors and owls in Finland in 1988). *Lintumies* **24**: 27–36.

Haapala, J., Korhonen, J. and Saurola, P. 1990. Petolintujen pesintä Suomessa 1989 (Summary: Breeding of raptors and owls in Finland in 1989). *Lintumies* **25**(3): 113–119.

Haapala, J., Korhonen, J. and Saurola, P. 1991. Petolintujen pesintä ja petolintukantojen seuranta Suomessa 1990 (Summary: Breeding and population trends of common raptors and owls in Finland in 1990). *Lintumies* **26**(1): 2–13.

Haapala, J., Lehtonen, J. and Saurola, P. 1992. Petolintujen pesintä ja petolintukantojen seuranta Suomessa 1991 (Summary: Breeding and population trends of common raptors and owls in Finland in 1991). *Lintumies* **27**(1): 2–13.

Haapala, J., Lehtonen, J. T., Korhonen, J. and Saurola, P. 1993. Petolintuvuosi 1992: Yltäkyllyydestä niukkuuteen (Summary: Breeding and population trends of common raptors and owls in Finland in 1992). *Linnut* **28**(1): 18–27.

Haapala, J., Korhonen, J. and Saurola, P. 1994. Petolintuvuosi 1993: Tyhjiä pesäpohjia ja höyhenpatjoja (Summary: Breeding and population trends of common raptors and owls in Finland in 1993). *Linnut* **29**(2): 20–24.

Haapala, J., Korhonen, J. and Saurola, P. 1995. Petolintuvuosi 1994: Nousuvuoden toiveikkuutta (Summary: Breeding and Population trends of common raptors and owls in Finland in 1994). *Linnut* **30**(2): 20–25.

Haapala, J., Korhonen, J. and Saurola, P. 1996a. Petolintuvuosi 1996: Pohjalta on hyvä ponnistaa (Summary: Breeding and population trends of common raptors and owls in Finland in 1996). *Linnut Yearbook* **1996**: 41–53.

Haapala, J., Korhonen, J. and Saurola, P. 1996b. Suomen petolintuvuosi 1995: Alamäen alkuliukua (Summary: Breeding and population trends of common raptors and owls in Finland in 1995). *Hippiäinen* **26**: 4–11.

Haapala, J., Korhonen, J. and Saurola, P. 1997. Petolintuvuosi 1997: Lamasta toipuminen käynnistyi hyvin (Summary: Breeding and population trends of common raptors and owls in Finland in 1997). *Linnut Yearbook* **1997**: 39–54.

Haartman, L. von 1968. The evolution of resident versus migratory habit in birds: some considerations. *Ornis Fenn.* **45**: 1–7.

Haartman, L. von 1971. Population dynamics. In: Farner, D. and King, J. R. (eds.), *Avian Biology.* Academic Press, New York, pp. 391–459.

Hagen, Y. 1952. *Rovfuglene og viltpleien.* Gyldendal Norsk Forlag, Oslo.

Haila, Y. 2002. A conceptual genealogy of fragmentation research: from island biogeography to landscape ecology. *Ecol. Appl.* **12**: 321–334.

Hakkarainen, H. 1994. *Reproductive effort, body size and their fitness consequences in Tengmalm's owl* (Aegolius funereus) *under cyclic food conditions.* PhD thesis, Department of Biology, University of Turku, Finland.

Hakkarainen, H. and Korpimäki, E. 1991. Reversed sexual size dimorphism in Tengmalm's owl: is small male size adaptive? *Oikos* **61**: 337–346.

Hakkarainen, H. and Korpimäki, E. 1993. The effect of female body size on clutch volume of Tengmalm's owls *Aegolius funereus* in varying food conditions. *Ornis Fenn.* **70**: 189–195.

Hakkarainen, H. and Korpimäki, E. 1994a. Does feeding effort of Tengmalm's owls reflect offspring survival prospects in cyclic food conditions? *Oecologia* **97**: 209–214.

Hakkarainen, H. and Korpimäki, E. 1994b. Environmental, parental and adaptive variation in egg size of Tengmalm's owls under fluctuating food conditions. *Oecologia* **98**: 362–368.

Hakkarainen, H. and Korpimäki, E. 1994c. Nest defence of Tengmalm's owls reflects offspring survival prospects under fluctuating food conditions. *Anim. Behav.* **48**: 843–849.

Hakkarainen, H. and Korpimäki, E. 1995. Contrasting phenotypic correlations in food provision of male Tengmalm's owls (*Aegolius funereus*) in a temporally heterogeneous environment. *Evol. Ecol.* **9**: 30–37.

Hakkarainen, H. and Korpimäki, E. 1996. Competitive and predatory interactions among raptors: an observational and experimental study. *Ecology* **77**: 1134–1142.

Hakkarainen, H. and Korpimäki, E. 1998. Why do territorial male Tengmalm's owls fail to obtain a mate? *Oecologia* **114**: 578–582.

Hakkarainen, H., Huhta, E., Lahti, K., Lundvall, P., Mappes, T., Tolonen, P. and Wiehn, J. 1996a. A test of male mating and hunting success in the kestrel: the advantages of smallness. *Behav. Ecol. Sociobiol.* **39**: 375–380.

Hakkarainen, H., Koivunen, V., Korpimäki, E. and Kurki, S. 1996b. Clear-cut areas and breeding success of Tengmalm's owls *Aegolius funereus*. *Wildl. Biol.* **3**: 253–258.

Hakkarainen, H., Korpimäki, E., Ryssy, J. and Vikström, S. 1996c. Low heritability in morphological characters of Tengmalm's owls: the role of cyclic food and laying date? *Evol. Ecol.* **10**: 207–219.

Hakkarainen, H., Koivunen, V. and Korpimäki, E. 1997a. Reproductive success and parental effort of Tengmalm's owls: effects of spatial and temporal variation in habitat quality. *Ecoscience* **4**: 35–42.

Hakkarainen, H., Korpimäki, E., Koivunen, V. and Kurki, S. 1997b. Boreal owl responses to forest management: a review. *J. Raptor Res.* **31**: 125–128.

Hakkarainen, H., Ilmonen, P., Koivunen, V. and Korpimäki, E. 1998. Blood parasites and nest defence behaviour of Tengmalm's owls. *Oecologia* **114**: 574–577.

Hakkarainen, H., Ilmonen, P., Koivunen, V. and Korpimäki, E. 2001. Experimental increase of predation risk induces breeding dispersal of Tengmalm's owl. *Oecologia* **126**: 355–359.

Hakkarainen, H., Korpimäki, E., Koivunen, V. and Ydenberg, R. 2002. Survival of male Tengmalm's owls under temporally varying food conditions. *Oecologia* **131**: 83–88.

Hakkarainen, H., Mykrä, S., Kurki, S., Korpimäki, E., Nikula, A. and Koivunen, V. 2003. Habitat composition as a determinant of reproductive success of Tengmalm's owls under fluctuating food conditions. *Oikos* **100**: 162–171.

Hakkarainen, H., Mykrä, S., Kurki, S., Tornberg, R. and Jungell, S. 2004. Competitive interactions among raptors in boreal forests. *Oecologia* **141**: 420–424.

Hakkarainen, H., Korpimäki, E., Laaksonen, T., Nikula, A. and Suorsa, P. 2008. Survival of male Tengmalm's owls increases with cover of old forest in their territory. *Oecologia* **155**: 479–486.

Halle, S. 1988. Avian predation upon a mixed community of common voles (*Microtus arvalis*) and wood mice (*Apodemus sylvaticus*). *Oecol. (Berl.)* **75**: 451–455.

Halle, S. and Lehmann, U. 1992. Cycle-correlated changes in the activity behaviour of field voles, *Microtus agrestis*. *Oikos* **64**: 489–497.

Halonen, M., Mappes, T., Meri, T. and Suhonen, J. 2007. Influence of snow cover on food hoarding in pygmy owls *Glaucidium passerinum*. *Ornis Fenn.* **84**: 105–111.

Hämäläinen, I. 1979. Viirupöllö ahdisti nauhuria. *Päijät-Hämeen Linnut* **10**: 78.

Hannula, H., Haapala, J. and Saurola, P. 2002. Petolintuvuosi 2001 myyrät harvojen herkkua (Summary: Breeding and population trends of common raptors and owls in Finland in 2001: a year with rodents almost absent). *Linnut Yearbook* **2001**: 15–25.

Hanski, I. and Gilbin, M. E. 1997. *Metapopulation Biology: Ecology, Genetics, and Evolution.* Academic Press, London.

Hanski, I. and Henttonen, H. 1996. Predation on competing rodent species: a simple explanation of complex patterns. *J. Anim. Ecol.* **65**: 220–232.

Hanski, I., Hansson, L. and Henttonen, H. 1991. Specialist predators, generalist predators, and the microtine rodent cycle. *J. Anim. Ecol.* **60**: 353–367.

Hanski, I., Turchin, P., Korpimäki, E. and Henttonen, H. 1993. Population oscillations of boreal rodents: regulation by mustelid predators leads to chaos. *Nature* **364**: 232–235.

Hansson, L. 1979. On the importance of landscape heterogeneity in northern regions for the breeding population densities of homeotherms: a general hypothesis. *Oikos* **33**: 182–189.

Hansson, L. 1982. Use of forest edges by Swedish mammals. *Fauna o. Flora* **77**: 301–308.

Hansson, L. 1983. Competition between rodents in successional stages of taiga forest: *Microtus agrestis* vs. *Clethrionomys glareolus*. *Oikos* **40**: 258–266.

Hansson, L. 1984. Winter reproduction of small mammals in relation to food conditions and population dynamics. *Bull. Carn. Mus. Nat. Hist.* **10**: 225–234.

Hansson, L. 1992. Landscape ecology of boreal forests. *Trends Ecol. Evol.* **7**: 299–302.

Hansson, L. and Henttonen, H. 1985. Gradients in density variations of small rodents: the importance of latitude and snow cover. *Oecol. (Berl.)* **67**: 394–402.

Hansson, L. and Henttonen, H. 1988. Rodent dynamics as community processes. *Trends Ecol. Evol.* **3**: 195–200.

Hardouin, L. A., Reby, D., Bavoux, C., Burneleau, G. and Bretagnolle, V. 2007. Communication of male quality in owl hoots. *Am. Nat.* **169**: 552–562.

Hardouin, L. A., Robert, D. and Bretagnolle, V. 2008. A dusk chorus effect in a nocturnal bird: support for mate and rival assessment functions. *Behav. Ecol. Sociobiol.* **62**: 1909–1918.

Hayward, G. D. 1997. Forest management and conservation of boreal owls in North America. *J. Raptor Res.* **31**: 114–124.

Hayward, G. D. and Garton, E. O. 1984. Roost habitat selection by three small forest owls. *Wilson Bull.* **96**: 690–692.

Hayward, G. D. and Garton, E. O. 1988. Resource partitioning among forest owls in the River of No Return Wilderness, Idaho. *Oecologia* **75**: 253–265.

Hayward, G. D. and Hayward, P. H. 1991. Body measurements of boreal owls in Idaho and a discriminant model to determine sex of live specimens. *Wilson Bull.* **103**: 497–500.

Hayward, G. D. and Hayward, P. H. 1993. Boreal owl *Aegolius funereus*. In: Poole, A. and Gill, F. (eds.), *The Birds of North America*. Academy of Natural Sciences and American Ornithologists Union, Philadelphia, PA, pp. 1–19.

Hayward, G. D., Hayward, P. H., Garton, E. O. and Escano, R. 1987. Revised breeding distribution of the boreal owl in the northern Rocky Mountains. *Condor* **89**: 431–432.

Hayward, G. D., Steinhorst, R. K. and Hayward, P. H. 1992. Monitoring boreal owl populations with nest boxes: sample size and cost. *J. Wildl. Manage.* **56**: 776–784.

Hayward, G. D., Hayward, P. H. and Garton, E. O. 1993. Ecology of boreal owls in the Northern Rocky Mountains, USA. *Wildl. Monogr.* **124**: 1–59.

Hedrick, A. V. and Temeles, E. J. 1989. The evolution of sexual size dimorphism in animals: hypotheses and tests. *Trends Evol. Ecol.* **4**: 136–138.

Heikura, K. 1984. The population dynamics and the influence of winter on the common shrew (*Sorex araneus* L.). *Carn. Mus. Nat. Hist. Spec. Publ.* **10**: 343–361.

Heino, A. 1985. *Pikkunisäkäsfaunan koostumuksesta ja päästäisten habitaatinvalinta- ja ravinto-suhteista Oulussa ja Oulangalla, Kuusamossa*. MSci thesis, Department of Zoology, University of Oulu, Finland.

Helle, P. 1986. Effects of forest succession and fragmentation on bird communities in boreal forests. *Acta Univ. Oul. Ser. A. Sci. Rer. Nat.* **178**: 1–41.

Henrioux, P. 2010. Etude d'une population de chouette de Tengmalm dans l'Ouest du Jura. *GERNOV* **2009**: 1–7.

Henttonen, H., Kaikusalo, A., Tast, J. and Viitala, J. 1977. Interspecific competition between small rodents in subarctic and boreal ecosystems. *Oikos* **29**: 581–590.

Henttonen, H., Oksanen, T., Jortikka, A. and Haukisalmi, V. 1987. How much do weasels shape microtine cycles in the northern Fennoscandian taiga? *Oikos* **50**: 353–365.

Henttonen, H., Haukisalmi, V., Kaikusalo, A., Korpimäki, E., Norrdahl, K. and Skarén, U. A. P. 1989. Long-term dynamics of the common shrew *Sorex araneus* in Finland. *Ann. Zool. Fenn.* **26**: 349–355.

Herren, V., Anderson, S. H. and Ruggiero, L. F. 1996. Boreal owl mating habitat in the northwestern United States. *J. Raptor Res.* **30**: 123–129.

Herrera, C. M. and Hiraldo, F. 1976. Food-niche and trophic relationships among European owls. *Ornis Scand.* **7**: 29–41.

Hildén, O. 1965. Habitat selection in birds: a review. *Ann. Zool. Fenn.* **2**: 53–75

Hildén, O. 1966. Changes in the bird fauna of Valassaaret, Gulf of Bothnia, during recent decades. *Ann. Zool. Fenn.* **3**: 245–269.

Hildén, O. 1975. Breeding system of Temminck's stint *Calidris temminckii*. *Ornis Fenn.* **52**: 117–147.

Hildén, O. 1978. Population dynamics in Temminck's stint *Calidris temminckii*. *Oikos* **30**: 17–28.

Hildén, O., Ulfvens, J., Pahtamaa, T. and Hästbacka, H. 1995. Changes in the archipelago bird populations of the Finnish Quark, Gulf of Bothnia, from 1957–69 to 1990–91. *Ornis Fenn.* **72**: 115–126.

Hinam, H. L. and St. Clair, C. C. 2008. High levels of habitat loss and fragmentation limit reproductive success by reducing home range size and provisioning rates of Northern saw-whet owls. *Biol. Conserv.* **141**: 524–535.

Hinde, C. A. and Kilner, R. M. 2007. Negotiations within the family over supply of parental care. *Proc. R. Soc. B* **274**: 53–60.

Hipkiss, T. 2002. Sexual size dimorphism in Tengmalm's owl (*Aegolius funereus*) on autumn migration. *J. Zool. Lond.* **257**: 281–285.

Hipkiss, T. and Hörnfeldt, B. 2004. High interannual variation in the hatching sex ratio of Tengmalm's owl broods during a vole cycle. *Popul. Ecol.* **46**: 263–268.

Hipkiss, T., Hörnfeldt, B., Eklund, U. and Berlin, S. 2002a. Year-dependent sex-biased mortality in supplementary-fed Tengmalm's owl nestlings. *J. Anim. Ecol.* **71**: 693–699.

Hipkiss, T., Hörnfeldt, B., Lundmark, Å., Norbäck, M. and Ellegren, H. 2002b. Sex ratio and age structure of nomadic Tengmalm's owls: a molecular approach. *J. Avian Biol.* **33**: 107–110.

Hirons, G. J., Hardy, A. R. and Stanley, P. I. 1984. Body weight, gonad development and moult in the tawny owl (*Strix aluco*). *J. Zool. Lond.* **202**: 145–164.

Hirshfield, M. F. and Tinkle, D. W. 1975. Natural selection and the evolution of reproductive effort. *Proc. Natl. Acad. Sci. USA* **72**: 2227–2231.

Hohtola, E., Pyörnilä, A. and Rintamäki, H. 1994. Fasting endurance and cold resistance without hypothermia in a small predatory bird: the metabolic strategy of Tengmalm's owl, *Aegolius funereus*. *J. Comp. Physiol. B* **164**: 430–437.

Holling, C. S. 1959. Some characteristics of simple types of predation and parasitism. *Can. Entomol.* **91**: 385–398.

Holmberg, T. 1979. Punkttaxering av pärluggla *Aegolius funereus*: en metodstudie. *Vår Fågelvärld* **38**: 237–244.

Holmberg, T. 1982. Breeding density and site tenacity of Tengmalm's owl, *Aegolius funereus*. *Vår Fågelvärld* **41**: 265–267.

Hongell, H. 1986. Keski-Pohjanmaan pöllöt 1985. *Ornis Botnica* **8**: 15–34.

Honkala, J. and Saurola, P. 2006. Petolintuvuosi 2005: monien ennätysten vuosi (Summary: Breeding and population trends of common raptors and owls in Finland in 2005). *Linnut Yearbook* **2005**: 9–22.

Honkala, J. and Saurola, P. 2007. Petolintuvuosi 2006: myyriä alkupaloiksi (Summary: Breeding and population trends of common raptors and owls in Finland in 2007). *Linnut Yearbook* **2006**: 54–67.

Honkala, J. and Saurola, P. 2008. Petolintuvuosi 2007 (Summary: Breeding and population trends of common raptors and owls in Finland in 2007). *Linnut Yearbook* **2007**: 36–51.

Honkala, J., Björklund, H. and Saurola, P. 2010. Petolintuvuosi 2009: monien ennätysten vuosi (Summary: Breeding and population trends of common raptors and owls in Finland in 2009). *Linnut Yearbook* **2009**: 78–89.

Hörnfeldt, B. and Eklund, U. 1990. The effect of food on laying date and clutch size in Tengmalm's owl. *Ibis* **132**: 395–406.

Hörnfeldt, B. and Nyholm, E. I. 1996. Breeding performance of Tengmalm's owl in a heavy metal pollution gradient. *J. Appl. Ecol.* **33**: 377–386.

Hörnfeldt, B., Carlsson, B.-G. and Nordström, L. L. 1988. Molt of primaries and age determination in Tengmalm's owl (*Aegolius funereus*). *Auk* **105**: 783–789.

Hörnfeldt, B., Carlsson, B.-G., Löfgren, O. and Eklund, U. 1990. Effects of cyclic food supply on breeding performance in Tengmalm's owl (*Aegolius funereus*). *Can. J. Zool.* **68**: 522–530.

Hörnfeldt, B., Hipkiss, T., Fridolfsson, A.-K., Eklund, U. and Ellegren, H. 2000. Sex ratio and fledging success of supplementary-fed Tengmalm's owl broods. *Molec. Ecol.* **9**: 187–192.

Hörnfeldt, B., Hipkiss, T. and Eklund, U. 2005. Fading out of vole and predator cycles? *Proc. R. Soc. Lond. B* **272**: 2045–2049.

Houston, A. I., Szekely, T. and McNamara, J. M. 2005. Conflict between parents over care. *Trends Ecol. Evol.* **20**: 33–38.

Huhta, E., Aho, T., Jäntti, A., Suorsa, P., Kuitunen, M., Nikula, A. and Hakkarainen, H. 2004. Forest fragmentation increases nest predation in the Eurasian treecreeper. *Conserv. Biol.* **18**: 148–155.

Huhtala, K., Korpimäki, E. and Pulliainen, E. 1987. Foraging activity and growth of nestlings in the hawk owl: adaptive strategies under northern conditions. In: Nero, R. W., Clark, R. J., Knapton, R. J. and Hamre, R. H. (eds.), *Biology and Conservation of Northern Forest Owls*. General Technical Report Rm-142. USDA Forest Service, Fort Collins, CO, pp. 152–156.

Huitu, O., Koivula, M., Korpimäki, E., Klemola, T. and Norrdahl, K. 2003a. Winter food supply limits growth of northern vole populations in the absence of predation. *Ecology* **84**: 2108–2118.

Huitu, O., Norrdahl, K. and Korpimäki, E. 2003b. Landscape effects on temporal and spatial properties of vole population fluctuations. *Oecologia* **135**: 209–220.

Huitu, O., Laaksonen, J., Klemola, T. and Korpimäki, E. 2008. Spatial dynamics of *Microtus* voles in continuous and fragmented agricultural landscapes. *Oecologia* **155**: 53–61.

Huitu, O., Kiljunen, N., Korpimäki, E., Koskela, E., Mappes, T., Pietiäinen, H., Pöysä, H. and Henttonen, H. 2009. Density-dependent vole damage in silviculture and associated economic losses at a nationwide scale. *Forest Ecol. Manage.* **258**: 1219–1224.

Hunter, D. B., Rohner, C. and Currie, D. C. 1997. Mortality in fledgling great horned owls from black fly hematophaga and leucocytozoonosis. *J. Wildl. Dis.* **33**: 486–491.

Hurme, E., Mönkkönen, M., Sippola, A.-L., Ylinen, H. and Pentinsaari, M. 2008. Role of the Siberian flying squirrel as an umbrella species for biodiversity in northern boreal forests. *Ecol. Ind.* **8**: 246–255.

Ilmonen, P., Hakkarainen, H., Koivunen, V., Korpimäki, E., Mullie, A. and Shutler, D. 1999. Parental effort and blood parasitism in Tengmalm's owl: effects of natural and experimental variation in food abundance. *Oikos* **86**: 79–86.

Ims, R. A., Henden, J.-A. and Killengreen, S. T. 2008. Collapsing population cycles. *Trends Ecol. Evol.* **23**: 79–86.

Itämies, J. and Korpimäki, E. 1987. Insect food of the kestrel, *Falco tinnunculus*, during breeding in western Finland. *Aquilo Ser. Zool.* **25**: 21–31.

Itämies, J. and Mikkola, H. 1972. The diet of honey buzzards *Pernis apivorus* in Finland. *Ornis Fenn.* **49**: 7–10.

Jacobsen, B. V. and Sonerud, G. A. 1987. Home range of Tengmalm's owl: a comparison between nocturnal hunting and diurnal roosting. In: Nero, R. W., Clark, R. J., Knapton, R. J. and Hamre, R. H. (eds.), *Biology and Conservation of Northern Forest Owls*. General Technical Report RM-142. USDA Forest Service, Fort Collins, CO, pp. 189–192.

Jacobsen, B. V. and Sonerud, G. A. 1993. Synchronous switch in diet and hunting habitat as a response to disappearance of snow cover in Tengmalm's owl *Aegolius funereus*. *Ornis Fenn.* **70**: 78–88.

Jäderholm, K. 1987. Diets of the Tengmalm's owl *Aegolius funereus* and the Ural owl *Strix uralensis* in central Finland. *Ornis Fenn.* **64**: 149–153.

Jaksic, F. M. and Braker, H. E. 1983. Food-niche relationships and guild structure of diurnal birds of prey: competition versus opportunism. *Can. J. Zool.* **61**: 2230–2241.

Jaksic, F. M. and Jiménez, J. E. 1986. Trophic structure and food-niche relationships of Nearctic and Neotropical raptors: an inferential approach. In: *Proceedings of the XIX International Ornithological Congress*, pp. 2336–2347.

Jamnaback, H. 1973. Recent developments in control of black flies. *Annu. Rev. Entomol.* **18**: 281–304.

Järvinen, A. 1989. Clutch-size variation in the pied flycatcher *Ficedula hypoleuca. Ibis* **131**: 572–577.

Järvinen, O. 1979. Geographical gradients of stability in European land bird communities. *Oecol. (Berl.)* **38**: 51–69.

Järvinen, O. and Väisänen, R. A. 1980. Quantitative biogeography of Finnish land birds as compared with regionality of other taxa. *Ann. Zool. Fenn.* **17**: 67–85.

Jedrzejewski, W., Jedrzejewska, B. and Szymura, L. 1995. Weasel population response, home range, and predation on rodents in a deciduous forest in Poland. *Ecology* **76**: 179–195.

Jedrzejewski, W., Jedrzejewska, B., Szymura, A. and Zub, K. 1996. Tawny owl (*Strix aluco*) predation in a pristine deciduous forest (Bialowieza National Park, Poland). *J. Anim. Ecol.* **65**: 105–120.

Jeschke, J. M. and Kokko, H. 2008. Mortality and other determinants of bird divorce rate. *Behav. Ecol. Sociobiol.* **62**: 1–9.

Jörlitschka, W. 1988. Untersuchungen zur Habitatstruktur von Revieren des Rauhfusskauzes *Aegolius funereus* im Nordschwarzwald. *Vogelwelt* **109**: 152–155.

Johnsgard, P. A. 1988. *North American Owls: Biology and Natural History.* Smithsonian Institution Press, Washington, DC.

Johnson, M. L. and Gaines, M. S. 1990. Evolution of dispersal: theoretical models and empirical tests using birds and mammals. *Annu. Rev. Ecol. Syst.* **21**: 449–480.

Johnsson, K., Nilsson, S. G. and Tjernberg, M. 1993. Characteristics and utilization of old black woodpecker *Dryocopus martius* holes by hole-nesting species. *Ibis* **135**: 410–414.

Jokinen, M., Kaikusalo, A. and Korpimäki, E. 1982. Breeding of owls in Finland in 1980. *Suomenselän Linnut* **17**: 15–22.

Jokinen, M., Kaikusalo, A., Korpimäki, E. and Pietiäinen, H. 1983. Breeding of owls in Finland in 1981. *Suomenselän Linnut* **18**: 135–142.

Jones, K. M., Ruxton, G. D. and Monaghan, P. 2002. Model parents: is full compensation for reduced partner nest attendance compatible with stable biparental care? *Behav. Ecol.* **13**: 838–843.

Joveniaux, A. and Durand, G. 1987. Gestion forestière et écologie des populations de chouette de Tengmalm (*Aegolius funereus*) dans l'est de la France. *Rev. Ecol. (Terre Vie)* **Suppl. 4**: 83–96.

Kacelnik, A. 1984. Central place foraging in starlings (*Sturnus vulgaris*). I. Patch residence time. *J. Anim. Ecol.* **53**: 283–299.

Kalela, O. 1962. On the fluctuations in the numbers of arctic and boreal small rodents as a problem of production biology. *Ann. Acad. Sci. Fenn. A IV* **66**: 1–38.

Källander, H. 1964. Invasion av pärluggla (*Aegolius funereus*) i Mellansverige 1958 samt något om artens förekomst i Sverige. *Vår Fågelvärld* **23**: 119–135.

Källander, H. and Smith, H. G. 1990. Food storing in birds: an evolutionary perspective. *Curr. Ornithol.* **7**: 147–209.

Kämpfer-Lauenstein, A. and Lederer, W. 1992. Bemerkenswerte Umsiedlungen von Rauhfusskauz (*Aegolius funereus*). *Vogelwarte* **36**: 236–237.

Kaplan, R. H. 1980. The implications of ovum size variability for offspring fitness and clutch size within several populations of salamanders (Ambystoma). *Evolution* **34**: 51–64.

Karell, P., Ahola, K., Karstinen, T., Zolei, A. and Brommer, J. E. 2009. Population dynamics in a cyclic environment: consequences of cyclic food abundance on tawny owl reproduction and survival. *J. Anim. Ecol.* **78**: 1050–1062.

Karell, P., Lehtosalo, N., Pietiäinen, H. and Brommer, J. E. 2010. Ural owl predation on field voles and bank voles by size, sex and reproductive state. *Ann. Zool. Fenn.* **47**: 90–98.

Keith, L. B., Todd, A. W., Brand, C. J., Adamcik, R. S. and Rusch, D. H. 1977. An analysis of predation during cyclic fluctuation of snowshoe hares. *Int. Congr. Game Biol.* **13**: 151–175.

Kellomäki, E. 1977. Food of the pygmy owl *Glaucidium passerinum* in the breeding season. *Ornis Fenn.* **54**: 1–29.

Kelly, E. J. and Kennedy, P. L. 1993. A dynamic state variable model of mate desertion in Cooper's hawks. *Ecology* **74**: 351–366.

Kenward, R. E. 2001. *A Manual for Wildlife Radio Tagging*. Academic Press, New York.

Kenward, R. E., Walls, S. S., Hodder, K. H., Pahkala, M., Freeman, S. N. and Simpson, V. R. 2000. The prevalence of non-breeders in raptor populations: evidence from rings, radio-tags and transect surveys. *Oikos* **91**: 271–279.

Keymer, A. E. and Read, A. F. 1991. Behavioural ecology: the impact of parasitism. In: Toft, C. A., Aeschlimann, A. and Bolis, L. (eds.), *Parasite–Host Associations: Coexistence or Conflict?* Oxford University Press, Oxford, pp. 37–61.

King, C. M. 1985. Interactions between woodland rodents and their predators. *Symp. Zool. Soc. Lond.* **55**: 219–247.

Klaus, S., Mikkola, H. and Wiesner, J. 1975. Aktivität und Ernährung des Rauhfusskauzes *Aegolius funereus* (L.) während der Fortpflanzungsperiode. *Zool. Jb. Syst.* **102**: 485–507.

Klemola, T., Korpimäki, E., Norrdahl, K., Tanhuanpää, M. and Koivula, M. 1999. Mobility and habitat utilization of small mustelids in relation to cyclically fluctuating prey abundances. *Ann. Zool. Fenn.* **36**: 75–82.

Klemola, T., Koivula, M., Korpimäki, E. and Norrdahl, K. 2000. Experimental tests of predation and food hypotheses for population cycles of voles. *Proc. R. Soc. Lond. B* **267**: 351–356.

Klemola, T., Tanhuanpää, M., Korpimäki, E. and Ruohomäki, K. 2002. Specialist and generalist natural enemies as an explanation for geographical gradients in population cycles of northern herbivores. *Oikos* **99**: 83–94.

Klomp, H. 1970. The determination of clutch size in birds: a review. *Ardea* **58**: 1–121.

Kloubec, B. and Pacenovsky, S. 1996. Vocal activity of Tengmalm's owl (*Aegolius funereus*) in southern Bohemia and eastern Slovakia: circadian and seasonal course, effects on intensity. *Buteo* **8**: 5–22.

Kloubec, B. and Vacik, R. 1990. Outline of food ecology of Tengmalm's owl (*Aegolius funereus* L.) in Czechoslovakia. *Tichodroma* **3**: 103–125.

Kochert, M. N., Steenhof, K. and Kennedy, P. L. 2010. In memoriam: Dr. Carl D. Marti 1944–2010. *J. Raptor Res.* **44**: 335–336.

Koivisto, E., Huitu, O. and Korpimäki, E. 2007. Smaller *Microtus* vole species competitively superior in the absence of predators. *Oikos* **116**: 156–162.

Koivula, M., Korpimäki, E. and Viitala, J. 1997. Do Tengmalm's owls see vole scent marks visible in ultraviolet light? *Anim. Behav.* **54**: 873–877.

Koivunen, V., Korpimäki, E. and Hakkarainen, H. 1996a. Differential avian predation on sex and size classes of small mammals: doomed surplus or dominant individuals? *Ann. Zool. Fenn.* **33**: 293–301.

Koivunen, V., Korpimäki, E., Hakkarainen, H. and Norrdahl, K. 1996b. Prey choice of Tengmalm's owls (*Aegolius funereus funereus*): preference for substandard individuals? *Can. J. Zool.* **74**: 816–823.

Koivunen, V., Korpimäki, E. and Hakkarainen, H. 1998a. Are mature female voles more susceptible than immature ones to avian predation? *Acta Oecol.* **19**: 389–393.

Koivunen, V., Korpimäki, E. and Hakkarainen, H. 1998b. Refuge sites of voles under owl predation risk: priority of dominant individuals? *Behav. Ecol.* **9**: 261–266.

König, C. 1968a. Kleiber (*Sitta europaea*) mauert Rauhfusskauz (*Aegolius funereus*) in Bruthöhle ein. *Orn. Mitt.* **20**: 10.

König, C. 1968b. Lautäusserungen von Rauhfusskauz (*Aegolius funereus*) und der Sperlingkauz (*Glaucidium passerinum*). *Beihefte der Vogelwelt* **1**: 115–138.

König, C. 1969. Sechsjährige Untersuchungen an einer Population des Rauhfusskauzes *Aegolius funereus*. *J. Ornithol.* **110**: 133–147.

König, C. and Weick, F. 2008. *Owls of the World*. Christopher Helm, London.

Kontiainen, P., Pietiäinen, H., Karell, P., Pihlaja, T. and Brommer, J. E. 2010. Hatching asynchrony is an individual property of female Ural owls which improves nestling survival. *Behav. Ecol.* **21**: 722–727.

Koopman, M. E., McDonald, D. B., Hayward, G. D., Eldegard, K., Sonerud, G. A. and Sermach, S. G. 2005. Genetic similarity among Eurasian subspecies of boreal owls *Aegolius funereus*. *J. Avian Biol.* **36**: 179–183.

Koopman, M. E., Hayward, G. D. and McDonald, D. B. 2007a. High connectivity and minimal genetic structure among North American boreal owl (*Aegolius funereus*) populations, regardless of habitat matrix. *Auk* **124**: 690–704.

Koopman, M. E., McDonald, D. B. and Hayward, G. D. 2007b. Microsatellite analysis reveals genetic monogamy among female boreal owls. *J. Raptor Res.* **41**: 314–318.

Korfanta, N. M. 2001. *Population genetics of migratory and resident burrowing owls* (Athene cunicularia) *elucidated by microsatellite DNA markers*. MSci thesis, University of Wyoming, Laramie, WY.

Korpimäki, E. 1972. Hiiri- ja helmipöllön pesinnästä ja ravinnosta samalla biotoopilla. *Suomenselän Linnut* **7**: 36–40.

Korpimäki, E. 1976. Eri kokoisiin koloihin ja pönttöihin pesivistä eläinlajeista. *Suomenselän Linnut* **11**: 102–104.

Korpimäki, E. 1978. Breeding biology of the starling *Sturnus vulgaris* in western Finland. *Ornis Fenn.* **55**: 93–104.

Korpimäki, E. 1981. On the ecology and biology of Tengmalm's owl (*Aegolius funereus*) in Southern Ostrobothnia and Suomenselkä western Finland. *Acta Univ. Oul. Ser. A. Sci. Rer. Nat.* **118**: 1–84.

Korpimäki, E. 1984. Clutch size and breeding success of Tengmalm's owl *Aegolius funereus* in natural cavities and nest-boxes. *Ornis Fenn.* **61**: 80–83.

Korpimäki, E. 1985a. Clutch size and breeding success in relation to nest-box size in Tengmalm's owl *Aegolius funereus*. *Holarct. Ecol.* **8**: 175–180.

Korpimäki, E. 1985b. Diet of the kestrel *Falco tinnunculus* in the breeding season. *Ornis Fenn.* **62**: 130–137.

Korpimäki, E. 1985c. Pöllöjen kannanvaihtelut 30 vuoden aikana sekä pöllöyhteisöjen rakenne Suomenselällä. *Suomenselän Linnut* **20**: 15–22.

Korpimäki, E. 1985d. Prey choice strategies of the kestrel *Falco tinnunculus* in relation to available small mammals and other Finnish birds of prey. *Ann. Zool. Fenn.* **22**: 91–104.

Korpimäki, E. 1985e. Rapid tracking of microtine populations by their avian predators: possible evidence for stabilizing predation. *Oikos* **45**: 281–284.

Korpimäki, E. 1986a. Diet variation, hunting habitat and reproductive output of the kestrel *Falco tinnunculus* in the light of the optimal diet theory. *Ornis Fenn.* **63**: 84–90.

Korpimäki, E. 1986b. Gradients in population fluctuations of Tengmalm's owl *Aegolius funereus* in Europe. *Oecol. (Berl.)* **69**: 195–201.

Korpimäki, E. 1986c. Niche relationships and life-history tactics of three sympatric *Strix* owl species in Finland. *Ornis Scand.* **17**: 126–132.

Korpimäki, E. 1986d. Reversed size dimorphism in birds of prey, especially in Tengmalm's owl *Aegolius funereus*: a test of the 'starvation hypothesis'. *Ornis Scand.* **17**: 309–315.

Korpimäki, E. 1986e. Seasonal changes in the food of Tengmalm's owl *Aegolius funereus* in western Finland. *Ann. Zool. Fenn.* **23**: 339–344.

Korpimäki, E. 1987a. Clutch size, breeding success and brood size experiments in Tengmalm's owl *Aegolius funereus*: a test of hypotheses. *Ornis Scand.* **18**: 277–284.

Korpimäki, E. 1987b. Composition of the owl communities in four areas in western Finland: importance of habitats and interspecific competition. *Acta Reg. Soc. Sci. Litt. Gothoburgensis. Zool.* **14**: 118–123.

Korpimäki, E. 1987c. Dietary shifts, niche relationships and reproductive output of coexisting kestrels and long-eared owls. *Oecol. (Berl.)* **74**: 277–285.

Korpimäki, E. 1987d. Prey caching of breeding Tengmalm's owls *Aegolius funereus* as a buffer against temporary food shortage. *Ibis* **129**: 499–510.

Korpimäki, E. 1987e. Selection for nest-hole shift and tactics of breeding dispersal in Tengmalm's owl *Aegolius funereus*. *J. Anim. Ecol.* **56**: 185–196.

Korpimäki, E. 1987f. Timing of breeding of Tengmalm's owl *Aegolius funereus* in relation to vole dynamics in western Finland. *Ibis* **129**: 58–68.

Korpimäki, E. 1987g. Wintering strategies of Tengmalm's owl *Aegolius funereus*. *Aquilo Ser. Zool.* **24**: 55–63.

Korpimäki, E. 1988a. Costs of reproduction and success of manipulated broods under varying food conditions in Tengmalm's owl. *J. Anim. Ecol.* **57**: 1027–1039.

Korpimäki, E. 1988b. Diet of breeding Tengmalm's owls *Aegolius funereus*: long-term changes and year-to-year variation under cyclic food conditions. *Ornis Fenn.* **65**: 21–30.

Korpimäki, E. 1988c. Effects of age on breeding performance of Tengmalm's owl *Aegolius funereus* in western Finland. *Ornis Scand.* **19**: 21–26.

Korpimäki, E. 1988d. Effects of territory quality on occupancy, breeding performance and breeding dispersal in Tengmalm's owl. *J. Anim. Ecol.* **57**: 97–108.

Korpimäki, E. 1988e. Factors promoting polygyny in European birds of prey: a hypothesis. *Oecologia* **77**: 278–285.

Korpimäki, E. 1989a. Breeding performance of Tengmalm's owl *Aegolius funereus*: effects of supplementary feeding in a peak vole year. *Ibis* **131**: 51–56.

Korpimäki, E. 1989b. Mating system and mate choice of Tengmalm's owls *Aegolius funereus*. *Ibis* **131**: 41–50.

Korpimäki, E. 1990a. Body mass of breeding Tengmalm's owls *Aegolius funereus*: seasonal, between-year, site and age-related variation. *Ornis Scand.* **21**: 169–178.

Korpimäki, E. 1990b. Low repeatability of laying date and clutch size in Tengmalm's owl: an adaptation to fluctuating food conditions. *Ornis Scand.* **21**: 282–286.

Korpimäki, E. 1991a. Lifetime reproductive success of male Tengmalm's owls. *Proc. Int. Ornithol. Congr.* **20**: 1528–1541.

Korpimäki, E. 1991b. Poor reproductive success of polygynously mated female Tengmalm's owls: are better options available? *Anim. Behav.* **41**: 37–47.

Korpimäki, E. 1992a. Diet composition, prey choice and breeding success of long-eared owls: effects of multiannual fluctuations in food abundance. *Can. J. Zool.* **70**: 2373–2381.

Korpimäki, E. 1992b. Fluctuating food abundance determines the lifetime reproductive success of male Tengmalm's owls. *J. Anim. Ecol.* **61**: 103–111.

Korpimäki, E. 1992c. Population dynamics of Fennoscandian owls in relation to wintering conditions and between-year fluctuations of food. In: Galbraith, C. A., Taylor, I. R. and Percival, S. (eds.), *The Ecology and Conservation of European Owls*. UK Nature Conservation, No. 5. Joint Nature Conservation Committee, Peterborough, UK, pp. 1–10.

Korpimäki, F. 1993a. Does nest-hole quality, poor breeding success or food depletion drive the breeding dispersal of Tengmalm's owls? *J. Anim. Ecol.* **62**: 606–613.

Korpimäki, E. 1993b. Helmipöllönaaraan pitkä puolisonetsintämatka. *Suomenselän Linnut* **28**: 130–131.

Korpimäki, E. 1994a. Nest predation may not explain poor reproductive success of polygynously mated female Tengmalm's owls. *J. Avian Biol.* **25**: 161–164.

Korpimäki, E. 1994b. Rapid or delayed tracking of multi-annual vole cycles by avian predators? *J. Anim. Ecol.* **63**: 619–628.

Korpimäki, E. 1997. Tengmalm's owl *Aegolius funereus*. In: Hagemeijer, E. J. M. and Blair, M. J. (eds.), *The EBCC Atlas of European Breeding Birds: Their Distribution and Abundance*. T & A D Poyser, London, pp. 420–421.

Korpimäki, E. and Hakkarainen, H. 1991. Fluctuating food supply affects the clutch size of Tengmalm's owl independent of laying date. *Oecologia* **85**: 543–552.

Korpimäki, E. and Hongell, H. 1986. Partial migration as an adaptation to nest-site scarcity and vole cycles in Tengmalm's owl *Aegolius funereus*. *Vår Fågelvärld* **Suppl. 11**: 85–92.

Korpimäki, E. and Huhtala, K. 1986. Nest visit frequencies and activity patterns of Ural owls *Strix uralensis*. *Ornis Fenn.* **63**: 42–46.

Korpimäki, E. and Krebs, C. J. 1996. Predation and population cycles of small mammals: a reassessment of the predation hypothesis. *BioSci.* **46**: 754–764.

Korpimäki, E. and Lagerström, M. 1988. Survival and natal dispersal of fledglings of Tengmalm's owl in relation to fluctuating food conditions and hatching date. *J. Anim. Ecol.* **57**: 433–441.

Korpimäki, E. and Marti, C. D. 1995. Geographical trends in the trophic characteristics of mammal- and bird-eating raptors in Europe and North America. *Auk* **112**: 1004–1023.

Korpimäki, E. and Norrdahl, K. 1989. Predation of Tengmalm's owls: numerical responses, functional responses and dampening impact on population fluctuations of voles. *Oikos* **54**: 154–164.

Korpimäki, E. and Norrdahl, K. 1991a. Do breeding nomadic avian predators dampen population fluctuations of small mammals? *Oikos* **62**: 195–208.

Korpimäki, E. and Norrdahl, K. 1991b. Numerical and functional responses of kestrels, short-eared owls, and long-eared owls to vole densities. *Ecology* **72**: 814–826.

Korpimäki, E. and Norrdahl, K. 1998. Experimental reduction of predators reverses the crash phase of small-rodent cycles. *Ecology* **76**: 2448–2455.

Korpimäki, E. and Rajala, E. 1985. Keidassoiden pesimälinnustosta Vaasan läänissä 1970-luvulla. *Suomenselän Linnut* **20**: 40–52.

Korpimäki, E. and Sulkava, S. 1987. Diet and breeding performance of Ural owls *Strix uralensis* under fluctuating food conditions. *Ornis Fenn.* **64**: 57–66.

Korpimäki, E. and Wiehn, J. 1998. Clutch size of kestrels: seasonal decline and experimental evidence for food limitation under fluctuating food conditions. *Oikos* **83**: 259–272.

Korpimäki, E., Lagerström, M. and Saurola, P. 1987. Field evidence for nomadism in Tengmalm's owl *Aegolius funereus*. *Ornis Scand.* **18**: 1–4.

Korpimäki, E., Huhtala, K. and Sulkava, S. 1990. Does the year-to-year variation in the diet of eagle and Ural owls support the alternative prey hypothesis? *Oikos* **58**: 47–54.

Korpimäki, E., Norrdahl, K. and Rinta-Jaskari, T. 1991. Responses of stoats and least weasels to fluctuating vole abundances: is the low phase of the vole cycle due to mustelid predation? *Oecologia* **88**: 552–561.

Korpimäki, E., Hakkarainen, H. and Bennett, G. F. 1993. Blood parasites and reproductive success of Tengmalm's owls: detrimental effects on females but not on males? *Funct. Ecol.* **7**: 420–426.

Korpimäki, E., Tolonen, P. and Valkama, J. 1994. Functional responses and load-size effect in central place foragers: data from the kestrel and some general comments. *Oikos* **69**: 504–510.

Korpimäki, E., Lahti, K., May, C. A., Parkin, D. T., Powell, G. P., Tolonen, P. and Wetton, J. H. 1996. Copulatory behaviour and paternity determined by DNA fingerprinting in kestrels: effects of cyclic food abundance. *Anim. Behav.* **51**: 945–955.

Korpimäki, E., Norrdahl, K., Klemola, T., Pettersen, T. and Stenseth, N. C. 2002. Dynamic effects of predators on cyclic voles: field experimentation and model extrapolation. *Proc. R. Soc. Lond. B* **269**: 991–997.

Korpimäki, E., Brown, P. R., Jacob, J. and Pech, R. P. 2004. The puzzles of population cycles and outbreaks of small mammals solved? *BioScience* **54**: 1071–1079.

Korpimäki, E., Norrdahl, K., Huitu, O. and Klemola, T. 2005a. Predator-induced synchrony in population oscillations of co-existing small mammal species. *Proc. R. Soc. B* **272**: 193–202.

Korpimäki, E., Oksanen, L., Oksanen, T., Klemola, T., Norrdahl, K. and Banks, P. B. 2005b. Vole cycles and predation in temperate and boreal zones of Europe. *J. Anim. Ecol.* **74**: 1150–1159.

Korpimäki, E., Hakkarainen, H., Laaksonen, T. and Vasko, V. 2008. Responses of owls and Eurasian kestrels to natural and human induced spatio-temporal variation. *Scottish Birds* **28**: 19–27.

Korpimäki, E., Hakkarainen, H., Laaksonen, T. and Vasko, V. 2009. Responses of owls and Eurasian kestrels to spatio-temporal variation of their main prey. *Ardea* **97**: 646–647.

Korpimäki, E., Salo, P. and Valkama, J. 2011. Sequential polyandry by brood desertion increases female fitness in a bird with obligatory bi-parental care. *Behav. Ecol. Sociobiol.* **65**: 1093–1102.

Kotler, B. P., Brown, J. S., Smith, R. J. and Wirtz II, W. O. 1988. The effects of morphology and body size on rates of owl predation on desert rodents. *Oikos* **53**: 145–152.

Krüger, O. 2002. Interactions between common buzzards *Buteo buteo* and goshawk *Accipiter gentilis*: trade-offs revealed by a field experiment. *Oikos* **96**: 441–452.

Krüger, O. 2005. The evolution of reversed sexual size dimorphism in hawks, falcons and owls: a comparative study. *Evol. Ecol.* **19**: 467–486.

Krüger, O. and Lindström, J. 2001. Lifetime reproductive success in common buzzard, *Buteo buteo*: from individual variation to population demography. *Oikos* **93**: 260–273.

Kuhk, R. 1949. Aus der Fortpflanzungsbiologie des Rauhfusskauzes *Aegolius funereus* (L.). In: Streseman, E. (ed.), *Ornithologie als Biologische Wissenschaft*. C. Winter, Heidelberg, pp. 171–182.

Kuhk, R. 1953. Lautäusserungen und jahreszeitliche Gesangtätigkeit des Rauhfusskauzes, *Aegolius funereus* (L.). *J. Ornithol.* **94**: 83–93.

Kuhk, R. 1969. Schlüpfen und Entwicklung der Nestjungen beim Rauhfusskauzes (*Aegolius funereus*). *Bonn. Zool. Beitr.* **20**: 141–150.

Kuhk, R. 1970. Die postembryonale Gewichtsentwicklung des Rauhfusskauzes (*Aegolius funereus*). *Beitr. Vogelkd.* **16**: 232–238.

Kuhlman, E. and Koskela, K. 1980. Lehto- ja helmipöllön pesintäaikaisesta ravinnosta. *Siipirikko* **7**: 46–50.

Kuno, E. 1987. Principles of predator–prey interaction in theoretical, experimental and natural population systems. *Adv. Ecol. Res.* **16**: 249–337.

Laaksonen, T. 2004. Hatching asynchrony as a bet-hedging strategy: an offspring diversity hypothesis. *Oikos* **104**: 616–620.

Laaksonen, T., Korpimäki, E. and Hakkarainen, H. 2002. Interactive effects of parental age and environmental variation on the breeding performance of Tengmalm's owls. *J. Anim. Ecol.* **71**: 23–31.

Laaksonen, T., Hakkarainen, H. and Korpimäki, E. 2004. Lifetime reproduction of a forest-dwelling owl increases with age and area of forests. *Proc. R. Soc. Lond. B* **271**: S461–S464.

Lack, D. 1946. Competition for food by birds of prey. *J. Anim. Ecol.* **15**: 123–129.

Lack, D. 1947. The significance of clutch size. I. Intraspecific variation. *Ibis* **89**: 302–352.

Lack, D. 1948. The significance of clutch-size. III. Some interspecific comparisons. *Ibis* **90**: 25–45.

Lack, D. 1954. *The Natural Regulation of Animal Numbers*. Oxford University Press, London.

Lack, D. 1966. *Population Studies of Birds*. Oxford University Press, Oxford.

Lack, D. 1968. *Ecological Adaptations for Breeding in Birds*. Methuen, London.

Lagerström, M. 1978. Helmipöllön ja viirupöllön yhteispesintä (Summary: Tengmalm's owl and Ural owl nesting simultaneously in the same box). *Ornis Fenn.* **55**: 183–184.

Lagerström, M. 1980. Helmipöllön populaatiodynamiikasta ja pesimisbiologiasta Pirkanmaalla 1961–1980. *Lintuviesti* **5**: 149–160.

Lagerström, M. 1982. The breeding record of the owls *Strix aluco, Strix uralensis* and *Aegolius funereus* in Pirkanmaa 1981. *Lintuviesti* **7**: 100–103.

Lagerström, M. 1983a. The nesting of owls in Pirkanmaa 1983. *Lintuviesti* **8**: 225–235.

Lagerström, M. 1983b. The nesting record of owls in Pirkanmaa in 1982. *Lintuviesti* **8**: 63–70.

Lagerström, M. 1985a. The nesting of owls in Pirkanmaa 1984. *Lintuviesti* **10**: 15–21.

Lagerström, M. 1985b. The occurrence and breeding of owls in Pirkanmaa, 1985. *Lintuviesti* **10**: 130–139.

Lagerström, M. 1991. Pirkanmaan pöllöillä hyvä nousuvuosi 1991. *Lintuviesti* **16**: 124–133.

Lagerström, M. and Häkkinen, I. 1978. Uneven sex ratio of voles in the food of *Aegolius funereus* and *Strix aluco*. *Ornis Fenn.* **55**: 149–153.

Lagerström, M. and Korpimäki, E. 1988. Helmipöllö. In: Pietiäinen, H., Ahola, K., Forsman, D., Haapala, J., Korpimäki, E., Lagerström, M. and Niiranen, S. (eds.), *Pöllöjen iän määrittäminen*. Helsingin yliopiston eläinmuseo, Helsinki, pp. 26–29.

Lambin, X., Elston, D. A., Petty, S. J. and MacKinnon, J. L. 1998. Spatial asynchrony and periodic travelling waves in cyclic populations of field voles. *Proc. R. Soc. Lond. B* **265**: 1491–1496.

Lambin, X., Petty, S. J. and MacKinnon, J. L. 2000. Cyclic dynamics in field vole populations and generalist predation. *J. Anim. Ecol.* **69**: 106–118.

Lambrechts, M. M., Adriaensen, F., Ardia, D. R., Artemyev, A. V., Atinzar, F., Babura, J., Barba, E., Bouvier, J.-C., Camprodon, J., Cooper, C. B., Dawson, R. D., Eens, M., Eeva, T., Faivre, B., Garamszegi, L. Z., Goodenough, A. E., Gosler, A. G., Grgoire, A., Griffith, S. C., Gustafsson, L., Johnson, L. S., Kania, W., Keis, O., Llambias, P. E., Mainwaring, M. C., Mänd, R., Massa, B., Mazgajski, T. D., Møller, A. P., Moreno, J., Naef-Daenzer, B., Nilsson, J.-A., Norte, A. C., Orell, M., Otter, K. A., Park, C. R., Perrins, C. M., Pinowski, J., Porkert, J., Potti, J., Reme, V., Richner, H., Rytkönen, S., Shiao, M.-T., Silverin, B., Slagsvold, T., Smith, H. G., Sorace, A., Stenning, M. J., Stewart, I., Thompson, C. F., Trk, J., Tryjanowski, P., van Noordwijk, A. J., Winkler, D. W. and Ziane, N. 2010. The design of artificial nestboxes for the study of secondary hole-nesting birds: a review of methodological inconsistencies and potential biases. *Acta Ornithol.* **45**: 1–26.

Lambrechts, M. M., Wiebe, K. L., Sunde, P., Solonen, T., Sergio, F., Roulin, A., Møller, A. P., Lopez, B. C., Fargallo, J. A., Exo, K.-M., DellOmo, G., Costantini, D., Charter, M., Butler, M. W.,. Bortolotti, G. R., Arlettaz, R. and Korpimäki, E. 2012. Nest-box design for the study of diurnal raptors and owls is still an overlooked point in ecological, evolutionary and conservation studies: a review. *J. Ornithol.* **153**: 23–34.

Lane, W. H., Andersen, D. E. and Nicholls, T. H. 1997. Habitat use and movements of breeding male boreal owls (*Aegolius funereus*) in Northeast Minnesota as determined by radio telemetry. In: Duncan, J. R., Johnson, D. H. and Nicholls, T. H. (eds.), *Biology and Conservation of Owls of the Northern Hemisphere.* General Technical Report NC-190. USDA Forest Service, St. Paul, MN, pp. 248–249.

Lane, W. H., Andersen, D. E. and Nicholls, T. H. 2001. Distribution, abundance, and habitat use of singing male boreal owls in northeast Minnesota. *J. Raptor Res.* **35**: 130–140.

Lauff, R. F. 2009. First nest records of the boreal owl *Aegolius funereus* in Nova Scotia, Canada. *Ardea* **97**: 497–502.

Lawless, S. G., Richison, G., Klatt, P. H. and Westneat, D. F. 1997. The mating strategies of Eastern screech owls: a genetic analysis. *Condor* **99**: 213–217.

Lawlor, L. R. 1980. Overlap, similarity, and competition coefficients. *Ecology* **61**: 245–251.

Lehikoinen, A. 2009. Pöllöt vaelsivat Haliakselle. *Tringa* **36**: 174–177.

Lehikoinen, A. and Vähätalo, A. 2000. Lintujen muuton ajoittuminen Hangon lintuasemalla vuosina 1979–1999. *Tringa* **27**: 150–226.

Lehikoinen, A., Ekroos, K., Jaatinen, K., Lehikoinen, P., Lindén, A., Piha, M., Vattulainen, A. and Vähätalo, A. 2008. Lintukantojen kehitys Hangon lintuaseman aineiston mukaan 1979–2009. *Tringa* **35**: 146–209.

Lehikoinen, A., Byholm, P., Ranta, E., Saurola, P., Valkama, J., Korpimäki, E., Pietiäinen, H. and Henttonen, H. 2009. Reproduction of the common buzzard at its northern range margin under climatic change. *Oikos* **118**: 829–836.

Lehikoinen, A., Ranta, E., Pietiäinen, H., Byholm, P., Saurola, P., Valkama, J., Huitu, O., Henttonen, H. and Korpimäki, E. 2011. The impact of climate and cyclic food abundance on the timing of breeding and brood size in four boreal owl species. *Oecologia* **165**: 349–355.

Lehtoranta, J. 1981. Helmipöllön pesintä asuinrakennuksessa. *Siipirikko* **8**: 62.

Levins, R. 1968. *Evolution in Changing Environments.* Princeton University Press, Princeton, NJ.

Libois, R. and Gailly, P. 1984. Sur le regime alimentaire de la Chouette de Tengmalm (*Aegolius funereus*) en Belgique. *Aves* **21**: 57–59.

Lillandt, B-G. 2009. Suupohjan kuukkeliprojekti 35 vuotta 1974–2009: tutkimus aidon metsäluonnon kuolemasta. *Hippiäinen* **39**: 5–13.

Lima, S. L. and Dill, L. M. 1990. Behavioural decisions made under the risk of predation: a review and prospectus. *Can. J. Zool.* **68**: 619–640.

Lindén, H. and Wikman, M. 1983. Goshawk predation on tetraonids: availability of prey and diet of the predator in the breeding season. *J. Anim. Ecol.* **52**: 953–968.

Lindhe, U. 1966. En undersökning av pärlugglans (*Aegolius funereus*) bytesval i SV Lapland. *Vår Fågelvärld* **25**: 40–48.

Lindström, E. 1988. Reproductive effort in the red fox, *Vulpes vulpes*, and future supply of a fluctuating prey. *Oikos* **52**: 115–119.

Lindström, J. 1999. Early development and fitness in birds and mammals. *Trends Ecol. Evol.* **14**: 343–348.

Linkhart, B. D. and Reynolds, R. T. 2006. Lifetime reproduction of flammulated owls in Colorado. *J. Raptor Res.* **40**: 29–37.

Linkola, P. and Myllymäki, A. 1969. Der Einfluss der Kleinsäugerfluktuationen auf das Brüten einiger Kleinsäugerfressenden Vögel im südlichen Häme, Mittelfinnland 1952–66. *Ornis Fenn.* **46**: 45–78.

Lochmiller, R. L. and Deerenberg, C. 2000. Trade-offs in evolutionary immunology: just what is the cost of immunity? *Oikos* **88**: 87–98.

Löfgren, O., Hörnfeldt, B. and Carlsson, B.-G. 1986. Site tenacity and nomadism in Tengmalm's owl (*Aegolius funereus* (L.)) in relation to cyclic food production. *Oecologia (Berl.)* **69**: 321–326.

Lohmus, A. 1999. Vole-induced regular fluctuations in the Estonian owl populations. *Ann. Zool. Fenn.* **36**: 167–178.

Longland, W. S. and Jenkins, S. H. 1987. Sex and age affect vulnerability of desert rodents to owl predation. *J. Mammal.* **68**: 746–754.

Lopez, B. C., Potrony, D., Lopez, A., Badosa, E., Bonada, A. and Salo, R. 2010. Nest-box use by boreal owls (*Aegolius funereus*) in the Pyrenees Mountains in Spain. *J. Raptor Res.* **44**: 40–49.

Lowe, V. P. W. 1980. Variation in digestion of prey by the tawny owl (*Strix aluco*). *J. Zool. Lond.* **192**: 283–293.

Lundberg, A. 1978. Beståndsuppskattning av slaguggla och pärluggla. *Anser*, **Suppl. 3**: 171–175.

Lundberg, A. 1979. Residency, migration and a compromise: adaptations to nest-site scarcity and food specialization in three Fennoscandian owl species. *Oecol. (Berl.)* **41**: 273–281.

Lundberg, A. 1986. Adaptive advantages of reversed sexual size dimorphism in European owls. *Ornis Scand.* **17**: 133–140.

Lundin, A. 1961. Sång och läten hos pärlugglan, *Aegolius funereus* L. *Fauna o. Flora* **56**: 95–128.

Maasikamäe, R. 1978. Vaatlusi karvasjalg-kaku (*Aegolius funereus*) pesä juures (Summary: Some observations on a nest of the Tengmalm's owl *Aegolius funereus*). *Loodusvatlusi* **1977**: 119–123.

MacArthur, R. H. and Levins, R. 1967. The limiting similarity, convergence, and divergence of coexisting species. *Am. Nat.* **101**: 377–385.

Magnhagen, C. 1991. Predation risk as a cost of reproduction. *Trends Ecol. Evol.* **6**: 183–186.

Magrath, R. D. 1990. Hatching asynchrony in altricial birds. *Biol. Rev.* **95**: 587–622.

Mammen, U. 1997. Bestandsentwicklung und Reproduktionsdynamik des Rauhfusskauzes (*Aegolius funereus*) in Deutschland. *Naturschutzreport* **13**: 30–39.

Mappes, T., Halonen, M., Suhonen, J. and Ylönen, H. 1993. Selective avian predation on a population of the field vole, *Microtus agrestis*: greater vulnerability of males and subordinates. *Ethol. Ecol. Evol.* **5**: 519–527.

Marine, R. and Dalmau, J. 2000. Habitat use by Tengmalm's owls *Aegolius funereus* in Andorra (eastern Pyrenees) during the breeding period. *Ardeola* **47**: 29–36.

Marks, J. S. and Marti, C. D. 1984. Feeding ecology of sympatric barn owls and long-eared owls in Idaho. *Ornis Scand.* **15**: 135–143.

Marks, J. S., Doremus, J. H. and Cannings, R. J. 1989. Polygyny in the Northern saw-whet owl. *Auk* **106**: 732–734.

Marks, J. S., Dickinson, J. L. and Haydock, J. 1999. Genetic monogamy in long-eared owls. *Condor* **101**: 854–859.

Marks, J. S., Dickinson, J. L. and Haydock, J. 2002. Serial polyandry and alloparenting in long-eared owls. *Condor* **104**: 202–204.

Marquiss, M. and Newton, I. 1982. A radio tracking study of the ranging behaviour and dispersion of European sparrowhawks *Accipiter nisus*. *J. Anim. Ecol.* **51**: 111–133.

Marti, C. D. 1987. Raptor food habits studies. In: Pendleton, B. A., Millsap, B. A., Cline, K. W. and Bird, D. M. (eds.), *Raptor Management Techniques Manual*. National Wildlife Federation, Washington, DC, pp. 67–80.

Marti, C. D. 1997. Lifetime reproductive success in barn owls near the limit of the species' range. *Auk* **114**: 581–592.

Marti, C. D. and Hogue, J. C. 1979. Selection of prey size in screech owls. *Auk* **96**: 319–327.

Marti, C. D., Korpimäki, E. and Jaksic, F. M. 1993a. Trophic structure of raptor communities: a three-continent comparison and synthesis. *Curr. Ornithol.* **10**: 47–137.

Marti, C. D., Steenhof, K., Kochert, M. N. and Marks, J. S. 1993b. Community trophic structure: the roles of diet, body size, and activity time in vertebrate predators. *Oikos* **67**: 6–18.

Marti, C. D., Bechard, M. and Jaksic, F. M. 2007. Food habits. In: Bird, D. M. and Bildstein, K. L. (eds.), *Raptor Research and Management Techniques*. Hancock House Publishers, Surrey, BC, Canada, pp. 129–149.

Martin, G. 1990. *Birds by Night*. T. & A. D. Poyser, London.

Martin, T. E. 1986. Competition in breeding birds: on the importance of considering processes at the level of the individual. *Curr. Ornithol.* **4**: 181–210.

Martin, T. E. 1987. Food as a limit on breeding birds: a life-history perspective. *Annu. Rev. Ecol. Syst.* **18**: 453–487.

Martin, T. E. 1995. Avian life history evolution in relation to nest sites, nest predation, and food. *Ecol. Monogr.* **65**: 101–127.

März, R. 1968. *Der Rauhfusskauz*. Die Neue Brehm-Bücherei, Wittenberg, Germany.

Masman, D., Dijkstra, C., Daan, S. and Bult, A. 1989. Energetic limitation of avian parental effort: field experiments in the kestrel (*Falco tinnunculus*). *J. Evol. Biol.* **2**: 435–455.

Massemin, S., Korpimäki, E. and Wiehn, J. 2000. Reversed sexual size dimorphism in raptors: evaluation of the hypotheses in kestrels breeding in a temporally changing environment. *Oecologia* **124**: 26–32.

May, R. M. 1975. Some notes on measuring the competition matrix. *Ecology* **56**: 737–741.

Meijer, T., Daan, S. and Dijkstra, C. 1988. Female condition and reproduction: effects of food manipulation in free-living and captive kestrels. *Ardea* **76**: 141–154.

Mendelsohn, J. M. 1983. Social behaviour and dispersion of the black-shouldered kite. *Ostrich* **54**: 1–18.

Mendelsohn, J. M. 1986. Sexual size dimorphism and roles in raptors: fat females, agile males. *Durban Mus. Novit.* **13**: 23, 321–336.

Merikallio, E. 1958. Finnish birds: their distribution and numbers. *Fauna Fenn.* **5**: 1–181.

Meyer, H., Gleixner, K. H. and Rudroff, S. 1998. Untersuchungen zu Populationsentwicklung, Brutbiologie und Verhalten des Rauhfusskauzes *Aegolius funereus* bei Hof und Münich. *Orn. Anz.* **37**: 81–101.

Meyer, W. 2003. Mit welchem Erfolg nutzt der Rauhfusskauz *Aegolius funereus* (L.) Naturhöhlen und Nistkästen zur Brut? *Vogelwelt* **124**: 325–331.

Meyer, W. 2010. Zum Wanderungsverhalten des Rauhfusskauzes *Aegolius funereus* auf der Grundlage von Beringungsergebnissen aus Thüringen. *Anz. Ver. Thüring. Ornithol.* **7**: 85–93.

Mezzavilla, F. and Lombardo, S. 1997. Biologia riproduttiva della civetta capogrosso (*Aegolius funereus*) nel bosco del Cansiglio. *Fauna* **4**: 101–114.

Mezzavilla, F., Lombardo, S. and Sperti, M. T. 1994. First data on biology and breeding success of Tengmalm's owl *Aegolius funereus* in Cansiglio. *Mus. Reg. Sci. Nat. Torino* **1994**: 325–334.

Mikkola, H. 1977. Pöllöjen esiintyminen ja ravinto Vilppulassa. *Suomenselän Linnut* **12**: 42–47.

Mikkola, H. 1983. *Owls of Europe*. T. & D. Poyser, Calton, UK.

Mikola, J. 1985. *Pöllöjen ravinnosta ja saalistuspaineesta Oulun Sanginjoella vuosina 1981–1984*. MSci thesis, Department of Zoology, University of Oulu, Finland.

Millon, A., Petty, S. J. and Lambin, X. 2010. Pulsed resources affect the timing of first breeding and lifetime reproductive success of tawny owls. *J. Anim. Ecol.* **79**: 426–435.

Møller, A. P. 1989a. Parasites, predators and nest boxes: facts and artefacts in nest box studies? *Oikos* **56**: 421–423.

Møller, A. P. 1989b. Frequency of extra-pair paternity in birds estimated from sex-differential heritability of tarsus length: reply to Lifjeld and Slagsvold's critique. *Oikos* **56**: 247–249.

Møller, A. P. 1994. Facts and artefacts in nest-box studies: implications for studies of birds of prey. *J. Raptor Res.* **28**: 143–148.

Møller, A. P., Arriero, E., Lobato, E. and Merino, S. 2009. A meta-analysis of parasite virulence in nestling birds. *Biol. Rev.* **84**: 567–588.

Mönkkönen, M., Helle, P. and Soppela, K. 1990. Numerical and behavioral responses of migrant passerines to experimental manipulation of resident tits (*Parus* spp.): heterospecific attraction in northern breeding bird communities? *Oecologia* **85**: 218–225.

Mönkkönen, M., Forsman, J. T. and Helle, P. 1996. Mixed-species foraging aggregations and heterospecific attraction in boreal bird communities. *Oikos* **77**: 127–136.

Mönkkönen, M., Helle, P., Niemi, G. J. and Montgomery, K. 1997. Heterospecific attraction affects community structure and migrant abundances in northern breeding bird communities *Can. J. Zool.* **75**: 2077–2083.

Montgomerie, R. D. and Weatherhead, P. J. 1988. Risks and rewards of nest defence by parent birds. *Q. Rev. Biol.* **63**: 167–187.

Moran, P. A. P. 1953. The statistical analysis of the Canadian lynx cycle. II. Synchronization and meteorology. *Aust. J. Zool.* **1**: 291–298.

Morosinotto, C., Thomson, R. L. and Korpimäki, E. 2010. Habitat selection as an antipredator behaviour in a multi-predator landscape: all enemies are not equal. *J. Anim. Ecol.* **79**: 327–333.

Morse, D. H. 1980. *Behavioral Mechanisms in Ecology*. Harvard University Press, Cambridge, MA.

Mossop, D. H. 1997. The importance of old growth forest refugia in the Yukon boreal forest to cavity-nesting owls. In: Duncan, J. R., Johnson, D. H. and Nicholls, T. H. (eds.), *Biology and Conservation of Owls of the Northern Hemisphere, Second International Symposium, February 5–9, 1997, Winnipeg, Manitoba, Canada*. General Technical Report NC-190. USDA Forest Service, North Central Forest Experimental Station, St. Paul, MN, pp. 584–586.

Mueller, H. C. 1986. The evolution of reversed sexual size dimorphism in owls: an empirical analysis of possible selective factors. *Wilson Bull.* **98**: 387–406.

Müller, W., Epplen, J. T. and Lubjuhn, T. 2001. Genetic paternity analyses in little owls (*Athene noctua*): does the high rate of paternal care select against extra-pair young? *J. Ornithol.* **142**: 195–203.

Murdoch, W. W. 1966. Population stability and life history phenomena. *Am. Nat.* **100**: 45–51.

Murdoch, W. W. and Oaten, A. 1975. Predation and population stability. *Adv. Ecol. Res.* **9**: 2–131.

Murphy, E. C. and Haukioja, E. 1986. Clutch size in nidicolous birds. In: Johnston, R. F. (ed.), *Current Ornithology*. Plenum Press, New York, pp. 141–180.

Murray, G. A. 1976. Geographic variation in the clutch size of seven owl species. *Auk* **93**: 602–613.

Myllymäki, A., Paasikallio, A. and Häkkinen, U. 1971. Analysis of a 'standard trapping' of *Microtus agrestis* (L.) with triple isotope marking outside the quadrat. *Ann. Zool. Fenn.* **8**: 22–34.

Mysterud, I. 1970. Hypotheses concerning characteristics and causes of population movements in Tengmalm's owl *Aegolius funereus* (L.). *Nytt Mag. Zool.* **18**: 49–74.

Naves, L. C., Cam, E. and Monnat, J. Y. 2007. Pair duration, breeding success and divorce in a long-lived seabird: benefits of mate fidelity? *Anim. Behav.* **73**: 433–444.

Negro, J. J., Villarroel, M., Tella, J. L., Kuhnlein, U., Hiraldo, F., Donazar, J. A. and Bird, D. M. 1996. DNA fingerprinting reveals a low incidence of extra-pair fertilizations in the lesser kestrel. *Anim. Behav.* **51**: 935–945.

Nelson, J. 1994. *Determinants of spacing behaviour, reproductive success and mating system in male field voles*, Microtus agrestis. PhD thesis, Department of Ecology, Lund University, Lund, Sweden.

Nelson, J. 1995. Intrasexual competition and spacing behaviour in male field voles, *Microtus agrestis*, under constant female density and spatial distribution. *Oikos* **73**: 9–14.

Newton, I. 1979. *Population Ecology of Raptors*. Poyser, Berkhamsted, UK.

Newton, I. 1985. Lifetime reproductive output of female sparrowhawks. *J. Anim. Ecol.* **54**: 241–253.

Newton, I. 1986. *The Sparrowhawk*. T. & D. Poyser, Calton, UK.

Newton, I. (ed.) 1989. *Lifetime Reproduction in Birds*. Academic Press, London.

Newton, I. 1992. Experiments on the limitation of bird numbers by territorial behaviour. *Biol. Rev.* **67**: 129–173.

Newton, I. 1998. *Population Limitation in Birds*. Academic Press, London.

Newton, I. and Marquiss, M. 1976. Occupancy and success of nesting territories in the European sparrowhawk. *Raptor Res.* **10**: 65–71.

Newton, I. and Marquiss, M. 1982. Fidelity to breeding area and mate in sparrowhawks *Accipiter nisus*. *J. Anim. Ecol.* **51**: 327–341.

Newton, I. and Marquiss, M. 1984. Seasonal trend in the breeding performance of sparrowhawks. *J. Anim. Ecol.* **53**: 809–829.

Newton, I. and Rothery, P. 1997. Senescence and reproductive value in sparrowhawks. *Ecology* **78**: 1000–1008.

Newton, I., Marquiss, M. and Moss, D. 1981. Age and breeding in sparrowhawks. *J. Anim. Ecol.* **50**: 839–854.

Newton, I., Bell, A. and Wyllie, I. 1982. Mortality of sparrowhawks and kestrels. *Brit. Birds* **75**: 195–204.

Newton, I., Marquiss, M. and Village, A. 1983. Weights, breeding, and survival in European sparrowhawks. *Auk* **100**: 344–354.

Newton, I., Wyllie, I. and Mearns, R. 1986. Spacing of sparrowhawks in relation to food supply. *J. Anim. Ecol.* **55**: 361–370.

Nielsen, O. 1999. Gyrfalcon predation on ptarmigan: numerical and functional responses. *J. Anim. Ecol.* **68**: 1034–1050.

Nilsson, I. N. 1981. Seasonal changes in food of the long-eared owl in southern Sweden. *Ornis Scand.* **12**: 216–223.

Nilsson, I. N. 1984. Prey weight, food overlap, and reproductive output of potentially competing long-eared and tawny owls. *Ornis Scand.* **15**: 176–182.

Nilsson, S. G., Johnsson, K. and Tjernberg, M. 1991. Is avoidance by black woodpeckers of old nests due to predation? *Anim. Behav.* **41**: 439–441.

Nisbet, I. C. T. 1973. Courtship-feeding, egg-size and breeding success in common terns. *Nature* **241**: 141–142.

Nol, E. and Smith, J. N. M. 1987. Effects of age and breeding experience on the seasonal reproductive success in the song sparrow. *J. Anim. Ecol.* **56**: 301–313.

Noordwijk, A. J. van 1984. Problems in the analysis of dispersal and a critique on its 'heritability' in the great tit. *J. Anim. Ecol.* **53**: 533–544.

Norberg, R. Å. 1964. Studier över pärlugglans (*Aegolius funereus*) ekologi och etologi. *Vår Fågelvärld* **23**: 228–244.

Norberg, R. Å. 1968. Physical factors in directional hearing in *Aegolius funereus* (Linné) (Strigiformes), with special reference to the significance of the asymmetry of the external ears. *Arkiv för Zoologi* **20**: 181–204.

Norberg, R. Å. 1970. Hunting technique of Tengmalm's owl *Aegolius funereus*. *Ornis Scand.* **1**: 51–64.

Norberg, R. Å. 1978. Skull asymmetry, ear structure and function, and auditory localization in Tengmalm's owl, *Aegolius funereus* (Linne). *Philos. Trans. R. Soc. Lond. B.* **282**: 325–410.

Norberg, R. Å. 1987. Evolution, structure, and ecology of northern forest owls. In: Nero, R. W., Clark, R. J., Knapton, R. J. and Hamre, R. H. (eds.), *Biology and Conservation of Northern Forest Owl: Symposium Proceedings, February 3–7, 1987, Winnipeg, Manitoba*. General Technical Report RM 142. USDA Forest Service, pp. 9–43.

Nordström, M. and Korpimäki, E. 2004. Effects of island isolation and feral mink removal on bird communities on small islands in the Baltic Sea. *J. Anim. Ecol.* **73**: 424–433.

Nordström, M., Högmander, J., Nummelin, J., Laine, J., Laanetu, N. and Korpimäki, E. 2002. Variable responses of waterfowl breeding populations to long-term removal of introduced American mink. *Ecography* **25**: 385–394.

Nordström, M., Högmander, J., Laine, J., Nummelin, J., Laanetu, N. and Korpimäki, E. 2003. Effects of feral mink removal on seabirds, waders and passerines on small islands of the Baltic Sea. *Biol. Conserv.* **109**: 359–368.

Norrdahl, K. and Korpimäki, E. 1993. Predation and interspecific competition in two *Microtus* voles. *Oikos* **67**: 149–158.

Norrdahl, K. and Korpimäki, E. 1995a. Effects of predator removal on vertebrate prey populations: birds of prey and small mammals. *Oecologia* **103**: 241–248.

Norrdahl, K. and Korpimäki, E. 1995b. Small carnivores and prey population dynamics in summer. *Ann. Zool. Fenn.* **32**: 163–169.

Norrdahl, K. and Korpimäki, E. 1996. Do nomadic avian predators synchronize population fluctuations of small mammals? A field experiment. *Oecologia* **107**: 478–483.

Norrdahl, K. and Korpimäki, E. 1998. Does mobility or sex of voles affect risk of predation by mammalian predators? *Ecology* **79**: 226–232.

Norrdahl, K. and Korpimäki, E. 2002a. Changes in individual quality during a 3-year population cycle of voles. *Oecologia* **130**: 239–249.

Norrdahl, K. and Korpimäki, E. 2002b. Changes in population structure and reproduction during a 3-year population cycle of voles. *Oikos* **96**: 331–345.

Nur, N. 1988. The consequences of brood size for breeding blue tits. III Measuring the costs of reproduction: survival, future fecundity, and differential dispersal. *Evolution* **42**: 351–362.

Nybo, J. and Sonerud, G. A. 1990. Seasonal changes in diet of hawk owls *Surnia ulula*. *Ornis Fenn.* **67**: 45–51.

Nyholm, E. 1970. Näädän elintavoista, saalistuksesta ja ravinnosta (Summary: On the ecology of the pine marten (*Martes martes*) in eastern and northern Finland). *Suomen Riista* **22**: 105–118.

O'Donnell, C. F. J. 2000. Cryptic local populations in a temperate rainforest bat *Chalinolobus tuberculatus* in New Zealand. *Anim. Conserv.* **3**: 287–297.

O'Donoghue, M., Boutin, S., Krebs, C. J. and Hofer, E. J. 1997. Numerical responses of coyotes and lynx to the snowshoe hare cycle. *Oikos* **80**: 150–162.

O'Donoghue, M., Boutin, S., Krebs, C. J., Murray, D. L. and Hofer, E. J. 1998a. Behavioural responses of coyotes and lynx to the snowshoe hare cycle. *Oikos* **82**: 169–183.

O'Donoghue, M., Boutin, S., Krebs, C. J., Zuleta, G., Murray, D. L. and Hofer, E. J. 1998b. Functional responses of coyotes and lynx to the snowshoe hare cycle. *Ecology* **79**: 1193–1208.

Oksanen, L. and Oksanen, T. 1992. Long-term microtine dynamics in north Fennoscandian tundra: the vole cycle and the lemming chaos. *Ecography* **15**: 226–236.

Oksanen, T. 1983. Prey caching in the hunting strategy of small mustelids. *Acta Zool. Fenn.* **174**: 197–199.

Oksanen, T. 1990. Exploitation ecosystems in heterogeneous habitat complexes. *Evol. Ecol.* **4**: 220–234.

Oksanen, T., Schneider, M., Rammul, Ü., Hambäck, P. and Aunapuu, M. 1999. Population fluctuations of voles in North Fennoscandian tundra: contrasting dynamics in adjacent areas with different habitat composition. *Oikos* **86**: 463–478.

Oksanen, T., Oksanen, L., Jedrzejewski, W., Jedrzejewska, B., Korpimäki, E. and Norrdahl, K. 2000. Predation and the dynamics of the bank vole, *Clethrionomys glareolus*. *Pol. J. Ecol.* **48** (Suppl.): 197–217.

Oksanen, T., Oksanen, L., Dahlgren, J. and Olofsson, J. 2008. Arctic lemmings, *Lemmus* spp. and *Dicrostonyx* spp.: integrating ecological and evolutionary perspectives. *Evol. Ecol. Res.* **10**: 415–434.

Olson, V. A., Liker, A., Freckleton, R. P. and Szekely, T. 2008. Parental conflict in birds: comparative analyses of offspring development, ecology and mating opportunities. *Proc. R. Soc. B* **275**: 301–307.

Orians, G. H. 1969. On the evolution of mating systems in birds and mammals. *Am. Nat.* **103**: 589–603.

Orians, G. H. and Wittenberger, J. F. 1991. Spatial and temporal scales in habitat selection. *Am. Nat.* **137**: S29–S49.

Oring, L. W. 1983. Avian polyandry. *Curr. Ornithol.* **3**: 309–351.

Östlund, S. 1984. Pärlugglerapporten. *Inf. Pärlugglegr.* **3**: 4–8.

Pakkala, H., Ojanen, M. and Tynjälä, M. 1993. On the autumn movements of the Tengmalm's owl (*Aegolius funereus*) at the Tauvo Bird Observatory, Northern Finland. *Ring* **16**: 70–76.

Palmer, D. A. 1986. *Habitat selection, movements and activity of boreal and saw-whet owls*. MSci thesis, Department of Fishery and Wildlife Biology, Colorado State University, Fort Collins, CO, USA.

Palmer, D. A. 1987. Annual, seasonal, and nightly activity variation in calling activity of boreal and northern saw-whet owls. In: Nero, R. W., Clark, R. J., Knapton, R. J. and Hamre, R. H. (eds.), *Biology and Conservation of Northern Forest Owls: Symposium Proceedings*. General Technical Report R-142. USDA Forest Service, Fort Collins, CO, pp. 162–168.

Palokangas, P., Alatalo, R. V. and Korpimäki, E. 1992. Female choice in the kestrel under different availability of mating options. *Anim. Behav.* **43**: 659–665.

Paradis, E., Baillie, S. R., Sutherland, W. J. and Gregory, R. D. 1998. Patterns of natal and breeding dispersal in birds. *J. Anim. Ecol.* **67**: 518–536.

Paradis, E., Baillie, S. R., Sutherland, W. J. and Gregory, R. D. 2000. Spatial synchrony in populations of birds: effects of habitat, population trend, and spatial scale. *Ecology* **81**: 2112–2125.

Parker, G. A., Royle, N. J. and Hartley, I. R. 2002. Intrafamiliar conflict and parental investment: a synthesis. *Philos. Trans. R. Soc. Lond. B.* **357**: 295–307.

Parsons, J. 1970. Relationship between egg-size and post-hatching chick mortality in the herring gull (*Larus argentatus*). *Nature* **228**: 1221–1222.

Pärt, T. and Gustafsson, L. 1989. Breeding dispersal in the collared flycatcher (*Ficedula albicollis*): possible causes and reproductive consequences. *J. Anim. Ecol.* **58**: 305–320.

Patthey, P., Chabloz, V. and Kunzlé, I. 2001. Corrélations entre le nombre de jeunes élevés, la date de ponte et la fructification du hétre *Fagus sylvatica* chez la chouette de Tengmalm *Aegolius funereus*. *Nos Oiseaux* **48**: 229–231.

Pearson, O. P. 1966. The prey of carnivores during one cycle of mouse abundance. *J. Anim. Ecol.* **35**: 217–233.

Perrins, C. M. 1965. Population fluctuations and clutch size in the great tit *Parus major* L. *J. Anim. Ecol.* **34**: 601–647.

Perrins, C. M. 1970. The timing of birds' breeding seasons. *Ibis* **112**: 242–255.

Peters, W. M. D. and Grubb, T. C. 1983. An experimental analysis of sex-specific foraging in the downy woodpecker, *Picoides pubescens*. *Ecology* **64**: 1437–1443.

Phillips, R. A., Dawson, D. A. and Ross, D. J. 2002. Mating patterns and reversed sexual size dimorphism in southern skuas (*Stercorarius skua lonnbergi*). *Auk* **119**: 858–863.

Pietiäinen, H. 1988. Breeding season quality, age, and the effects of experience on the reproductive success of the Ural owl (*Strix uralensis*). *Auk* **105**: 316–324.

Pietiäinen, H. 1989. Seasonal and individual variation in the production of offspring in the Ural owl *Strix uralensis*. *J. Anim. Ecol.* **58**: 905–920.

Pilastro, A., Biddau, L., Marin, G. and Mingozzi, T. 2001. Female brood desertion increases with number of available mates in the rock sparrow. *J. Avian Biol.* **32**: 68–72.

Pilson, D. and Rausher, M. D. 1988. Clutch size adjustment by a swallowtail butterfly. *Nature* **333**: 361–363.

Pokorny, J. 2000. The diet of the Tengmalm's owl (*Aegolius funereus*) in north Bohemian mountain areas damaged by immissions. *Buteo* **11**: 107–114.

Pokorny, J., Kloubec, B. and Obusch, J. 2003. Comparison of Tengmalm's owl *Aegolius funereus* diet in several Czech mountain areas. *Vogelwelt* **124**: 313–323.

Polis, G. A., and Holt, R. D. 1992. Intraguild predation: the dynamics of complex trophic interactions. *Trends Ecol. Evol.* **7**: 151–154.

Polis, G. A., Myers, C. A. and Holt, R. D. 1989. The ecology and evolution of intraguild predation: potential competitors that eat each other. *Annu. Rev. Ecol. Syst.* **20**: 297–330.

Postupalsky, S. 1989. Osprey. In: Newton, I. (ed.), *Lifetime Reproduction in Birds*. Academic Press, London, pp. 297–313.

Pouttu, P. 1985. Palokärjen (Dryocopus m. martius) pesimäbiologiasta Etelä-Hämeessä vv. 1977–1984. *Kanta-Hämeen Linnut* **9**: 40–50.

Pöysä, H. and Pöysä, S. 2002. Nest-site limitation and density dependence of reproductive output in the common goldeneye *Bucephala clangula*: implications for the management of cavity-nesting birds. *J. Appl. Ecol.* **39**: 502–510.

Prodon, R., Alamany, O., Garcia-Ferre, D., Canut, J., Novoa, C. and Dejaifve, P.-A. 1990. L'aire de distribution pyrénéenne de la Chouette de Tengmalm *Aegolius funereus*. *Alauda* **58**: 233–243.

Prugh, L. R., Hodges, K. E., Sinclair, A. R. E. and Brashares, J. S. 2008. Effect of habitat area and isolation on fragmented animal populations. *Proc. Natl. Acad. Sci. USA* **105**: 20770–20775.

Pulliainen, E. and Keränen, J. 1979. Composition and function of beard lichen stores accumulated by bank voles, *Clethrionomys glareolus* Schreb. *Aquilo Ser. Zool.* **19**: 73–76.

Pulliainen, E. and Ollinmäki, P. 1996. A long-term study of the winter food niche of the pine marten *Martes marte*s in northern boreal Finland. *Acta Theriol.* **41**: 337–352.

Pykal, J. and Kloubec, B. 1994. Feeding ecology of Tengmalm's owl *Aegolius funereus* in the Sumava National Park, Czechoslovakia. In: Meyburg, B. U. and Chancellor, R. D. (eds.), *Raptor Conservation Today*. WWGBP/The Pica Press, pp. 537–541.

Pyke, G. H. 1984. Optimal foraging theory: a critical review. *Annu. Rev. Ecol. Syst.* **15**: 523–575.

Pyke, G. H., Pulliam, H. R. and Charnov, E. L. 1977. Optimal foraging: a selective review of theory and tests. *Q. Rev. Biol.* **52**: 137–154.

Rabenhold, K. N. 1993. Latitudinal gradients in avian species diversity and the role of long-distance migration. *Curr. Ornithol.* **10**: 247–274.

Råberg, L., Nilsson, J.-Å., Ilmonen, P., Stjerman, M. and Hasselquist, D. 2000. The cost of immune response: vaccination reduces parental effort. *Ecology Letters* **3**: 382–386.

Raczynski, J. and Ruprecht, A. L. 1974. The effect of digestion on the osteological composition of owl pellets. *Acta Ornithol.* **14**: 25–38.

Ralls, K. 1976. Mammals in which females are larger than males. *Q. Rev. Biol.* **51**: 245–276.

Randla, T. 1976. *Eesti Röövlinnud*. Valgus, Tallinn.

Ranta, E., Kaitala, V., Lindström, J. and Lindén, H. 1995. Synchrony in population dynamics. *Proc. R. Soc. Lond. B* **262**: 113–118.

Rassi, P., Hyvärinen, E., Juslen, A. and Mannerkoski, I. (eds.) 2010. *The Red List of Finnish Species 2010*. Special Publication, Ministry of Environment, Finland.

Ratcliffe, N., Furness, R. W. and Hamer, K. C. 1998. The interactive effects of age and food supply on the breeding ecology of great skuas. *J. Anim. Ecol.* **67**: 853–862.

Ravussin, P.-A. 1991. Biologie de reproduction de la Chouette de Tengmalm, *Aegolius funereus*, dans le Jura vaudois (Suisse). In: Juillard, M. et al (eds.), *Rapaces Nocturnes*. Actes du 30e Colloque Interregional d'Ornithologie, Porrentruy, pp. 201–216.

Ravussin, P.-A. 2004. Dem Rauhfusskauz auf der Spur: Kleine Eule mit grossen Geheimnissen. *Ornis* **2**4: 16–19.

Ravussin, P.-A., Trolliet, D., Willenegger, L. and Béguin, D. 1993. Observations sur les fluctuations d'une population de Chouette de Tengmalm (*Aegolius funereus*) dans le Jura vaudois (Suisse). *Nos Oiseaux* **42**: 127–142.

Ravussin, P.-A., Walder, P., Henrioux, P., Chabloz, V. and Menétrey, Y. 1994. Répartition de Chouette de Tengmalm (*Aegolius funereus*) dans les sites naturels du Jura vaudois (Suisse). *Nos Oiseaux* **42**: 245–260.

Ravussin, P.-A., Trolliet, D., Béguin, D., Willenegger, L. and Matalon, G. 2001a. Observations et remarques sur la biologie de la Chouette de Tengmalm *Aegolius funereus* dans le massif du Jura suite à l'invasion du printemps 2000. *Nos Oiseaux* suppl. **5**: 235–246.

Ravussin, P.-A., Trolliet, D., Willenegger, L., Béguin, D. and Matalon, G. 2001b. Choix du site nidification chez la Chouette de Tengmalm *Aegolius funereus*: influence des nichoirs. *Nos Oiseaux* suppl. **5**: 41–51.

Ravussin, P.-A., Trolliet, D., Metraux, V., Gorgerat, V. and Roch, J. 2008. Saison 2008 chez la Chouette de Tengmalm. *GOBE* **2008**: 1–5.

Ravussin, P.-A., Trolliet, D., Metraux, V., Longchamp, L., Clemencon, F. and Roch, J. 2010. Saison 2009 chez la Chouette de Tengmalm. *GOBE* **2009**: 1–5.

Redpath, S. M. 1995. Habitat fragmentation and the individual: tawny owls *Strix aluco* in woodland patches. *J. Anim. Ecol.* **64**: 652–661.

Redpath, S. M. and Thirgood, S. 1999. Numerical and functional responses in generalist predators: hen harriers and peregrines. *J. Anim. Ecol.* **68**: 879–892.

Redpath, S. M., Leckie, F. M., Arroyo, B., Amar, A. and Thirgood, S. J. 2006. Compensating for the costs of polygyny in hen harriers *Circus cyaneus*. *Behav. Ecol. Sociobiol.* **60**: 386–391.

Reif, V. and Tornberg, R. 2006. Using time-lapse digital video recording for a nesting study of birds of prey. *Eur. J. Wildl. Res.* **52**: 251–258.

Reif, V., Tornberg, R., Jungell, S. and Korpimäki, E. 2001. Diet variation of common buzzards in Finland supports the alternative prey hypothesis. *Ecography* **24**: 267–274.

Reif, V., Jungell, S., Korpimäki, E., Tornberg, R. and Mykrä, S. 2004. Numerical response of common buzzards and predation rate of main and alternative prey under fluctuating food conditions. *Ann. Zool. Fenn.* **41**: 599–607.

Reynolds, J. D. 1996. Animal breeding systems. *Trends Ecol. Evol.* **11**: 68–72.

Reynolds, R. T. 1972. Sexual dimorphism in accipiter hawks: a new hypothesis. *Condor* **74**: 191–197.

Rice, W. R. 1983. Sensory modality: an example of its effects on optimal foraging behavior. *Ecology* **64**: 403–406.

Richner, H. 1989. Habitat-specific growth and fitness in carrion crows (*Corvus corone corone*). *J. Anim. Ecol.* **58**: 427–440.

Ritter, F. and Zienert, W. 1972. Bemerkungen zum Schutz des Rauhfusskauzes (*Aegolius funereus*). *Landschaftspfl. Naturschutz Thür.* **9**: 12–17.

Ritter, F., Heidrich, M. and Zienert, W. 1978. Statistische Daten zur Brutbiologie Thüringer Rauhfusskäuze, *Aegolius funereus* (L.). *Thüring. Orn. Mitt.* **24**: 37–45.

Roberge, J.-M. and Angelstam, P. 2006. Indicator species among resident forest birds: a cross-regional evaluation in northern Europe. *Biol. Conserv.* **130**: 134–147.

Roberge, J.-M., Mikusinski, G. and Svensson, S. 2008. The white-backed woodpecker: umbrella species for forest conservation planning? *Biodiv. Conserv.* **17**: 2479–2494.

Rodenhouse, N. L., Sherry, T. W. and Holmes, R. T. 1997. Site-dependent regulation of population size: a new synthesis. *Ecology* **78**: 2025–2042.

Roff, D. A. 1992. *The Evolution of Life Histories: Theory and Analysis*. Chapman & Hall, New York.

Roff, D. A. 2002. *Life History Evolution*. Sinauer Associates, Sunderland, MA.

Rohner, C. 1995. Great horned owls and snowshoe hares: what causes the time lag in the numerical response of predators to cyclic prey? *Oikos* **74**: 61–68.

Rohner, C. 1996. The numerical response of great horned owls to the snowshoe hare cycle: consequences of non-territorial 'floaters' on demography. *J. Anim. Ecol.* **65**: 359–370.

Rohner, C. and Hunter, D. B. 1996. First-year survival of great horned owls during a peak and decline of the snowshoe hare cycle. *Can. J. Zool.* **74**: 1092–1097.

Rohner, C. and Krebs, C. J. 1996. Owl predation on snowshoe hares: consequences of antipredator behaviour. *Oecologia* **108**: 303–310.

Rohner, C., Smith, J. N. M., Stroman, J., Joyce, M., Doyle, F. I. and Boonstra, R. 1995. Northern hawk owls in the Nearctic boreal forest: prey selection and population consequences of multiple prey cycles. *Condor* **97**: 208–220.

Rohner, C., Krebs, C. J., Hunter, D. B. and Currie, D. C. 2000. Roost site selection of great-horned owls in relation to black fly activity: an anti-parasite behavior? *Condor* **102**: 950–955.

Rohner, C., Doyle, F. I. and Smith, J. N. M. 2001. Great horned owls. In: Krebs, C. J., Boutin, S. and Boonstra, R. (eds.), *Ecosystem Dynamics of the Boreal Forest: The Kluane Project*. Oxford University Press, Oxford, pp. 340–376.

Ronce, O. 2007. How does it feel to be like a rolling stone? Ten questions about dispersal evolution. *Annu. Rev. Ecol. Evol. Syst.* **38**: 231–253.

Roulin, A. 2002. Offspring desertion by double-brooded female barn owls (*Tyto alba*). *Auk* **119**: 515–519.

Roulin, A. 2004. The function of food stores in bird nests: observations and experiments in the barn owl *Tyto alba*. *Ardea* **92**: 69–78.

Roulin, A., Müller, W., Sasvari, L., Dijsktra, C., Ducrest, A.-L., Riols, C., Wink, M. and Lubjuhn, T. 2004. Extra-pair paternity, testes size and testosterone level in relation to colour polymorphism in the barn owl *Tyto alba*. *J. Avian Biol.* **35**: 492–500.

Rowe, L., Ludwig, D. and Schluter, D. 1994. Time, condition, and the seasonal decline of avian clutch size. *Am. Nat.* **143**: 698–722.

Rudat, V., Kühlke, D., Meyer, W. and Wiesner, J. 1979. Zur Nistökologie von Schwarzspecht (*Dryocopus martius* (L.)), Rauhfusskauz (*Aegolius funereus* (L.)) und Hohltaube (*Columba oenas* L.). *Zool. Jb. Syst.* **106**: 295–310.

Ryder, R. A., Palmer, D. A. and Rawinski, J. J. 1987. Distribution and status of the boreal owl in Colorado. In: Nero, R. W., Clark, R. J., Knapton, R. J. and Hamre, R. H. (eds.), *Biology and Conservation of Northern Forest Owls*. General Technical Report RM-142. USDA Forest Service, Fort Collins, CO, pp. 169–174.

Ryszkowski, L., Goszczynski, J. and Truszkowski, J. 1973. Trophic relationships of the common vole in cultivated fields. *Acta Theriol.* **18**: 125–165.

Saether, B.-E. 1990. Age-specific variation in reproductive performance of birds. *Curr. Ornithol.* **7**: 251–283.

Sala, O. E., Chapin, F. S., Armesto, J. J., Berlow, E., Bloomfield, J., Dirzo, R., Huber-Sanwald, E., Huenneke, L. F., Jackson, R. B., Kinzig, A., Leemans, R., Lodge, D. M., Mooney, H. A., Oesterheld, M., Poff, N. L., Sykes, M. T., Walker, B. H., Walker, M. and Wall, D. H. 2000. Global biodiversity scenarios for the year 2100. *Science* **287**: 1770–1774.

Saladin, V., Ritschard, M., Roulin, A., Bize, P. and Richner, H. 2007. Analysis of genetic parentage in the tawny owl (*Strix aluco*) reveals extra-pair paternity is low. *J. Ornithol.* **148**: 113–116.

Salamolard, M., Butet, A., Leroux, A. and Bretagnolle, V. 2000. Responses of an avian predator to variations in prey density at a temperate latitude. *Ecology* **81**: 2428–2441.

Salo, P., Korpimäki, E., Banks, P. B., Nordström, M. and Dickman, C. R. 2007. Alien predators are more dangerous than native predators to prey populations. *Proc. R. Soc. B* **274**: 1237–1243.

Salo, P., Nordström, M., Thomson, R. L. and Korpimäki, E. 2008. Risk induced by a native top predator reduces alien mink movements. *J. Anim. Ecol.* **77**: 1092–1098.

Salo, P., Banks, P. B., Dickman, C. R. and Korpimäki, E. 2010. Predator manipulation experiments: impacts on populations of terrestrial vertebrate prey. *Ecol. Monogr.* **80**: 531–546.

Santangeli, A., Hakkarainen, H., Laaksonen, T. and Korpimäki, E. 2012. Home range size is determined by habitat composition but feeding rate by food availability in male Tengmalm's owls. *Anim. Behav.* **83**: 1118–1123.

Saurola, P. 1979. Autumn movements of Tengmalm's owl in Finland. *Lintumies* **14**: 104–110.

Saurola, P. 1985. Finnish birds of prey: status and population changes. *Ornis Fenn.* **62**: 64–72.

Saurola, P. 1986. The raptor grid: an attempt to monitor Finnish raptors and owls. *Vår Fågelvärld* Suppl. **11**: 187–190.

Saurola, P. 1987. Mate and nest-site fidelity in Ural and tawny owls. In: Nero, R. W., Clark, R. J., Knapton, R. J. and Hamre, R. H. (eds.), *Biology and Conservation of Northern Forest Owls*. General Technical Report RM-142. USDA Forest Service, Fort Collins, CO, pp. 81–86.

Saurola, P. 1989. Ural owl. In: Newton, I. (ed.), *Lifetime Reproduction in Birds*. Academic Press, London, pp. 327–345.

Saurola, P. 2002. Natal dispersal distances of Finnish owls: results from ringing. In: Newton, I., Kavanagh, R., Olsen, J. and Taylor, I. (eds.), *Ecology and Conservation of Owls*. CSIRO Publishing, Collingwood, Australia, pp. 42–55.

Saurola, P. 2008. Monitoring birds of prey in Finland: a summary of methods, trends, and statistical power. *Ambio* **37**: 413–419.

Saurola, P. 2009. Bad and good news: population changes of Finnish owls during 1982–2007. *Ardea* **97**: 469–482.

Schäffer, N., Mertel, A. and Rost, R. 1991. Siedlungsdichte, Bruterfolg und Brutverluste des Rauhfusskauzes *Aegolius funereus* in Nordostbayern. *Vogelwelt* **112**: 216–225.

Schaffer, W. M. 1974. Optimal reproductive effort in fluctuating environments. *Am. Nat.* **108**: 783–790.

Schelper, W. 1972. *Die Biologie des Rauhfusskauzes* Aegolius funereus *(L.)*. PhD thesis, Georg-August-Universität zu Göttingen, Germany.

Schelper, W. 1989. Zur Brutbiologie, Ernährung und Populationsdynamik des Rauhfusskauzes *Aegolius funereus* im Kaufunger Wald (Südniedersachsen). *Vogelkd. Ber. Niedersachsen* **21**: 33–53.

Scherzinger, W. 1971. Beobachtungen zur Jugendentwicklung einigen Eulen (Strigidae). *Z. Tierpsychol.* **28**: 494–504.

Scheuren, F. 1970. Übersicht über die Bruten des Rauhfusskauzes (*Aegolius funereus*) in den belgischen Ostkantonen Eupen-Malmedy-St. Vith. *Charadrius* **6**: 1–6.

Schmiegelow, F. K. A. and Mönkkönen, M. 2002. Habitat loss and fragmentation in dynamic landscapes: avian perspectives from the boreal forest. *Ecol. Appl.* **12**: 375–389.

Schoener, T, W, 1983. Field experiments on interspecific competition. *Am. Nat.* **122**: 240–285.

Schwerdtfeger, O. 1984. Verhalten und Populationsdynamik des Rauhfusskauzes (*Aegolius funereus*). *Vogelwarte* **32**: 183–200.

Schwerdtfeger, O. 1988. Analyse der Depotbeute in den Bruthöhlen des Rauhfusskauzes (*Aegolius funereus*). *Vogelwelt* **109**: 176–181.

Schwerdtfeger, O. 1990. Modell zur Dispersiondynamik des Rauhfusskauzes (*Aegolius funereus*). In: *Proceedings of International 100 DO-G Meeting: Current Topics in Avian Biology. Bonn, 1988.* pp. 241–247.

Schwerdtfeger, O. 1991. Alterstruktur und Populationsdynamik beim Rauhfusskauz. *Populationsökol. Greifvogel u. Eulenarten* **2**: 493–506.

Schwerdtfeger, O. 1993. Ein Invasionjahr des Rauhfusskauzes (*Aegolius funereus*) im Harz: eine populationsökologische Analyse und ihre Konsequenzen für den Artenschutz. *Ökol. Vögel* **15**: 121–136.

Schwerdtfeger, O. 1996. Wie optimiert der Rauhfusskauz (*Aegolius funereus*) seine Reproduktionsrate? *Populationsökol. Greifvogel u. Eulenarten* **3**: 365–376.

Schwerdtfeger, O. 1997. Höhlennutzung und lokale Dispersionsdynamik beim Rauhfusskauz (*Aegolius funereus*). *Naturschutzreport* **13**: 50–60.

Schwerdtfeger, O. 2000. Entwicklung und Lebenserwartung junger Rauhfusskäuze *Aegolius funereus*. *Populationsökol. Greifvogel u. Eulenarten* **4**: 505–516.

Schwerdtfeger, O. 2008. Ist der Rauhfusskauz *Aegolius funereus* ein echter Harzer? Über die Bedeutung einer lokalen Population. *Vogelkd. Ber. Niedersachsen* **40**: 247–253.

Seaman, D. E. and Powell, R. A. 1996. An evaluation of the accuracy of kernel density estimators from home range analysis. *Ecology* **77**: 2075–2085.

Seaman, D. E., Millspaugh, J. J., Kernohan, B. J., Brundige, G. C., Raedeke, K. J. and Gitzen, R. A. 1999. Effects of sample size on kernel home range estimates. *J. Wildl. Manage.* **63**: 739–747.

Selonen, V., Hanski, I. K. and Stevens, P. C. 2001. Space use of the Siberian flying squirrel *Pteromys volans* in fragmented forest landscapes. *Ecography* **24**: 588–601.

Selonen, V., Sulkava, P., Sulkava, R., Sulkava, S. and Korpimäki, E. 2010. Decline of flying and red squirrels in boreal forests revealed by long-term diet analyses of avian predators. *Anim. Cons.* **13**: 579–585.

Sergio, F. and Hiraldo, F. 2008. Intraguild predation in raptor assemblages: a review. *Ibis* **150**: 132–145.

Sergio, F. and Newton, I. 2003. Occupancy as a measure of territory quality. *J. Anim. Ecol.* **72**: 857–865.

Sergio, F., Newton, I. and Marchesi, L. 2005. Top predators and biodiversity. *Nature* **436**: 192.

Sergio, F., Newton, I., Marchesi, L. and Pedrini, P. 2006. Ecologically justified charisma: preservation of top predators delivers biodiversity conservation. *J. Appl. Ecol.* **43**: 1049–1055.

Sergio, F., Caro, T., Brown, D., Clucas, B., Hunter, J., Ketchum, J., McHugh, K. and Hiraldo, F. 2008a. Top predators as conservation tools: ecological rationale, assumptions, and efficacy. *Annu. Rev. Ecol. Evol. Syst.* **39**: 1–19.

Sergio, F., Newton, I. and Marchesi, L. 2008b. Top predators and biodiversity: much debate, few data. *J. Anim. Ecol.* **45**: 992–999.

Shine, R. 1988. The evolution of large body size in females: a critique of Darwin's 'fecundity advantage' model. *Am. Nat.* **131**: 124–131.

Sidorovich, V., Ivanovsky, V. V. and Adamovich, S. 2003. Food niche and dietary overlap in owls of northern Belarus. *Vogelwelt* **124**: 271–279.

Sih, A. 1993. Effects of ecological interactions on forager diets: competition, predation risk, parasitism and prey behaviour. In: Hughes, R. N. (ed.), *Diet Selection*. Blackwell Scientific Publications, Cambridge, pp. 182–211.

Sih, A. and Christensen, B. 2001. Optimal diet theory: when does it work, and when and why does it fail? *Anim. Behav.* **61**: 379–390.

Sih, A., Crowley, P., McPeek, M., Petranka, J. and Strohmeier, K. 1985. Predation, competition and prey communities: a review of field experiments. *Annu. Rev. Ecol. Syst.* **16**: 269–311.

Siivonen, L. 1948. Structure of short-cycling fluctuations in numbers of mammals and birds in the Northern Hemisphere. *Pap. Game Res.* **1**: 1–116.

Siivonen, L. 1954. On the short-term fluctuations in numbers of tetraonids. *Pap. Game Res.* **13**: 1–10.

Siivonen, L. 1957. The problem of the short-term fluctuations in numbers of tetraonids in Europe. *Pap. Game Res.* **19**: 1–44.

Siivonen, L. 1974. *Pohjolan nisäkkäät*. Otava, Helsinki.

Simberloff, D. 1998. Flagships, umbrellas, and keystones: is single-species management passé in the landscape era? *Biol. Conserv.* **83**: 247–257.

Simmons, R. E. 1988. Food and the deceptive acquisition of mates by polygynous male harriers. *Behav. Ecol. Sociobiol.* **23**: 83–92.

Simmons, R. E., Barnard, P., MacWhirter, B. and Hansen, G. L. 1986a. The influence of microtines on polygyny, productivity, age, and provisioning of breeding hen harriers: a 5-year study. *Can. J. Zool.* **64**: 2447–2456.

Simmons, R. E., Smith, P. C. and MacWhirter, R. B. 1986b. Hierarchies among northern harrier (*Circus cyaneus*) harems and the costs of polygyny. *J. Anim. Ecol.* **55**: 755–771.

Singleton, G. R., Brown, P. R., Pech, R. P., Jacob, J., Davis, S., Mutze, G. J. and Krebs, C. J. 2005. One hundred years of eruptions of house mice in Australia: a natural biological curio. *Biol. J. Linn. Soc.* **84**: 617–627.

Slagsvold, T. 1986. Nest site settlement by the pied flycatcher: does the female choose her mate for the quality of his house or himself? *Ornis Scand.* **17**: 210–220.

Slagsvold, T., Lifjeld, J. T., Stenmark, G. and Breiehagen, T. 1988. On the cost of searching for a mate in female pied flycatchers *Ficedula hypoleuca*. *Anim. Behav.* **36**: 433–442.

Smith, C. C. and Fretwell, S. D. 1974. The optimal balance between size and number of offspring. *Am. Nat.* **108**: 499–506.

Smith, D. G. and Hiestand, E. 1990. Alloparenting at an Eastern screech-owl nest. *Condor* **92**: 246–247.

Smith, J. N. M. 1981. Does high fecundity reduce survival in song-sparrows. *Evolution* **35**: 1142–1148.

Solantie, R. 1975. The areal distribution of winter precipitation and snow depth in March in Finland. *Ilmat. Lait. Tied.* **28**: 1–66.

Solantie, R. 1977. On the persistence of snow cover in Finland. *Ilmat. Lait. Tutk.* **60**: 1–68.

Solantie, R. 2000. *Snow Depth on January 15th and March 15th in Finland 1991–98, and its Implications for Soil Frost and Forest Ecology*. Meteorological Publication No. 34. Finnish Meteorological Institute, Helsinki.

Solantie, R., Drebs, A., Hellsten, E. and Saurio, P. 1996. *Timing and Duration of Snow Cover in Finland during 1961–1993*. Meteorological Publications No. 34. Finnish Meteorological Institute, Helsinki.

Solheim, R. 1983. Breeding frequency of Tengmalm's owl *Aegolius funereus* in three localities in 1974–78. In: *Proceedings of Third Nordic Congress in Ornithology 1981*, pp. 79–84.

Solheim, R. 1984. Caching behaviour, prey choice and surplus killing by pygmy owls *Glaucidium passerinum* during winter, a functional response of a generalist predator. *Ann. Zool. Fenn.* **21**: 301–308.

Solomon, M. E. 1949. The natural control of animal population. *J. Anim. Ecol.* **18**: 1–35.

Solonen, T. 2004. Are vole-eating owls affected by mild winters in southern Finland? *Ornis Fenn.* **81**: 65–74.

Solonen, T., Tiainen, J., Korpimäki, E. and Saurola, P. 1991. Dynamics of Finnish starling *Sturnus vulgaris* populations in recent decades. *Ornis Fenn.* **68**: 158–169.

Sonerud, G. A. 1985a. Nest hole shift in Tengmalm's owl *Aegolius funereus* as defence against nest predation involving long-term memory in the predator. *J. Anim. Ecol.* **54**: 179–192.

Sonerud, G. A. 1985b. Risk of nest predation in three species of hole nesting owls: influence on choice of nesting habitat and incubation behaviour. *Ornis Scand.* **16**: 261–269.

Sonerud, G. A. 1986. Effect of snow cover on seasonal changes in diet, habitat, and regional distribution of raptors that prey on small mammals in boreal zones of Fennoscandia. *Holarct. Ecol.* **9**: 33–47.

Sonerud, G. A. 1989. Reduced predation by pine martens on nests of Tengmalm's owl in relocated boxes. *Anim. Behav.* **37**: 332–334.

Sonerud, G. A. 1992a. Functional responses of birds of prey: biases due to the load-size effect in central place foragers. *Oikos* **63**: 223–232.

Sonerud, G. A. 1992b. Nest predation may make the 'deception hypothesis' unnecessary to explain polygyny in the Tengmalm's owl. *Anim. Behav.* **43**: 871–874.

Sonerud, G. A. 1993. Reduced predation by nest box relocation: differential effect on Tengmalm's owl nests and artificial nests. *Ornis Scand.* **24**: 249–253.

Sonerud, G. A., Solheim, R. and Jacobsen, B. V. 1986. Home range use and habitat selection during hunting in a male Tengmalm's owl *Aegolius funereus*. *Fauna Norv. Ser. C, Cinclus* **9**: 100–106.

Sonerud, G. A., Nybo, J. O., Fjeld, P. E. and Knoff, C. 1987. A case of bigyny in the hawk owl *Surnia ulula*: spacing of nests and allocation of male feeding effort. *Ornis Fenn.* **64**: 144–148.

Sonerud, G. A., Solheim, R. and Prestrud, K. 1988. Dispersal of Tengmalm's owl *Aegolius funereus* in relation to prey availability and nesting success. *Ornis Scand.* **19**: 175–181.

Sorbi, S. 1993. Recherche sur les cavitiés naturelles propices à la chouette de Tengmalm (*Aegolius funereus*) en Haute Ardenne (Summary: Research on natural cavities favourable to the Tengmalm's owl (*Aegolius funereus*) in High Ardenne. *Aves* **30**: 81–93.

Sorbi, S. 1995. La chouette de Tengmalm (*Aegolius funereus*) en Belgique: Synthèse et mise à jour du stat (Summary: The Tengmalm's owl (*Aegolius funereus*) in Belgium: synthesis and updating). *Aves* **32**: 101–132.

Sorbi, S. 2003. Étendue et utilisation du domaine vital chez la chouette de Tengmalm *Aegolius funereus* en Haute-Ardenne Belge: suivi par radio-pistage (Summary: Size and use of Tengmalm's owl *Aegolius funereus* home range in the high Belgian Ardennes: results from radio-tracking). *Alauda* **71**: 215–220.

Southern, H. N. and Lowe, V. P. 1968. The pattern of distribution of prey and predation in tawny owl territories. *J. Anim. Ecol.* **37**: 75–97.

Sperti, M. T., Mezzavilla, F. and Lombardo, S. 1991. Nidification de la chouette de Tengmalm, *Aegolius funereus*, dans la foret de Cansiglio (nord-est de l'Italie). In: Juillard, M. (ed.), *Rapaces Nocturnes*. Nos Oiseaux, Porrentruy, Switzerland, pp. 247–254.

Stahlecker, D. W. and Duncan, R. B. 1996. The boreal owl at the southern terminus of the Rocky Mountains: undocumented longtime resident or recent arrival? *J. Raptor Res.* **98**: 153–161.

Stearns, S. C. 1976. Life history tactics: a review of the ideas. *Q. Rev. Biol.* **51**: 3–47.

Stearns, S. C. 1977. The evolution of life history traits: a critique of the theory and a review of the data. *Annu. Rev. Ecol. Syst.* **8**: 145–171.

Stearns, S. C. 1989. Trade-offs in life-history evolution. *Funct. Ecol.* **3**: 259–268.

Stearns, S. C. 1992. *The Evolution of Life Histories*. Oxford University Press, Oxford.

Steenhof, K. and Kochert, M. N. 1985. Dietary shifts of sympatric buteos during a prey decline. *Oecol. (Berl.)* **66**: 6–16.

Steenhof, K. and Kochert, M. N. 1988. Dietary responses of three raptor species to changing prey densities in a natural environment. *J. Anim. Ecol.* **57**: 37–48.

Stenmark, G., Slagsvold, T. and Lifjeld, J. T. 1988. Polygyny in the pied flycatcher, *Ficedula hypoleuca*: a test of the deception hypothesis. *Anim. Behav.* **36**: 1646–1657.

Stephens, D. W. and Krebs, J. R. 1986. *Foraging Theory*. Princeton University Press, Princeton, NJ.

Stephens, D. W., Brown, J. S. and Ydenberg, R. C. 2007. *Foraging. Behavior and Ecology*. University of Chicago Press, Chicago, I.L.

Stoleson, S. H. and Beissinger, S. R. 1995. Hatching asynchrony and the onset of incubation in birds, revisited: when is the critical period? *Curr. Ornithol.* **12**: 191–270.

Stoleson, S. H. and Beissinger, S. R. 1997. Hatching asynchrony, brood reduction, and food limitation in a neotropical parrot. *Ecol. Monogr.* **67**: 131–154.

Strann, K. B., Yoccoz, N. G. and Ims, R. A. 2002. Is the heart of Fennoscandian rodent cycle still beating? A 14-year study of small mammals and Tengmalm's owls in northern Norway. *Ecography* **25**: 81–87.

Suhonen, J., Halonen, M., Mappes, T. and Korpimäki, E. 2007. Interspecific competition limits larders of pygmy owls. *J. Avian Biol.* **38**: 630–634.

Sulkava, P. 1964. Varpushaukan pesimäaikaisesta käyttäytymisestä ja ravinnosta (Summary: On the behaviour and food habits of the sparrowhawk (*Accipiter nisus*) during the nesting season). *Suomen Riista* **17**: 93–105.

Sulkava, P. 1965. Vorkommen und Nahrung der Waldohreule, *Asio otus* (L.), in Ilmajoki (EP) in den Jahren 1955–1963. *Aquilo Ser. Zool.* **2**: 41–47.

Sulkava, P. 1972. *Varpushaukan*, Accipiter nisus *L., pesimisbiologiasta ja pesimisaikaisesta ravinnosta.* Phil. Lic. thesis, Department of Zoology, University of Helsinki, Finland.

Sulkava, P. and Sulkava, S. 1971. Die nistzeitliche Nahrung des Rauhfusskauzes *Aegolius funereus* in Finnland 1958–67. *Ornis Fenn.* **48**: 117–124.

Sulkava, S. 1964. Zur Nahrungsbiologie des Habichts, *Accipiter g. gentilis* (L.). *Aquilo Ser. Zool.* **3**: 1–103.

Sulkava, S. 1966. Huuhkajan pesimisaikaisesta ravinnosta Suomessa (Summary: Feeding habits of the eagle owl (*Bubo bubo*) in Finland). *Suomen Riista* **18**: 145–156.

Sulkava, S. 1968. A study on the food of the peregrine, *Falco p. peregrinus* Tunstall, in Finland. *Aquilo Ser. Zool.* **6**: 18–31.

Sulkava, S. and Huhtala, K. 1997. The great gray owl (*Strix nebulosa*) in the changing forest environment of Northern Europe. *J. Raptor Res.* **31**: 151–159.

Sulkava, S., Huhtala, K. and Rajala, P. 1984. Diet and breeding success of the golden eagle in Finland 1958–82. *Ann. Zool. Fenn.* **21**: 283–286.

Sulkava, S., Tornberg, R. and Koivusaari, J. 1997. Diet of the white-tailed eagle *Haliaeetus albicilla* in Finland. *Ornis Fenn.* **74**: 65–78.

Sulkava, S., Huhtala, K., Rajala, P. and Tornberg, R. 1998. Changes in the diet of the golden eagle *Aquila chrysaetos* and small game populations in Finland in 1957–96. *Ornis Fenn.* **76**: 1–16.

Sulkava, S., Lokki, H. and Koivu, J. 2008. Huuhkajan pesimäaikainen ravinto Hämeessä 1994–2006 (Summary: The diet of the eagle owl (*Bubo bubo*) during the nesting season in Häme (southern Finland)). *Suomen Riista* **54**: 83–94.

Sunde, P. 2005. Predators control post-fledging mortality in tawny owls, *Strix aluco. Oikos* **110**: 461–472.

Sunde, P., Bolstad, M. S. and Desfor, K. B. 2003. Diurnal exposure as a risk sensitive behaviour in tawny owls *Strix aluco. J. Avian Biol.* **34**: 409–418.

Sundell, J., Huitu, O., Henttonen, H., Kaikusalo, A., Korpimäki, E., Pietiäinen, H., Saurola, P. and Hanski, I. 2004. Large-scale spatial dynamics of vole populations in Finland revealed by the breeding success of vole-eating avian predators. *J. Anim. Ecol.* **73**: 167–178.

Suopajärvi, M. 2001. Syksyn 2000 helmipöllöt. *Sirri* **26**: 7.

Suopajärvi, M. 2006. Helmipöllöjen syyspyynnit 2005. *Sirri* **31**: 8.

Suopajärvi, P. and Suopajärvi, M. 2003. Mittaaminen kannattaa aina. Kanahaukan, tuulihaukan, lapinpöllön ja helmipöllön poikasten iänmääritys siiven pituudesta. *Sirri* **28**: 54–59.

Suorsa, P., Huhta, E., Nikula, A., Nikinmaa, M., Jäntti, A., Helle, H. and Hakkarainen, H. 2003. Forest management is associated with physiological stress in an old-growth forest passerine. *Proc. R. Soc. Lond. B* **270**: 963–969.

Suorsa, P., Huhta, E., Jäntti, A., Nikula, A., Helle, H., Kuitunen, M., Koivunen, V. and Hakkarainen, H. 2005. Thresholds in selection of breeding habitat by the Eurasian treecreeper (*Certhia familiaris*). *Biol. Conserv.* **121**: 443–452.

Sutherland, W. J. 1996. *From Individual Behaviour to Population Ecology.* Oxford University Press, Oxford.

Sydeman, W. J., Penniman, J. F., Penniman, T. M., Pyle, P. and Ainley, D. G. 1991. Breeding performance in the western gull: effects of parental age, timing of breeding and year in relation to food availability. *J. Anim. Ecol.* **60**: 135–149.

Sykkö, M. and Vikström, S. 1987. Helmipöllön vaelluskäyttäytymisestä. *Lintumies* **22**: 232–237.

Szekely, T., Webb, J. N., Houston, A. I. and McNamara, J. N. 1996. An evolutionary approach to offspring desertion in birds. In: Nolan, V. and Ketterson, E. D. (eds.), *Current Ornithology*. Plenum Press, New York, pp. 265–324.

Taivalmäki, J.-P., Haapala, J. and Saurola, P. 1998. Petolintuvuosi 1998: vaisu sadekesä, myyrälaikut harvassa (Summary: Breeding and population trends of common raptors and owls in Finland in 1998). *Linnut Yearbook* **1998**: 38–53.

Taivalmäki, J.-P., Haapala, J. and Saurola, P. 2001. Petolintuvuosi 2000: tuulihaukka ja varpuspöllö menestyivät (Summary: Breeding and population trends of common raptors and owls in Finland in 2000). *Linnut Yearbook* **2000**: 44–54.

Taylor, I. 1994. *Barn Owls: Predator–Prey Relationships and Conservation*. Cambridge University Press, Cambridge.

Taylor, I. R. 2009. How owls select their prey: a study of barn owls *Tyto alba* and their small mammal prey. *Ardea* **97**: 635–644.

Taylor, R. J. 1984. *Predation*. Chapman and Hall, London.

Temeles, E. J. 1985. Sexual size dimorphism of bird-eating hawks: the effect of prey vulnerability. *Am. Nat.* **125**: 485–499.

Temple, S. A. 1987. Do predators always capture substandard individuals disproportionately from prey populations? *Ecology* **68**: 669–674.

Tjernberg, M., Johnsson, K. and Nilsson, S. G. 1993. Density variation and breeding success of the black woodpecker *Dryocopus martius* in relation to forest fragmentation. *Ornis Fenn.* **70**: 155–162.

Tkadlec, E. and Stenseth, N. C. 2001. A new geographical gradient in vole population dynamics. *Proc. R. Soc. Lond. B* **268**: 1547–1552.

Toft, C. A. 1991. Current theory of host parasite interactions. In: Loye, J. E. and Zuk, M. (eds.), *Bird–Parasite Interactions*. Oxford University Press, Oxford, pp. 3–15.

Tomé, R., Santos, N., Cardia, P., Ferrand, N. and Korpimäki, E. 2005. Factors affecting the prevalence of blood parasites of little owls *Athene noctua* in southern Portugal. *Ornis Fenn.* **82**: 63–72.

Tomppo, E., Henttonen, H., Korhonen, K., Aarnio, A., Ahola, A., Ihalainen, A., Heikkinen, J. and Tuomainen, T. 1999. Forest resources and their development in central Finland, 1967–1996. *Metsätieteen aikakauskirja Folia Forestalia* **2B**/1999: 309–388 (in Finnish).

Törmälehto, H. and Korpimäki, E. 1978. Helmipöllön (*Aegolius funereus*) poikanen selvinnyt lentoon hiiripöllön (*Surnia ulula*) pesästä. *Aureola* **3**: 36–37.

Tornberg, R., Korpimäki, E., Jungell, S. and Reif, V. 2005. Delayed numerical response of goshawks to population fluctuations of forest grouse. *Oikos* **111**: 408–415.

Tornberg, R., Korpimäki, E. and Byholm, P. 2006. Ecology of the Northern goshawk in Fennoscandia. *Stud. Avian Biol.* **31**: 141–157.

Tornberg, R., Mönkkönen, M. and Pahkala, M. 1999. Changes in diet and morphology of Finnish goshawks from the 1960s to 1990s. *Oecologia* **121**: 369–376.

Trivers, R. L. 1972. Parental investment and sexual selection. In: Campbell, B. (ed.), *Sexual Selection and the Descent of Man*. Heinemann, Chicago, IL, pp. 136–179.

Tuomi, J., Hakala, T. and Haukioja, E. 1983. Alternative concepts of reproductive effort, costs of reproduction, and selection in life-history evolution. *Am. Zool.* **23**: 25–34.

Ulfstrand, S. and Högstedt, G. 1976. Hur många fåglar häckar i Sverige? *Anser* **15**: 1–32.

Uttendörfer, O. 1952. *Neue Ergebnisse über die Ernährung der Greifvögel und Eulen*. Eugen Ulmer, Stuttgart.

Valkama, J. and Korpimäki, E. 1999. Nestbox characteristics, habitat quality and reproductive success of Eurasian kestrels. *Bird Stud*. **46**: 81–88.

Valkama, J., Korpimäki, E., Holm, A. and Hakkarainen, H. 2002. Hatching asynchrony and brood reduction in Tengmalm's owl *Aegolius funereus*: the role of temporal and spatial variation in food abundance. *Oecologia* **133**: 334–341.

Valkama, J., Korpimäki, E., Arroyo, B., Beja, P., Bretagnolle, V., Bro, E., Kenward, R., Manosa, S., Redpath, S. M., Thirgood, S. and Vinuela, J. 2005. Birds of prey as limiting factors of gamebird populations in Europe: a review. *Biol. Rev*. **80**: 171–203.

Verner, J. 1964. Evolution of polygamy in the long-billed marsh wren. *Evolution* **18**: 252–261.

Viitala, J., Korpimäki, E., Palokangas, P. and Koivula, M. 1995. Attraction of kestrels to vole scent marks visible in ultraviolet light. *Nature* **373**: 425–427.

Vikström, S. 1988. Helmipöllön (*Aegolius funereus*) syysvaellus Kokkolassa 1988. *Ornis Botnica* **10**: 14–26.

Vikström, S. 2009. Keski-pohjalaisen helmipöllökannan tulevaisuudennäkymät. *Rengastajan Vuosikirja* **2009**: 27–28.

Village, A. 1982. The diet of the kestrels in relation to vole abundance. *Bird Stud*. **29**: 129–138.

Village, A. 1985. Spring arrival times and assortative mating of kestrels in south Scotland. *J. Anim. Ecol*. **54**: 857–868.

Village, A. 1986. Breeding performance of kestrels at Eskdalemuir, south Scotland. *J. Zool. Lond*. **208**: 367–378.

Viñuela, J. (1999) Sibling aggression, hatching asynchrony, and nestling mortality in the black kite (*Milvus migrans*). *Behav. Ecol. Sociobiol*. **45**: 3345.

Virkkala, R., Rajasärkkä, A., Väisänen, R. A., Vickholm, M. and Virolainen, E. 1994. Conservation value of nature reserves: do hole-nesting birds prefer protected forests in southern Finland? *Ann. Zool. Fenn*. **31**: 173–186.

Viro, P. and Sulkava, S. 1985. Food of the bank vole in northern Finnish spruce forests. *Acta Theriol*. **30**: 259–266.

Vrezec, A. 2003. Breeding density and altitudinal distribution of the Ural, tawny, and boreal owls in North Dinaric Alps (central Slovenia). *J. Raptor Res*. **37**: 55–62.

Wahlstedt, J. 1959. Ugglornas spelvanor. *Fauna o. Flora* **3–4**: 81–112.

Wallin, K. and Andersson, M. 1981. Adult nomadism in Tengmalm's owl *Aegolius funereus*. *Ornis Scand*. **14**: 63–65.

Weatherhead, P. J. and Robertson, R. J. 1979. Offspring quality and the polygyny threshold: 'the sexy son hypothesis'. *Am. Nat*. **113**: 201–208.

Wellicome, T. I. 2005. Hatching asynchrony in burrowing owls is influenced by clutch size and hatching success but not by food. *Oecologia* **142**: 326–334.

Whitman, J. S. 2008. Post-fledgling estimation of annual productivity in boreal owls based on prey detritus mass. *J. Raptor Res*. **42**: 58–60.

Whitman, J. S. 2009. Diet and prey consumption rates of nesting boreal owls, *Aegolius funereus*, in Alaska. *Can. Field Nat*. **123**: 112–116.

Whitman, J. S. 2010. Boreal owl breeding ecology in interior Alaska. Manuscript.

Wiebe, K. L. 2005. Asymmetric costs favor female desertion in the facultatively polyandrous northern flicker (*Colaptes auratus*): a removal experiment. *Behav. Ecol. Sociobiol*. **57**: 429–437.

Wiebe, K. L. 2010. Negotiation of parental care when the stakes are high: experimental handicapping of one partner during incubation leads to short-term generosity. *J. Anim. Ecol.* **79**: 63–70.

Wiebe, K. L. and Bortolotti, G. R. 1994. Food supply and hatching spans of birds: energy constraints or facultative manipulation? *Ecology* **75**: 813–823.

Wiebe, K. L., Korpimäki, E. and Wiehn, J. 1998. Hatching asynchrony in Eurasian kestrels in relation to the abundance and predictability of cyclic prey. *J. Anim. Ecol.* **67**: 908–917.

Wiehn, J. and Korpimäki, E. 1998. Resource levels, reproduction and resistance to haematozoan infections. *Proc. R. Soc. Lond. B* **265**: 1197–1201.

Wiehn, J., Korpimäki, E. and Pen, I. 1999. Haematozoan infections in the Eurasian kestrel: effects of fluctuating food supply and experimental manipulation of parental effort. *Oikos* **84**: 87–98.

Wiehn, J., Ilmonen, P., Korpimäki, E., Pahkala, M. and Wiebe, K. L. 2000. Hatching asynchrony in the Eurasian kestrel *Falco tinnunculus*: an experimental test of the brood reduction hypothesis. *J. Anim. Ecol.* **69**: 85–95.

Wiens, J. A. 1989. *The Ecology of Bird Communities. Volume 2. Processes and Variations.* Cambridge University Press, Cambridge.

Wijnandts, H. 1984. Ecological energetics of the long-eared owl (*Asio otus*). *Ardea* **72**: 1–92.

Wiklund, C. G. 1996. Determinants of dispersal in breeding merlins (*Falco columbarius*). *Ecology* **77**: 1920–1927.

Wiktander, U., Olsson, O. and Nilsson, S. G. 2000. Parental care and social mating system in the lesser spotted woodpecker *Dendrocopos minor*. *J. Avian Biol.* **31**: 447–456.

Williams, G. C. 1966. Natural selection, the costs of reproduction, and a refinement of Lack's principle. *Am. Nat.* **100**: 687–690.

Winkler, D. W. and Wilkinson, G. S. 1988. Parental effort in birds and mammals: theory and measurement. *Oxford Surv. Evol. Biol.* **5**: 185–214.

Wirdheim, A. 2008. Invasion av pärlugglor. *Vår Fågelvärld* **7**: 23–25.

Wittenberger, J. F. 1976. The ecological factors selecting for polygyny in altricial birds. *Am. Nat.* **110**: 779–799.

Wittenberger, J. F. and Tilson, R. L. 1980. The evolution of monogamy: hypotheses and evidence. *Annu. Rev. Ecol. Syst.* **11**: 197–232.

Woo, P. T. K. 1970. The hematocrit centrifuge technique for the detection of trypanosomes in blood. *Can. J. Zool.* **47**: 921–923.

Ydenberg, R. C. 1987. Nomadic predators and geographical synchrony in microtine population cycles. *Oikos* **50**: 270–272.

Zanette, L. 2000. Fragment size and the demography of an area-sensitive songbird. *J. Anim. Ecol.* **69**: 458–470.

Zang, H. and Kunze, P. 1978. Zur Ernährung des Rauhfusskauzes (*Aegolius funereus*) im Harz mit einer Bemerkung zur Gefährdung durch Eichhörnchen (*Sciurus vulgaris*). *Vogelkd. Ber. Niedersachsen* **2**: 41–44.

Zang, H. and Kunze, P. 1985. Zum Ansiedlungsverhalten des Rauhfusskauzes (*Aegolius funereus*) in einem suboptimalen Habitat des Harzes. *Vogelwarte* **106**: 264–267.

Zárybnická, M. 2008. Circadian activity of the Tengmalm's owl (*Aegolius funereus*) in the Krusne hory Mts.: the effect of different parental roles. *Sylvia* **44**: 51–61.

Zárybnická, M. 2009a. Activity patterns of male Tengmalm's owls, *Aegolius funereus* under varying food conditions. *Folia Zool.* **58**: 104–112.

Zárybnická, M. 2009b. Parental investment of female Tengmalm's owls *Aegolius funereus*: correlation with varying food abundance and reproductive success. *Acta Ornithol.* **44**: 81–88.

Zárybnická, M., Sedlacek, O. and Korpimäki, E. 2009. Do Tengmalm's owls alter parental feeding effort under varying conditions of main prey availability? *J. Ornithol.* **150**: 231–237.

Zárybnická, M., Riegert, J. and Stastny, K. 2011. Diet composition in the Tengmalm's owl *Aegolius funereus*: a comparison of camera surveillance and pellet analysis. *Ornis Fenn.* **88**: 147–153.

Zhou, Y. B., Newman, C., Xu, W. T., Buesching, C. D., Zalewski, A., Kaneko, Y., Macdonald, D. W. and Xie, Z. Q. 2011. Biogeographical variation in the diet of Holarctic martens (genus *Martes*, Mammalia: Carnivora: Mustelidae): adaptive foraging in generalists. *J. Biogeograph.* **38**: 137–147.

Index

Printed in the United States
By Bookmasters